P9-AOA-777

TECHNOLOGY *and YOU*

757
BOEING 757
MODEL 78

TECHNOLOGY and YOU

David L. Goetsch
John A. Nelson

DELMAR PUBLISHERS INC.®

Dedication

To a good friend, colleague, and favorite teacher, C. M. "Duque" Wilson

COVER PHOTO: Courtesy of NASA

Delmar Staff
Executive Editor: Mark W. Huth
Associate Editor: Jonathan Plant
Managing Editor: Barbara A. Christie
Production Editor: Eleanor Isenhart
Design Coordinator: Susan Mathews

For information, address Delmar Publishers Inc.,
2 Computer Drive West, Box 15-015
Albany, New York 12212-5015

Printed in the United States of America
Published simultaneously in Canada
by Nelson Canada,
A Division of International Thomson Limited

10 9 8 7 6 5 4 3 2

Library of Congress Cataloging in Publication Data

Goetsch, David L.
Technology and you.
Includes index.
1. Technology. 2. Industrial arts. I. Nelson, John A. II. Title.
T47.G63 1987 600 86–19751
ISBN 0–8273–2662–9

CONTENTS

PART TWO
THE PHYSICAL TECHNOLOGIES

PART FIVE
ISSUES AND OUTLOOKS IN TECHNOLOGY

PREFACE

Technology and You was written to provide an easy to understand, well illustrated book on technology for junior high school students. Rapid change has become a normal part of our everyday lives. Since most change results from advances in technology, it is important for young people to become "technologically literate."

In the past, literacy depended only on reading, writing, speaking, listening, and math skills. Now a literate person is one who has these skills *and* knows what technology is, the good and bad effects of technology, and how to make sure that the effects are as good as possible.

Technological literacy is the purpose of this book. This is a broader concept than those (such as woodworking, metal shop, and power shop) covered in a traditional industrial arts setting. Whereas industrial arts courses have traditionally been limited to only a part of the student body, technology education courses are for all students.

NOTE ON TECHNOLOGY LEARNING ACTIVITIES

A basic principle of teaching and learning is that people learn by *doing*. This principle was an important consideration in all phases of developing *Technology and You*. Hands-on learning is promoted through special "Technology Learning Activities" which have been included at the end of each chapter. In addition, two supplements, *Technology Activity Guides One and Two*, have been developed to go with the book. The technology learning activities in this book were not meant to be heavy hands-on activities. Rather, they were meant to bridge the gap between classroom learning and the more involved hands-on activities contained in *Technology Activity Guides One and Two.*

SPECIAL FEATURES OF THE BOOK

Technology and You was especially designed to meet a new and emerging need for all students which the International Technology Education Association feels is a critical need. It was especially designed to bridge the gap between the traditional industrial arts concept and the more recent technology education concepts such as those formulated in the Jackson Mill's Study. Foremost among these is the concept of the system, which allows students to analyze and understand each area of technology, and to see how they all fit together.

In developing *Technology and You*, the authors talked with many experienced teachers in the field of technology education and industrial arts. We also studied the curriculum guides and course syllabi from several states that are developing technology education curricula. The input from these sources led to a number of special features of *Technology and You*.

- The text is written in a simple conversational style that is easy to read and understand. The subject matter is technology, but steps have been taken to ensure that the text is not overly technical in style. When difficult terms are used, they are immediately explained within the text.
- Each chapter includes numerous everyday examples to illustrate the various technological concepts presented. Students, even those with no technological background, can easily relate to these examples.
- The entire book is heavily illustrated with photographs and drawings that directly support the reading material. Each illustration has a caption that relates to the text material. Photographs and drawings should be emphasized. Together with their captions, they are important learning materials.
- *Technology and You* is divided into five parts. Part 1 is an overview of technology. Students learn what technology is, what the buzzword "high technology" means, how high technology relates to the broader concept of technology, what the seven resources of technology are, what technological systems are, and how technology is divided into such areas as the physical, information/communication, and biotechnologies. Part 2 explains the physical technologies: construction, energy and power, manufacturing, and transportation. Part 3 explains the information/communication technologies: information processing, graphic communication, and electronic communication. Part 4 explains the biotechnologies: agriculture/food processing and health care/medicine. Part 5 deals with technology and the future. In this part, the problems of technology and how they can be solved are discussed. Future developments and your role in them are also discussed.
- The various key words associated with technology, such as CADD, CAM, NC, CNC, DNC, and robotics are covered. Rather than being treated as individual concepts, they are worked into the technology areas where they apply. A glossary is also included to further reinforce the learning of technology terms.

NOTE ON ILLUSTRATIONS

People tend to be visual learners. This means we like drawings and photographs to go along with what we read. For this reason,

Technology and You contains many drawings and photographs. Most of these were prepared especially for this book. Each figure has a caption that amplifies the illustration and relates it to the reading material.

After researching the various memory aids which help young people learn, the authors selected the circular diagram as one of the ways to reinforce important concepts. Systems of technology are frequently presented by means of these diagrams.

A NOTE FOR TEACHERS

Technology education will represent change for most teachers. The authors are teachers. We understand that even the most positive changes can be uncomfortable. For this reason, we designed *Technology and You* based upon your stated wants and needs, as determined through nationally circulated surveys.

If technology education means change for you, we are confident that *Technology and You* will help make that change more comfortable. The special features outlined in the last section will help. In addition, a comprehensive Instructor's Guide is available which will help you further. It contains answers to the questions in the text, helpful suggestions, proven teaching strategies, and a variety of supplemental teaching/learning materials.

The setting for technology education classes will not only be the shop or lab. However, the basic teaching and learning approach is still hands-on. We feel this book bridges the gap between the old and new by including the best of both.

ACKNOWLEDGMENTS

The authors wish to acknowledge those people without whose help this book would not have been completed. Deborah M. Goetsch traveled around the country taking the photographs needed to supplement the text material. Dr. Don McDaniel, Director of the Health Performance Center of Sacred Heart Hospital in Pensacola, Florida gave especially invaluable input in health care and medical technology.

We also thank the many professionals in our field who gave of their time and expertise to review the manuscript. These educators made *Technology and You* a better book. They are Lynn Barrier, Paul H. Cunningham, Henry Heinlein, and Ken Swartz.

Acknowledgments for Part Opening Photos

Part I

Top	Courtesy of Boeing Company
Center left	Courtesy of Compugraphic Corporation
Bottom left	Courtesy of General Motors Corporation
Bottom right	Courtesy of Sears, Roebuck and Co.

Part II

Top left	Courtesy of Commodore Electronics Limited
Top right	Courtesy Philadelphia Electric Company
Bottom left	Courtesy Chrysler Corporation
Bottom right	Courtesy of Bethlehem Steel

Part III

Top left	Courtesy of Sperry Corp.
Top right	Photo by Deborah M. Goetsch
Center right	Courtesy of GTE Communication Systems
Bottom	Courtesy of Diagram Corporation

Part IV

Top left	Permission by Weyerhauser Corporation
Top right	Photo by Robert C. Atwood
Bottom left	Photo by Deborah M. Goetsch
Bottom right	Photo by Robert C. Atwood

Part V

Top left	Courtesy of Mentor Graphics Corp.
Top right	Courtesy of Lockheed—California, A Division of Lockheed Corp.
Bottom left	Courtesy Pratt and Whitney Group of United Technologies Corporation
Bottom right	Courtesy of Cincinnati Milacron

Acknowledgments for Color Inserts

Air Traffic Control Systems
All photos Courtesy of the F.A.A.

Cars and Trucks: The New Look
(1) Courtesy of Ford Motor Company
(2) Courtesy of Ford Motor Company
(3) Courtesy of General Motors Corporation
(4) Courtesy of Freuhauf Corporation

Computers Plus Design and Drafting Equal CADD
(1) Courtesy of Keuffel and Esser Company, Parsippany, NJ 07054
(2) Courtesy of Auto-trol Technology Corporation
(3) Reprinted with permission from Computervision Corporation
(4) Courtesy of Texas Instruments, Inc.

The Changing World of Construction
(1) Courtesy of the Turner Corporation
(2) Courtesy of Kajima Corporation
(3) Courtesy of Japan Railway Construction Public Corporation

Tapping Alternative Sources of Energy
(1) Courtesy of PG and E
(2) Courtesy of Southern California Edison
(3) Courtesy of Department of Energy

Industry: The New Look
(1) Courtesy of Ford Motor Company
(2) Courtesy of Cincinnati Milacron
(3) Courtesy of EMCO MAIER

Airplanes: The Biggest and the Fastest
(1) Courtesy of Lockheed—California Company
(2) Courtesy of Lockheed Corporation
(3) Courtesy of British Airways
(4) Courtesy of Boeing Company

The Navy's Nuclear Fleet
All photos Courtesy of the U.S. Navy

New Technology for Railroads
(1) Photo by Jeffrey Plant
(2) Courtesy of Santa Fe Railway
(3) Courtesy of Sperry Corporation
(4) Courtesy of General Motors Corporation

The Computer Revolution
(1) Courtesy of Apple Computer, Inc.
(2) Courtesy of Bank of America NT and SA
(3) Courtesy of AT and T Information Systems
(4) Courtesy of James Balog/Black Star

Computers Come in All Sizes
(1) Courtesy of Cray Research Inc.
(2) Courtesy of Prime Computer, Inc.
(3) Courtesy of Panasonic Industrial Co.
(4) Courtesy of Hewlett-Packard, Inc.

Communication Breakthroughs
(1) Courtesy of NASA
(2) Courtesy of Sperry Corporation
(3) Courtesy of Texas Instruments

High Technology Down on the Farm
(1) Courtesy of Sperry Corporation
(2) Courtesy of U.S. Dept. of Agriculture
(3) Courtesy of NASA
(4) Courtesy of U.S. Dept. of Agriculture

Solving Waste Disposal Problems
(1) Photo by Jonathan Plant
(2) Courtesy of Aluminum Corporation of America
(3) Courtesy of Sperry Corporation

Mines on the Moon
Both photos Courtesy of NASA

Space Stations
Photo Courtesy of Boeing Aerospace Company

NOTICE TO THE READER

INTRODUCTION

PEOPLE CREATE TECHNOLOGY

In this book you will learn about technology as a concept and about the many ways technology has enhanced the quality of life for people. You will also learn about the harmful effects of technology. However, before you begin, you must understand one important fact. Whether the effects of it are good or bad, *people create technology*.

Technological development is now and has always been the result of people trying to improve the quality of their lives or trying to extend their horizons. Therefore, the best way to begin a study of technology is to examine the profiles of the people who have been responsible for some of the most important technological developments.

There are many people, famous and unknown, who have made important contributions to technological development over the years. Eleven are profiled in this Introduction.

1. Thomas Edison
2. Alexander Graham Bell
3. Guglielmo Marconi
4. James Watt
5. Benjamin Franklin
6. Johannes Gutenberg
7. George Washington Carver
8. Orville and Wilbur Wright
9. Samuel F. B. Morse
10. Eli Whitney
11. Grace Murray Hopper

Timeline of technological developments.

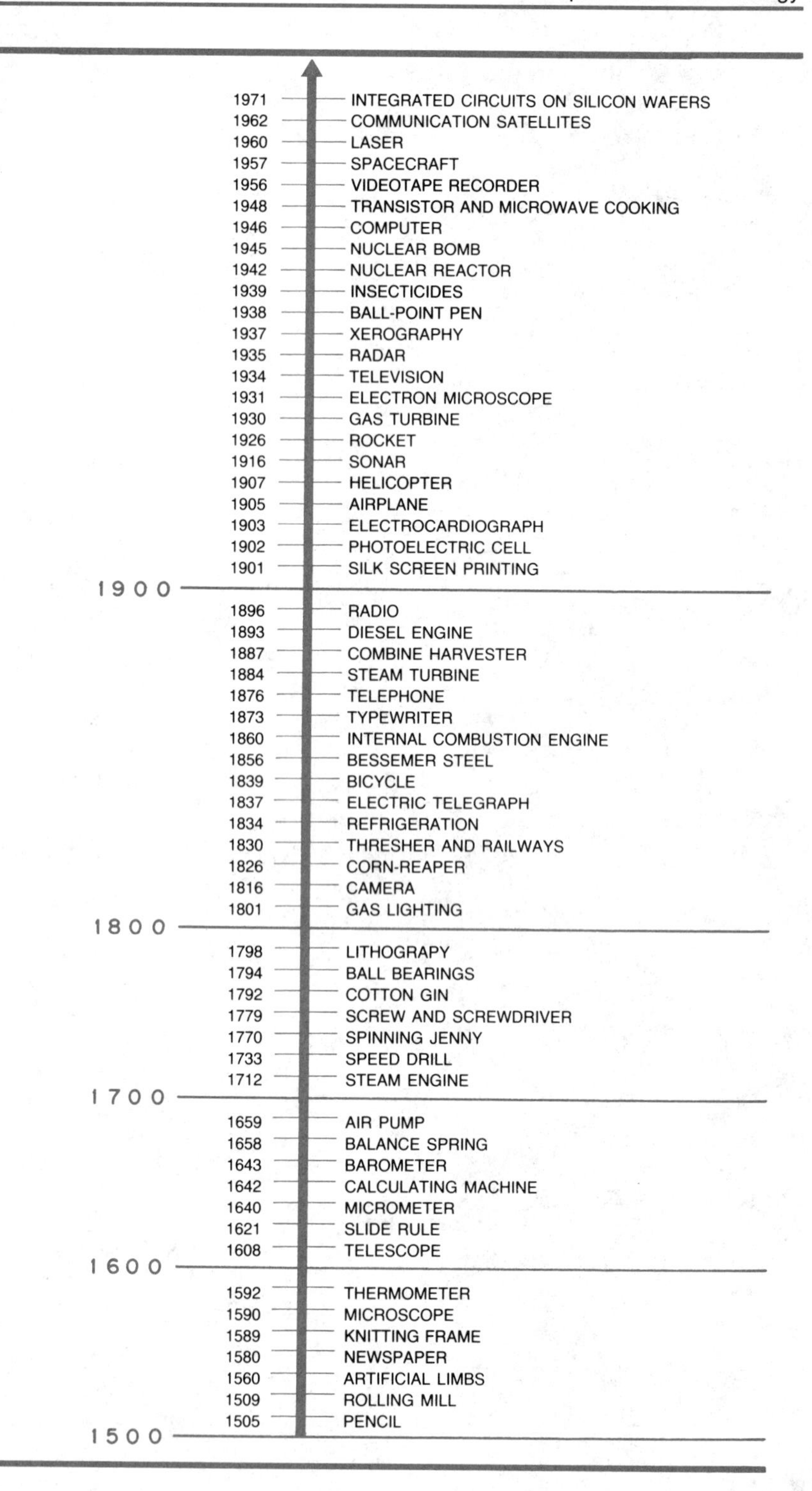
1971 INTEGRATED CIRCUITS ON SILICON WAFERS
1962 COMMUNICATION SATELLITES
1960 LASER
1957 SPACECRAFT
1956 VIDEOTAPE RECORDER
1948 TRANSISTOR AND MICROWAVE COOKING
1946 COMPUTER
1945 NUCLEAR BOMB
1942 NUCLEAR REACTOR
1939 INSECTICIDES
1938 BALL-POINT PEN
1937 XEROGRAPHY
1935 RADAR
1934 TELEVISION
1931 ELECTRON MICROSCOPE
1930 GAS TURBINE
1926 ROCKET
1916 SONAR
1907 HELICOPTER
1905 AIRPLANE
1903 ELECTROCARDIOGRAPH
1902 PHOTOELECTRIC CELL
1901 SILK SCREEN PRINTING
1900
1896 RADIO
1893 DIESEL ENGINE
1887 COMBINE HARVESTER
1884 STEAM TURBINE
1876 TELEPHONE
1873 TYPEWRITER
1860 INTERNAL COMBUSTION ENGINE
1856 BESSEMER STEEL
1839 BICYCLE
1837 ELECTRIC TELEGRAPH
1834 REFRIGERATION
1830 THRESHER AND RAILWAYS
1826 CORN-REAPER
1816 CAMERA
1801 GAS LIGHTING
1800
1798 LITHOGRAPY
1794 BALL BEARINGS
1792 COTTON GIN
1779 SCREW AND SCREWDRIVER
1770 SPINNING JENNY
1733 SPEED DRILL
1712 STEAM ENGINE
1700
1659 AIR PUMP
1658 BALANCE SPRING
1643 BAROMETER
1642 CALCULATING MACHINE
1640 MICROMETER
1621 SLIDE RULE
1608 TELESCOPE
1600
1592 THERMOMETER
1590 MICROSCOPE
1589 KNITTING FRAME
1580 NEWSPAPER
1560 ARTIFICIAL LIMBS
1509 ROLLING MILL
1505 PENCIL
1500

THOMAS EDISON

Thomas Edison was one of the most productive inventors in the history of the world. His most famous invention is the incandescent light bulb, invented in 1879. He also invented the phonograph. This 1877 invention was the forerunner of modern stereo systems. If you like to play records and tapes, you can thank Thomas Edison.

Edison was involved in the development of a number of other inventions, including the typewriter, telephone, electric generator, and motion pictures. In 1882, in New York City, he established the world's first electric power station. Edison's inventions represented major advances in energy/power and communications technology.

ALEXANDER GRAHAM BELL

Like Edison, Bell was a prolific inventor. His most famous invention is the telephone. He produced the world's first successful telephone in 1875. Bell's assistant, Thomas A. Watson, was the first person to hear the human voice transmitted by telephone.

In addition to the telephone, Bell made improvements to the phonograph that Thomas Edison had invented. He also developed special tools for use in medical and health care. Bell's inventions represented major advances in communication and medical/health care technology.

Thomas Edison. *Courtesy Historical Picture Service, Chicago.*

Alexander Graham Bell. *Courtesy Historical Picture Service, Chicago.*

GUGLIELMO MARCONI

Guglielmo Marconi is credited with developing one of the most important inventions in the history of communications technology—the radio. At the time of its invention, Marconi's device was known as a "wireless telegraph." It was not used in the way that our modern radios are used (i.e., for music, news, sports, and weather). Rather, it was used to send and receive coded messages.

These messages were sent through the air without the aid of telegraph wires. Instead the message was carried by radio waves. Marconi sent the first wireless broadcast in 1895. It was transmitted only about a mile. However, in 1901 he succeeded in sending a message across the Atlantic Ocean. Like the earlier telegraph, Marconi's "wireless" sent messages in code that had to be translated at the receiving end.

By improving on Marconi's device, other inventors were able to send actual voice transmissions rather than coded electrical impulses. The first voice transmission by radio took place in 1906. Marconi's wireless telegraph represented an important development in communications technology.

JAMES WATT

James Watt is credited with building the first practical steam engine. Several other inventors had developed their versions of the steamengine before Watt. However, they were all too cumbersome or inefficient to be of any practical use.

Guglielmo Marconi. *Courtesy Historical Picture Service, Chicago.*

James Watt. *Courtesy Historical Picture Service, Chicago.*

Watt made several improvements to the steam engine and made it a practical source of power for pumps, boats, trains, and electric generators. He so improved the system of gears that the speed of the steam engine was doubled. He also made changes to keep the steam hotter, which cut down on the amount of reheating needed. Watt's steam engine was a major development in transportation and energy/power technology.

BENJAMIN FRANKLIN

Benjamin Franklin is best known as one of this country's founding fathers. He was one of the authors of the Declaration of Independence. However, Franklin was also an inventor.

His most important discovery was that lightning is a form of electricity. With this discovery, he was able to invent the lightning rod to protect buildings and other structures from lightning. Franklin also invented bifocal glasses.

JOHANNES GUTENBERG

Johannes Gutenberg invented printing with movable type. The Gutenberg press was the first printing press to use movable type instead of hand-carved wooden plates. Before his invention, the words for each page of a book to be printed had to be carved into the face of a flat wooden plate. The plate then served as a stamp in making as many copies of the book as needed. This, of course, was not much better than copying the book by hand.

Benjamin Franklin. *Courtesy Historical Picture Service, Chicago.*

Johannes Gutenberg. *Courtesy Historical Picture Service, Chicago.*

Movable type was made up of blocks, each containing a backwards impression of one letter, number, or symbol. By putting the blocks together and arranging them as necessary, a "typesetter" could spell the words and make sentences and paragraphs as needed. The blocks could then be rearranged to form the words for another page.

The Gutenberg press revolutionized the printing industry. Because it made the printing of books easier and faster, movable type printing caused the use of books to increase rapidly. Gutenberg's invention was a major development in graphic communications technology.

GEORGE WASHINGTON CARVER

George Washington Carver spent his entire life working to improve agricultural technology. He is especially well known for his work with the peanut. During Carver's early life (the mid-1800s), the peanut was not a popular crop. But Carver knew it had a lot of potential. Because the peanut is a legume, it would help restore nutrients to wornout soil. It could also be used for many other things.

To convince skeptical farmers to plant peanuts, Carver first had to find enough uses for the peanut to make it a profitable cash crop. He did. Carver is credited with developing over 300 separate uses for the peanut. One is peanut butter! Carver's work with the peanut was instrumental in making the South more than a one-crop region. This is considered a major advance in agricultural technology.

ORVILLE AND WILBUR WRIGHT

Orville and Wilbur Wright are considered the fathers of modern aviation. In 1903 at Kitty Hawk, North Carolina, an airplane they designed and built became the first motorized manned aircraft to fly. Orville flew the airplane for a total of 12 seconds. Their work led to our modern system of air transportation. The Wright brothers' invention is considered by many to be the most important development in the history of air transportation technology.

SAMUEL F. B. MORSE

Samuel F. B. Morse is credited with inventing the telegraph in 1837. Morse's invention was a device that transmitted coded messages over telegraph wires. The messages were sent in the form of electrical impulses. Morse sent the first long-distance telegraph message in 1844. It went from Washington, D.C. to Baltimore, Maryland.

Morse also developed a code, which is named after him. Morse code is a system of dots and dashes that can be used to form letters and words. A dash was a long burst of electricity, and the dot was a short burst. Morse's invention is considered one of the most important developments in the history of communications technology.

George Washington Carver. *Courtesy Historical Picture Service, Chicago.*

Orville and Wilbur Wright. *Courtesy Historical Picture Service, Chicago.*

Samuel F. B. Morse. *Courtesy Historical Picture Service, Chicago.*

ELI WHITNEY

Eli Whitney is credited with developing two of the most important inventions in the history of agricultural and manufacturing technology: the cotton gin and the concept of interchangeable parts. The cotton gin (short for engine) was a machine that separated cotton seeds from balls of cotton. Until the cotton gin was invented, seeds had to be separated by hand. This process was slow and difficult and involved much hard labor. As a result, cotton was not a profitable crop, and few farmers planted it.

Whitney's invention solved this problem to such an extent that cotton eventually became the most important crop in the South. Along with rice and peanuts, cotton became a staple of the southern farm.

Whitney is also credited with developing the concept of interchangeable parts—one of the essential elements of mass production. Mass production involves manufacturing a large amount of product in a short time. Before mass production, all manufacturing was done on an individual basis. For example, if 100 rifles were needed, each rifle would be manufactured separately. No two of the rifles would be exactly alike. If a part wore out, it couldn't be replaced; it had to be remade especially for that rifle.

Whitney solved this problem with his concept of interchangeable parts. With this system, for example, if 100 rifles were needed, then 100 barrels, 100 stocks, 100 triggers, and so forth, would be manufactured. Each part would be exactly like the other 99 parts, or within an acceptable tolerance. After all of the interchangeable parts were manufactured, they could be assembled on an assembly line.

This concept revolutionized the manufacturing industry because it reduced the time and cost of manufacturing products and introduced the concept of spare parts. If a part wears out, it can be replaced by

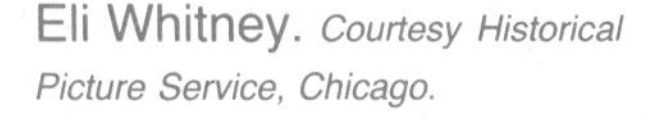

Eli Whitney. *Courtesy Historical Picture Service, Chicago.*

another part that is just like it. The cotton gin is considered one of the most important developments in the history of agricultural technology. The concept of interchangeable parts is considered one of the most important developments in the history of manufacturing technology.

GRACE MURRAY HOPPER

Grace Murray Hopper is now Admiral Hopper of the United States Navy. She is a true pioneer in the field of computers. In fact, she is one of the most successful women in the history of the computer field. Along with several others, she is responsible for the development of the COBOL computer language, one of the most widely used languages.

In 1959, the Department of Defense held a conference on data systems languages. As part of this conference a special committee, called the CODASYL, was formed. The purpose of the committee was to develop a new computer language for business applications. Grace Hopper was an important member of this committee.

The final result of the CODASYL committee's work was a new computer language called COBOL. COBOL is the acronym for *CO*mmon *B*usiness *O*riented *L*anguage. It is still one of the most widely used computer languages.

Grace Murray Hopper.
Courtesy of Sperry Corporation.

Many other important people could be profiled here. Others who are unknown are just as important or even more important. For example, who invented fire or the wheel? If these people were known, they would surely rank in importance with the people we have mentioned. However, the purpose of this introduction is not to list everyone who has made important contributions to technological development. Its purpose is to illustrate the point that, for better or worse, *people create technology*.

BOEING 767
767
N767BA

ASEA
20

PART ONE

OVERVIEW OF TECHNOLOGY

CHAPTER 1

WHAT IS TECHNOLOGY?

OBJECTIVES

After completing this chapter, you should be able to

- define technology.
- apply criteria in deciding if something is an example of technology.
- explain how technology causes change.
- explain the ways in which technology affects people.

Have you ever heard someone say, "This is the age of technology"? It is easy to see what they mean. Evidence of technology is all around us. In your house you have furniture, appliances, food, clothing, a television, a telephone, a stereo, and maybe a VCR. Your parents probably have a car or a truck. Maybe you have a bicycle. All of these things are products of technology.

DEFINITION OF TECHNOLOGY

How would you answer the question, "What is technology?" Do you know how to define this concept? It's not an easy thing to do. Many people use the term "technology" in many different ways. One way you can learn to understand this concept is by comparing it with a cake. Both a cake and technology are made up of separate ingredients. Each ingredient has its own identity, but when mixed together they form something else.

What are the ingredients needed to make a cake? You need flour, butter, eggs, sugar, and milk. Each of these ingredients has its own identity. But when they are mixed together in the right amounts and baked, they become a cake. What are the ingredients needed to make technology? You need tools, resources, processes, and people. Each of these ingredients also has its own identity. But when taken together to solve a problem or to extend our capabilities, they become technology.

Now you know the ingredients of technology: tools, resources, processes, and people. You also know the purposes of technology: to solve problems or to extend our capabilities. Let's use these things we know to see if we can define technology.

Technology is people using tools, resources, and processes to solve problems or to extend their capabilities. What do you think of this definition? Did yours agree with ours? We will talk more about tools, resources, and processes in Chapter 2.

For now, using this definition, let's look at a test case and decide if it is an example of technology.

BAKING BREAD

When Mr. Smith, the baker, and his assistant make a loaf of bread, is that an example of technology? To decide, let's test the question against our definition.

STEP 1 Are people involved? Yes, the baker, Mr. Smith, and his assistant are people.

STEP 2 Are tools involved? Yes, Mr. Smith and his assistant use such tools as measuring cups, mixing bowls, spoons, a baking pan, and an oven.

STEP 3 Are resources involved? Yes, Mr. Smith relies on heat as a resource and such materials as flour, water, eggs, salt, and baking soda.

STEP 4 Are processes involved? Yes, mixing and baking are both processes.

STEP 5 Does baking bread solve a problem or extend our capabilities? Yes, baking bread helps solve the problem of hunger for the family that buys the bread.

In all five steps of the test we were able to answer "yes," so baking bread is an example of technology.

Using this definition and this test, you can decide if anything is an example of technology. Try a few more by yourself now. Is building a house an example of technology? Is pulling a tooth an example of technology? Is riding a bicycle an example of technology? The answer is "yes" to all three questions. Do you see why?

Can you think of some examples that are ***not*** technology? Is a river

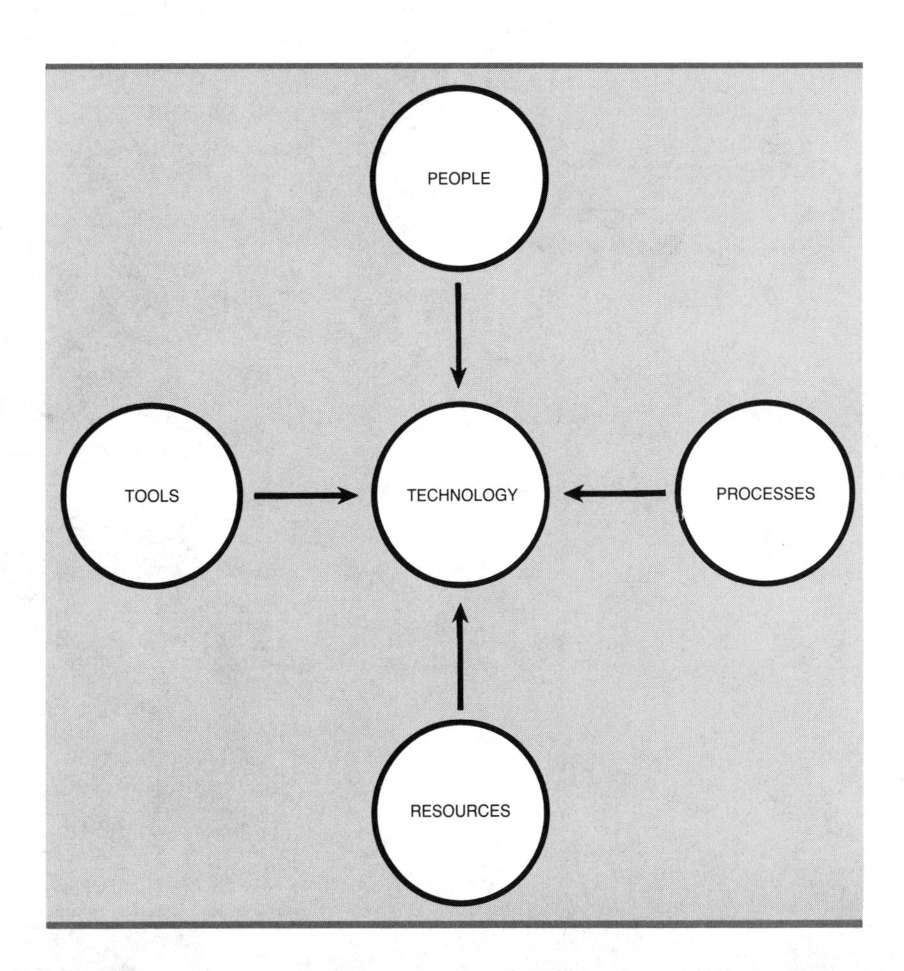

The elements of technology.

technology? Is a tree technology? Is a lake technology? By using the definition and test we used earlier, you would decide that these are not examples of technology. Rivers, trees, and lakes are examples of nature, not of technology. However, think carefully before answering no. The river might have been made by hand or by machine. The trees might have been grown scientifically by using special fertilizers and chemicals. The lake might have been created when a river was damned. If each of these were the case, the river, tree, and lake would be examples of technology.

We now know the definition of technology, and we have looked at some examples of technology. One more thing we can do to help understand technology even better is look at the history of technology.

We often hear the term "high technology." In fact, it has become a buzzword. A *buzzword* is a word that becomes a fad. Actually "high technology" is an unfortunate term because, like so many buzzwords, it doesn't have a clear, universal definition.

We hear the term used in so many different ways that it can be difficult to pinpoint what the user means. However, it seems that what most people mean is technology that relies in a fundamental way on state-of-the-art electronics. Of course, what is state-of-the-art changes constantly. Right now it refers to technological developments such as

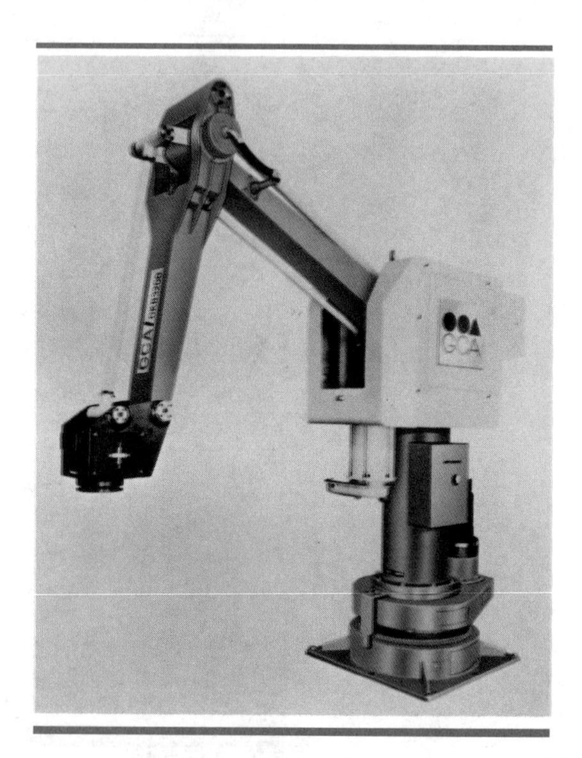

Industrial robots are an example of high technology. *Courtesy of GCA.*

The personal computer is an example of high technology. *Courtesy of Radio Shack, A Tandy Corporation.*

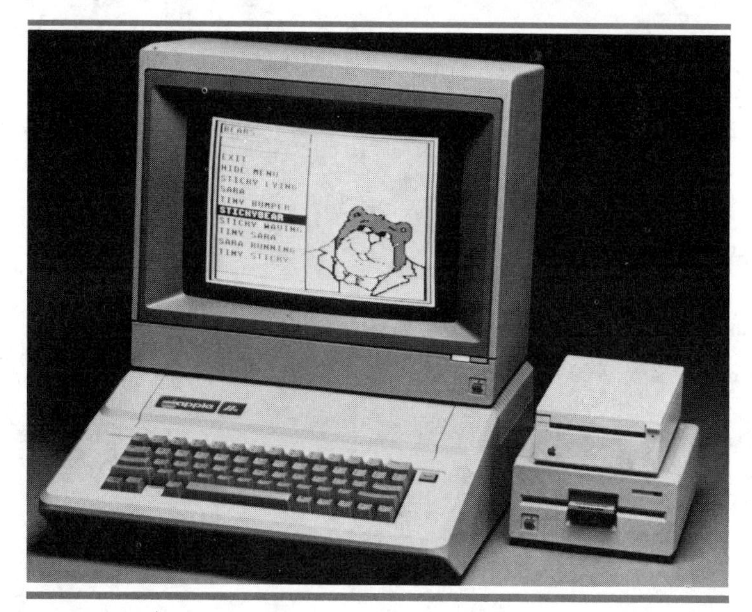

High-technology devices such as the computer are used on the job, in the home, and even in the classroom. *Courtesy of Apple Computers.*

integrated circuits, computers, robots, lasers, computer-aided design and drafting (CADD), computer-assisted manufacturing (CAM), and telecommunication devices.

A better term would be "advanced technology." This term would clearly refer to the most advanced developments in all technology areas, including construction, manufacturing, graphic communications, health care and medicine, transportation, energy and power, electronic communication, and agricultural technology.

HISTORY OF TECHNOLOGY

Technology has been around ever since humans first picked up rocks or sticks to kill wild game or protect themselves. Can you apply our test to this example? If a stick was used to kill an animal, the stick was a tool, the tree it came from was a resource, swinging the stick and striking the animal was a process, and hunger was a problem. It works!

In the earliest times, there weren't many examples of technology. But as time passed, the need to solve problems or to extend **capabilities** caused people to rely more and more on technology. Technology grew so much over the years that now we are surrounded by it and we are dependent on it.

Cars can be traced to the invention of the wheel and axle thousands of years ago. NASA's space shuttle can be traced to the Wright brothers' flight at Kitty Hawk. Solar energy systems can be traced to primitive people warming themselves in the sun. The newspaper, magazine, television, and telephone can be traced to the crude drawings found on the walls of caves. Our houses can be traced to the most primitive thatched huts.

Try to imagine how life would be without all of the different examples of technology that we have come to depend on. How do you get to school in the morning? Do you ride a bus, a car, or a bike? Do you live in a house, an apartment, or a condominium? Do you like to watch television, play video games, listen to the radio, or play records

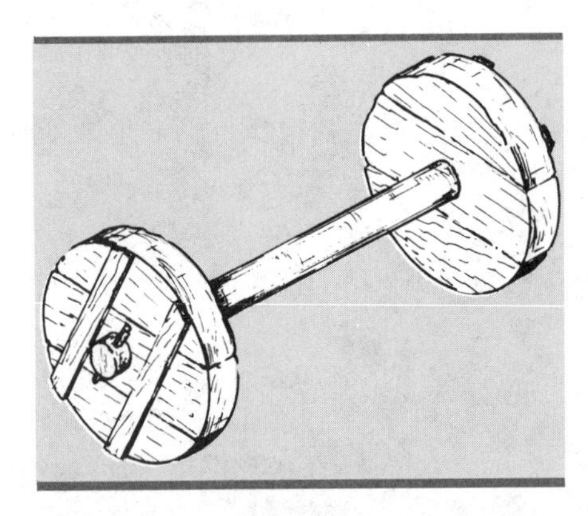

The invention of the wheel and axle led to the development of modern land transportation.

CARS AND TRUCKS: THE NEW LOOK

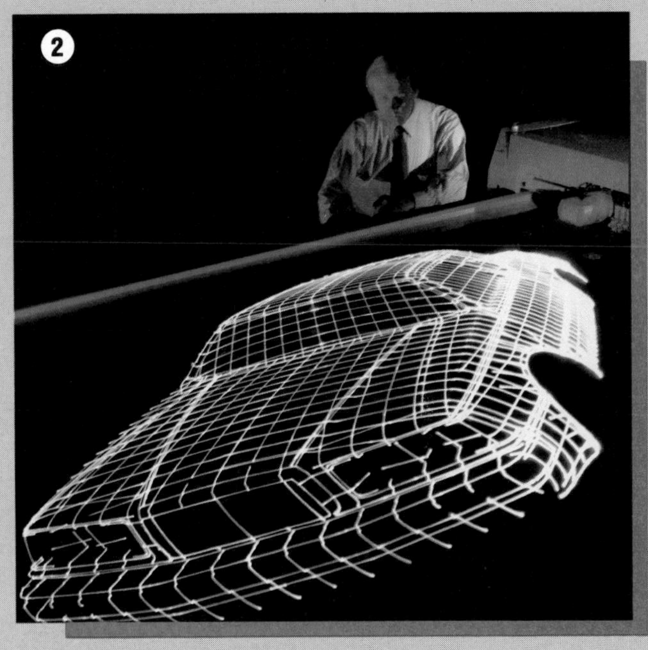

Engineers and technologists are constantly working to make our vehicles safer, more comfortable and more fuel efficient. Extensive research goes into each new design. (1) Full-sized clay models are often made to test a new design. Highly skilled workers are needed to create this kind of model from drawings and plans. (2) Computer-aided drafting and design (CADD) systems are also used extensively in today's auto industry. The results of this new technology are vehicles like (3) the Aero 2000 and (4) the FEV 2000. These represent the state of the art in mileage, good handling and comfort for drivers.

COMPUTERS PLUS DESIGN AND DRAFTING EQUAL CADD

(1) Not long ago designs for new products were drawn up by hand. While the skills of technical drawing and drafting are still very important, more and more design work is being done with computers using CADD systems. (2) Notice the three-dimensional image on the screen of this compact CADD system. (3) Complex electronic circuitry can be designed using CADD technology. (4) These students are using a CADD system at their school.

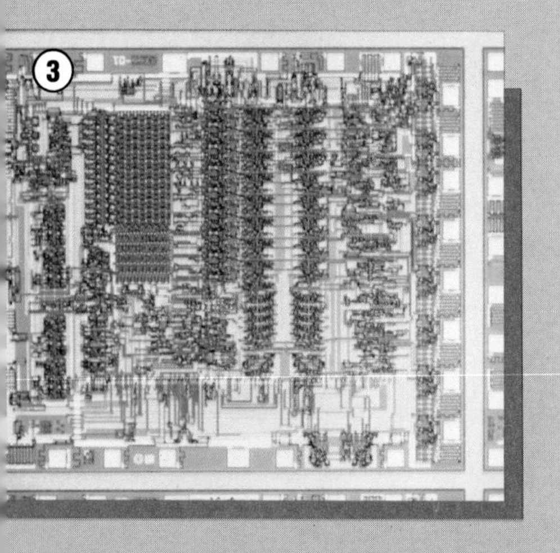

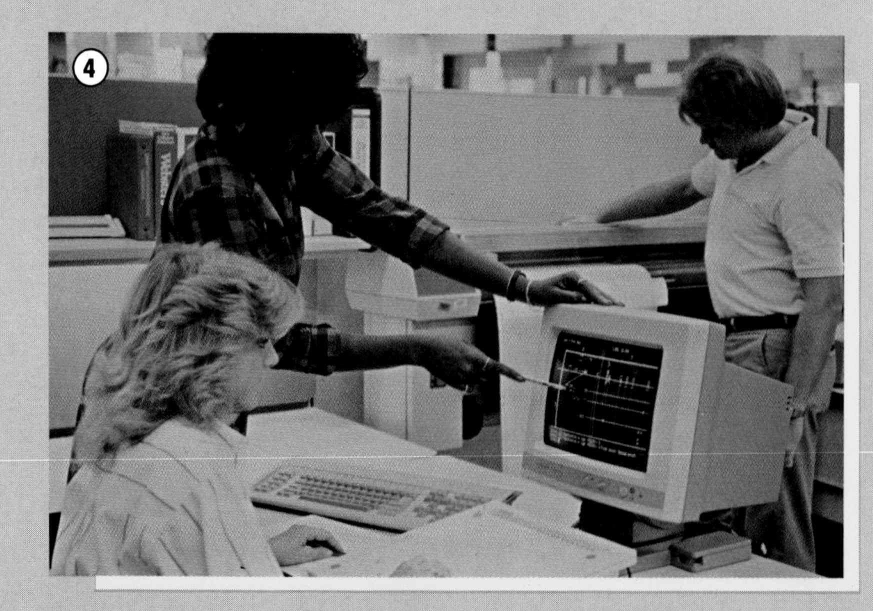

This new concept car uses satellite navigation. *Courtesy of Chrysler Corporation.*

NASA's space shuttle. *Courtesy of Spizzirri, Inc.*

School buses transport thousands of students to school.
Photo by Deborah M. Goetsch.

and tapes? Do you like to read? What is your favorite food? All of these questions concern technologies that can be traced back to long ago.

It is easy to see that technology is always changing. Technology causes change, and it is caused by change. You should know about change and what technology has to do with it.

TECHNOLOGY AND CHANGE

Technology never stands still. It is always changing. Sometimes it changes quickly, and sometimes it changes slowly, but it always changes. First we have a problem or a need, so we develop technology to solve the problem. The new technology causes **change**. Once the change has occurred, the whole cycle starts over again. Let's look at an example of this cycle using the field of **transportation**.

Walking was the first means of transportation. To be able to travel farther and faster, we learned how to ride horses. The switch to horses caused change. People could go places they couldn't go and do things they couldn't do before. However, people still wanted to go faster and farther than horses could go, so the automobile and the train were invented. These inventions caused even more changes. But after getting used to traveling by car and train, people still wanted to go farther and faster, which led to the invention of the airplane.

Change is part of a three-step cycle.

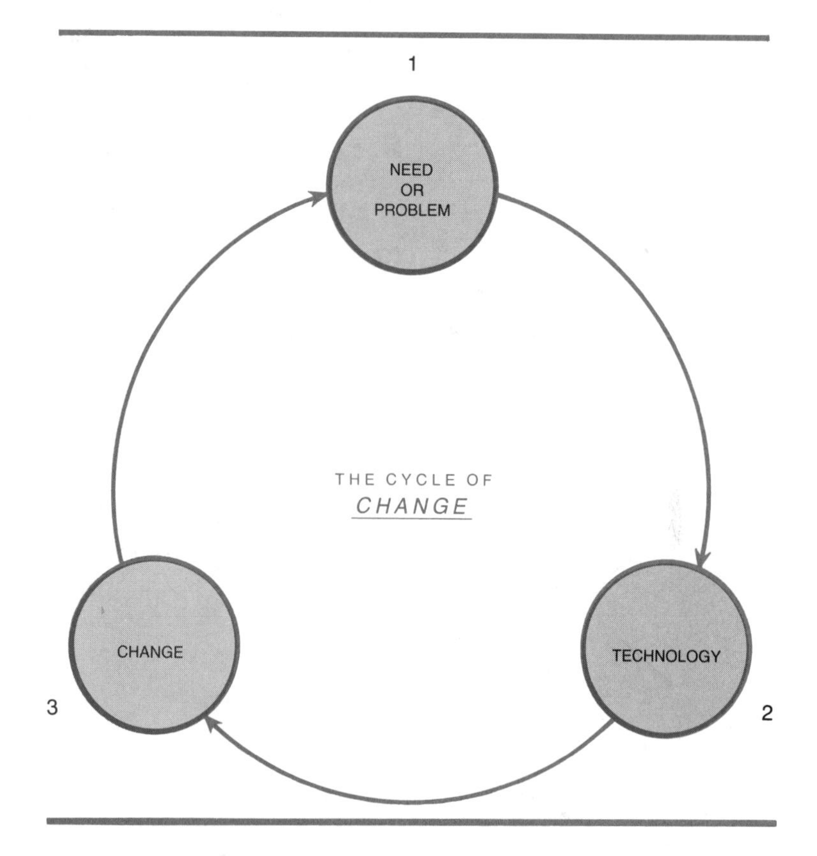

Early cars such as this 1905 Oldsmobile Model B Standard Runabout were a big improvement over horses. *Courtesy of Spizzirri, Inc.*

With jet airliners, we can fly around the world in just hours—New York to London takes just 7 hours! The airplane has caused even more changes. But wouldn't you know it: people still want to go farther and faster. So now we have sent men to the moon and are flying through space in NASA's space shuttle. Looking at changes is a good way to understand the changing nature of technology. Let's look at some other examples of how technology has changed over the years.

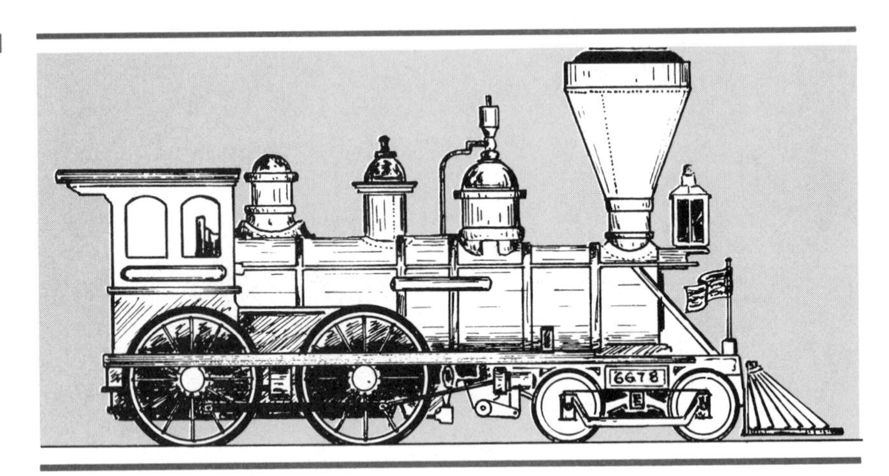

The train allowed people to travel farther and faster than horses.

The Boeing 747 jumbo jet is one of the most widely used of modern commercial airliners.
Courtesy of Spizzirri, Inc.

Some Changes in Communications Technology

In the earliest times, people communicated by using hand signs and making grunting noises. Try this with a friend. Try to talk to a friend by using only hand signs. How would you say, "Let's go to a movie," without using words? It's not easy, is it? This is why languages were invented. It is also why people around the world speak different languages. Talking is called spoken **communication**.

At the same time people were learning to talk to each other, they were also learning to communicate graphically. Graphic communication means communicating by writing or by drawing pictures or doing both. This book is a form of graphic communication because it uses words and pictures to communicate with you.

When people began to talk to each other, they found that they sometimes needed to talk with someone who was not in hearing distance. This was the beginning of the problem of talking across distance. They found that by yelling one person could communicate with another a short distance away. But what if you needed the help of a neighbor who was a mile away?

At first, the problem of communicating across distance was solved in different ways. You've probably heard that the American Indians used smoke signals to communicate across distance. Signal flags and runners were used in the military. Written letters sent by pony express were another early way of communicating across distance.

Smoke signals and flags were hard to understand. Runners and the pony express were slow and unreliable. People wanted to be able to communicate across distance fast and accurately. This need led to the invention of the telegraph. Using a telegraph, people could send coded messages over long distances. The code of dots and dashes was invented by Samuel F. B. Morse and has come to be called *Morse code*.

Telegraph communication was much better than signal flags and runners, but there were problems with it too. The person sending the message and the person receiving it had to understand the code. Before the message could be sent, it had to be translated into code. Before it could be read at the other end, it had to be translated back into words.

Soon people began to feel the need to transmit the human voice across distance. This need led to the invention of the telephone. Now, thanks to communications satellites and special fiber-optic cables, we can talk to people overseas as if they were right next door.

The videophone allows people to view information on the screen and talk at the same time. *Courtesy of A T & T.*

Graphic communication has also undergone rapid change over the years. It began as crude drawings on the walls of caves. People then developed written languages using words made of letters of an alphabet. Have you ever seen any alphabet other than ours? There are lots of them.

Graphic communication was also changed by the continual development of better materials. Writing materials went from stone tablets, to quills and leather parchment, to pencils and pens on paper. Think of all the different writing materials we have now: pencils, fountain pens, ballpoint pens, felt-tip pens, chalk, magic markers, crayons, and many different types and colors of paper.

About 1450, mass communication came into being with the invention of the Gutenberg press. Until then, books had to be copied by hand or carved into wooden blocks, as we mentioned earlier. Suppose someone wanted a copy of one of your books and you had to sit down with a pen and a stack of paper and copy every page in longhand. You wouldn't give away many copies of your book would you? This is how things were until the Gutenberg press was invented.

The press allowed books to be printed one page at a time. All of the words for an entire page were arranged on the press with movable letters. This process came to be known as *typesetting*. The letters on the press were coated with ink and pressed against the paper, so one entire page could be printed. If 10 copies of a book were needed, the pressman would make 10 copies of page 1, 10 of page 2, 10 of page 3, and so on. Then the pages would be sorted and bound together to make 10 books.

The *Voyager* satellite was launched into space to collect information and collect photographs of the planet Jupiter. *Courtesy of Spizzirri, Inc.*

Over the years the press and various other printing processes were continually changed as people demanded faster, better results. Think of the various forms of mass communication we have today, and try to imagine how it would be to live without them. We have television and radio for transmitting the spoken word. We have computers for receiving, storing, processing, and sending the written word. And we

Computers are used in many different ways, including graphic communication. *Courtesy of Compugraphic Corporation.*

have copiers that can copy, reduce, and enlarge almost any document in a matter of seconds.

Other examples of how technology has continually changed over the years can be found in the area of health care. Let's take a look at some of the changes in health care over the years.

Some Changes in Health Care Technology

In the earliest of times there was very little **health care**. Much of what existed was based on superstition. In those days, any serious injury or disease usually meant death. During the fourteenth century, the black plague killed millions of people. At the time, doctors didn't even know what caused it. When they finally learned that the plague was caused by fleas who jumped from infected rats to people, they were able to bring the disease under control.

Over the years health care technology has progressed from an age of ignorance and superstition, in which sick people were cut and bled to rid their bodies of supposed poisons and evil spirits, to an age of sophistication. Today an artificial heart can be substituted for the real thing, and delicate microsurgery can be performed with a laser beam.

Have you been in a hospital lately? Have you noticed all of the electronic equipment there? A nurse may take your temperature with a digital thermometer. Because of advances in health care, killing and disabling diseases such as diphtheria, typhoid, rickets, scarlet fever, polio, and tuberculosis are almost unheard of today. In less than 500 years, the average life span of people in advanced countries has increased from under 40 years to over 70 years.

Another way to view the changes in technology that have occurred over the years is to look at the changes that have taken place in tools, resources, and processes.

Some Changes in Tools

It is easy to see the changes that have taken place in **tools** over the years. Farm tools make a good example. People have always relied on tools for planting and harvesting crops. An early planting tool was the stick. The stick was used to make holes in the ground into which seeds were put. Next came the wooden plow, then the stone-edged plow, and then the metal plow. Farmers now use complex metal plows and disks pulled by tractors.

The earliest harvesting tool was the human hand. The problems with this tool are obvious. They led to the invention of the sickle. The sickle was a half-moon-shaped metal blade with a wooden handle. It was a vast improvement over the human hand, but it still required people to stoop over all day while harvesting crops. This problem led to the invention of the scythe. The scythe was a long, slightly curved metal blade with a very long wooden handle. It allowed a person to harvest grain crops without stooping and bending.

One of the most important inventions for harvesting crops was Cyrus McCormick's reaping machine. Pulled by animals, the reaper could harvest in hours what used to take days. This invention led to the many mechanized farm machines we have today.

Some Changes in Resources

There are many different types of **resources**. Two that make good examples of the changes that have taken place over the years are materials and **energy/power**. The earliest tools were made of stone or wood. Then metal was discovered, first bronze and then iron. Then it was learned that iron could be mixed with other materials to make

Crude early tools led to modern power tools such as this Shop Smith. *Courtesy of Shop Smith.*

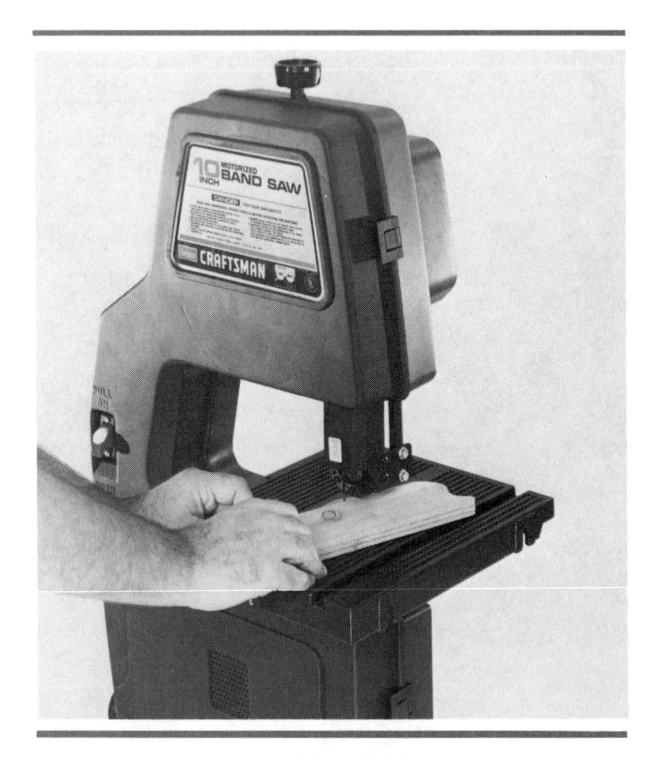

This 10-inch motorized bandsaw is a modern version of the crude saws of earlier days. *Courtesy of Sears.*

Steel is one of the most widely used building materials. *Courtesy of Bethlehem Steel Corporation.*

steel, an even stronger metal. Now we have many different grades of steel and many other mixes of metals, called *alloys*, that are used every day.

In addition to steel, we have developed materials such as concrete and plastic. Concrete is a mix of cement, water, sand, and rocks. By adding steel bars, strands, or mesh to concrete, we can strengthen it so that it can be used on jobs that normally require steel.

Not very long ago, no one had even heard of plastic. Now it is one of the most frequently used materials. How many things can you name that are made out of plastic?

There have also been many changes in the energy/power resources we use in our lives. At first, people used wood to make heat and light. Then they used coal, then whale oil, and then fossil fuels such as oil and natural gas. Now look at all of the different sources of energy/power we use. In addition to fossil fuels, we use solar energy, nuclear energy, geothermal energy, and even energy produced by burning garbage.

Some Changes in Processes

A **process** is a set of operations or procedures that people use to do a job. A process can be very simple or extremely complex. Cooking is a process that converts raw food into edible food. Drilling is a process that creates holes. There are so many different processes used by

The diesel engine converts fuel energy into power. *Courtesy of Yanmar Diesel Engine Co., LTD.*

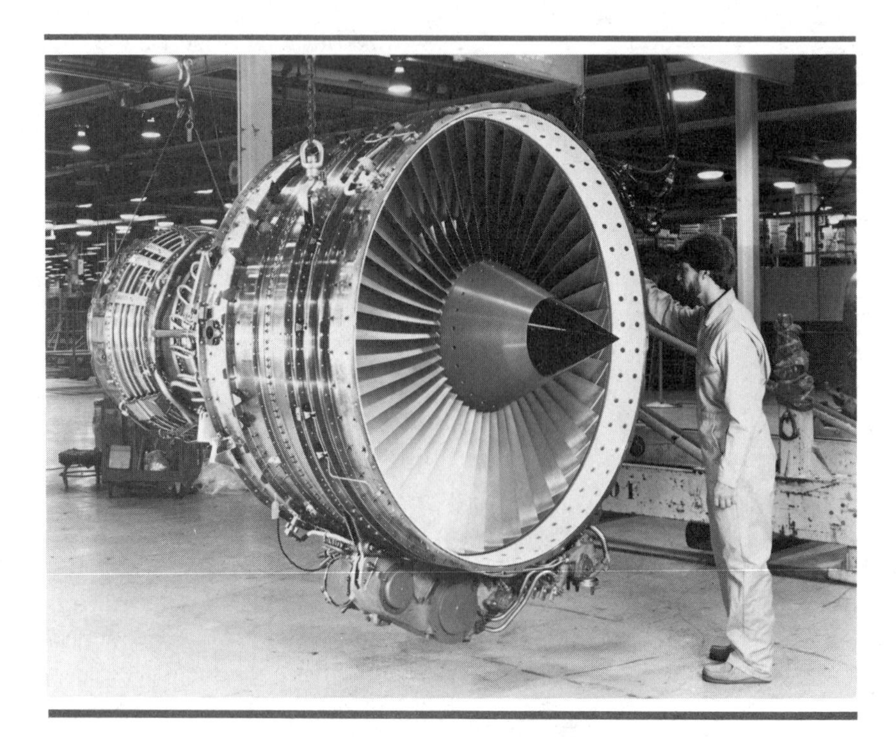

The jet engine converts the energy of jet fuel into power. *Courtesy of General Electric.*

people that it is necessary to select an example to illustrate the changes that have occurred. Let's look at machining as a manufacturing process and see how it has changed over the years.

In the early days of machining, the raw stock was loaded onto the machine by a person. That person operated the machine and did the desired task, which might have been drilling, turning, milling, or some other task. Then the same person unloaded the machine and repeated the process until the desired number of workpieces had been made.

Through the years, machines and materials being machined changed a lot, but the process changed very little. However, in modern machine shops this process has changed. In the most modern shops, raw stock might be loaded onto a machine by a robot. The machining operation might be controlled by a computer, and the machine might be unloaded by a robot. The workpiece, in some cases, is completed without ever being touched by human hands.

Can you think of other processes that have changed over the years? How about the way we milk a cow? Have you ever seen a milking machine? How about the way we wash a car? Have you ever ridden in a car as it went through a car wash?

Now that you know something about tools, materials, and processes, you need to learn more about the most important ingredient in technology—people.

In modern manufacturing plants, robots do work that people used to do. *Courtesy of General Motors.*

PEOPLE AND TECHNOLOGY

You already know that technology helps us solve our problems or extend our capabilities. These are the planned purposes of technology. There are also some unplanned effects of technology.

Thanks to technology, people live longer now than they used to. In the United States, the most advanced country in terms of technology, the life expectancy of a person is 70 years. In some of the less developed countries of the world, the life expectancy is less than 40 years. People live longer in the United States and other advanced countries because through technology the food, clothing, shelter, transportation, and health care are better.

Thanks to technology people in advanced countries have more **leisure** time. Did you know that not all countries have weekends? Some countries don't have 8-hour workdays either. In some countries people work from sunup to sunset seven days a week. The reason is that without technology they have to rely on people to do the work, and people can't always work as fast as machines. But in advanced countries technology helps us do our work faster and, in some cases, does the work for us.

Let's look at an example. Suppose you wrote a 10-page report and gave it to your teacher, who liked it so much that she asked to have five more copies of it by tomorrow. It will take you all night to copy the report five times by hand. But, thanks to technology, all you have to do is find a copying machine and make five copies. The whole task will take less than 15 minutes. In almost no time you can be finished and still have plenty of time left to spend with your friends after school.

But a student in an underdeveloped country isn't as lucky. He doesn't have a copying machine. So after school he has to sit down with a pencil and a stack of paper and spend the entire evening copying the report five times. Unlike you, he won't have any leisure time.

We even depend on technology during our leisure time. People in advanced countries spend much of their leisure time watching television, listening to the radio or tapes, playing video games, or playing sports. All of these activities involve technology. What are some other things you like to do? Do they involve technology? Bicycling, three-wheeling, roller skating, dancing, ice skating, model building, reading, football, baseball, basketball, and soccer all involve technology. Do you see why?

Technology has made working much easier for us. We have already seen this in the previous sections. Let's see if we can think of some other ways that technology makes work easier. How would house painters do their job without a ladder? How would secretaries do their job without a typewriter or word processor? How would carpenters do their job without hammers and saws? How would plumbers do their job without wrenches? How would you rake the yard without a rake?

So far we have discussed the good aspects of technology, and there are many. But there is also a bad side to technology. The same technology that brought us modern factories, automobiles, and jet airplanes also brought us **pollution**. In some large cities that have a lot of cars, the air sometimes becomes dangerous to breathe. This kind of air pollution is called *smog*.

In some cases, robots that have made manufacturing plants more productive have, at least temporarily, put people out of work. The automobile has made us very mobile, but automobile accidents kill thousands of people each year. Insecticides and chemicals have improved farm **production**, but they have also caused air, water, and ground pollution. Technology allows countries to protect themselves from foreign invaders, but it has also given several countries the ability to destroy the entire world in minutes.

You can see from these examples that technology has had a very good effect on the world, but if used in the wrong ways it can also have very bad effects. It is up to people to make sure that technology is used for good. It is one of the main reasons for reading this book. When you have finished this book, you should have a better understanding of technology and of how to use it for good.

SUMMARY

Technology is people using tools, resources, and processes to solve problems or extend their capabilities. Technology has existed as long as people have existed. Technology is always changing. It never stands still.

In the field of transportation, we have land, water, and air travel. Land travel changed over the years from walking to riding animals to animals pulling wagons and coaches to driving automobiles. Water travel grew from swimming to riding rafts and canoes to sailing vessels to steam-powered ships to diesel-powered ships to the nuclear-powered vessels of today. Air transportation changed more rapidly than any other form of travel. Beginning with lighter-than-air balloons, it went to dirigibles, propeller airplanes, jet aircraft, and, finally, to spacecraft such as NASA's space shuttle.

In the field of communication, technology has changed from hand signs, grunting, and crude drawings on cave walls to television, telephone, and computer.

In the field of health care, technology has changed from an age of superstition to an age of heart transplants and laser surgery. The tools, resources, and processes used by people have also changed continually over the years.

The effects of technology on people can be both good and bad. Some of the good effects are longer, healthier lives, easier work, and more leisure time. Some of the bad effects are pollution, accidental deaths, and the potential for world destruction. One of the main reasons for studying technology is to learn how to make sure that its effects on people will be good.

KEY WORDS

Technology
Tools
Processes
Transportation
Health care
Energy/power
Leisure
Change
Resources
Capabilities
Communication
Production
Pollution

SOME QUESTIONS TO TRY

1. What is technology?
2. What are the four ingredients of technology?
3. Is a bicycle an example of technology? Explain why.
4. Give five examples of technology in your daily life.
5. How does technology cause change?
6. Trace the history of land transportation from the earliest times to now.
7. Trace the history of water transportation from the earliest times to now.
8. Trace the history of air transportation from the earliest times to now.
9. Trace the history of communication from the earliest times to now.
10. Give three examples of the good effects of technology.
11. Give three examples of the bad effects of technology.
12. Explain why it is important to study technology.

TECHNOLOGY LEARNING ACTIVITIES

1. Make a list of every type of transportation technology you have ever used.
2. Make a list of every type of communications technology you have ever used.
3. Visit a museum and look at the changes in tools that have taken place over the years.
4. Go to the library and write a report on one of the following topics:
   - History of air travel
   - History of land travel
   - History of water travel
5. Select one of the following technology areas and make a chart that shows the most important inventions in it from the earliest times until now:
   - Transportation technology
   - Construction technology
   - Agricultural technology
   - Health care technology
   - Manufacturing technology

CHAPTER 2

ELEMENTS OF TECHNOLOGY

OBJECTIVES

After completing this chapter, you should be able to

- define tools, mechanisms, and machines.
- explain how people use tools, mechanisms, and machines.
- explain what resources are.
- explain how people use resources.
- explain what processes are.
- explain how people use processes.

In Chapter 1 you learned that three of the most important ingredients of technology are tools, resources, and processes. From now on, instead of calling these things ingredients, we will call them elements. When you see or hear the term elements of technology, think of three things: tools, resources, and processes.

TOOLS

What do you think of when you hear the word "tools"? Most people think of hammers, saws, screwdrivers, and wrenches. You probably don't think of pulleys, gears, wheels, or computers. However, in the language of technology, all of these things are tools.

In the language of technology, we sometimes use the word "tools" in a general way to mean tools, mechanisms, or machines. It is like using the word "fruit" in a general way to mean apples, oranges, and bananas. You need to know about tools, mechanisms, machines, and systems and how they are used.

Tools

We use tools often. In fact, we use tools so much that we don't even think about it. We just take them for granted—that is, until we need a tool and can't find it. That is when we realize how important tools are.

We differ in many ways from animals. One important way is in our ability to make and use tools. If you are playing baseball and the ball lands on the roof of your house, you know how to use the ladder to get the ball down. The ladder is a tool to help extend our reach. Most animals wouldn't know how to use a ladder.

Let's see if we can define "tool." It will help to begin with what we know. We know that a hammer, saw, and wrench are tools. A hammer drives nails, a saw cuts lumber, and a wrench turns nuts and bolts. Each tool has a different task, but they really all serve the same purpose. They help people do things they couldn't do with their bare hands. This is the key to our definition.

A **tool** is a device or object that allows people to perform tasks they

People use tools such as this modular tool pouch system so much that they take tools for granted. *Courtesy of Sears.*

Imagine trying to cut lumber without modern tools such as these saw blades. *Courtesy of Delta Machinery.*

couldn't perform with their bare hands. Tools increase our ability to do work. Try to imagine driving a nail without a hammer, cutting a board without a saw, or loosening a rusty bolt without a wrench. Now that you know what a tool is, let's look at how tools work.

How Tools Work All tools work differently, of course. You don't use a saw in the same way that you use a hammer. However, in certain basic ways, there are similarities in the way that all tools work. These similarities come from the scientific **principles** that the tools are based on.

Most tools are based on the principle of **leverage**. Have you ever heard of a lever? A lever looks something like a seesaw. It has four elements: the *lever*, *fulcrum*, *load*, and *force*. A lever helps people lift more weight than they normally could. A good example of a tool that is based on leverage is a bumper jack for a car. Without a jack, how would you lift a car to change the tire? It would take a very strong

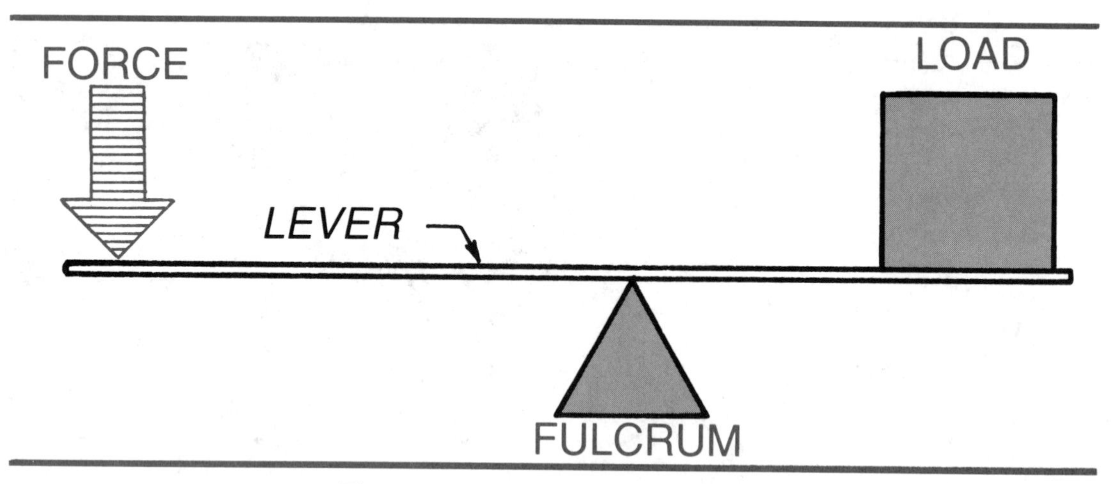

The principal of leverage.

person to lift a car without a jack. But even a small person can lift one with a jack, because the jack gives the person leverage.

Many of the tools we use are also based on leverage. Hammers, wrenches, balance scales, and clawed nail pullers are all based on leverage. Do you see why? With a hammer, the lever is the handle, the fulcrum is your hand, the load is the head of the hammer, and the force is you. You need to know two other scientific principles that tools are based on: rotation and inclination. They sound complicated, but they aren't. Both are spinoffs of leverage.

The principle of **rotation** comes from spinning the lever around the fulcrum. A wheel turning on an axle demonstrates the principle of rotation. Since it is much easier to roll something than to pull or drag it, the wheel is an important tool. The wheel is a tool based on the principle of rotation.

What other tools are based on rotation? Can you name any? In the shop there is the drill. Painters use the paint roller, and bakers use the rolling pin. These tools operate on the principle of rotation.

The principle of **inclination** comes from leaning the lever on the fulcrum to form a wedge. Have you ever ridden your bicycle up a ramp or seen a car in the movies drive up a ramp and jump over some other cars? The ramp is a wedge. Many tools are based on the principle of the wedge (inclination). The saw, planer, and paint scraper are all based on the wedge. If you look closely at the teeth on a saw, you will see that they have wedge-shaped edges. The blade on a planer and a paint scraper are also wedge shaped. Can you make any other tools that are based on the wedge? How about a can opener or a carpet knife?

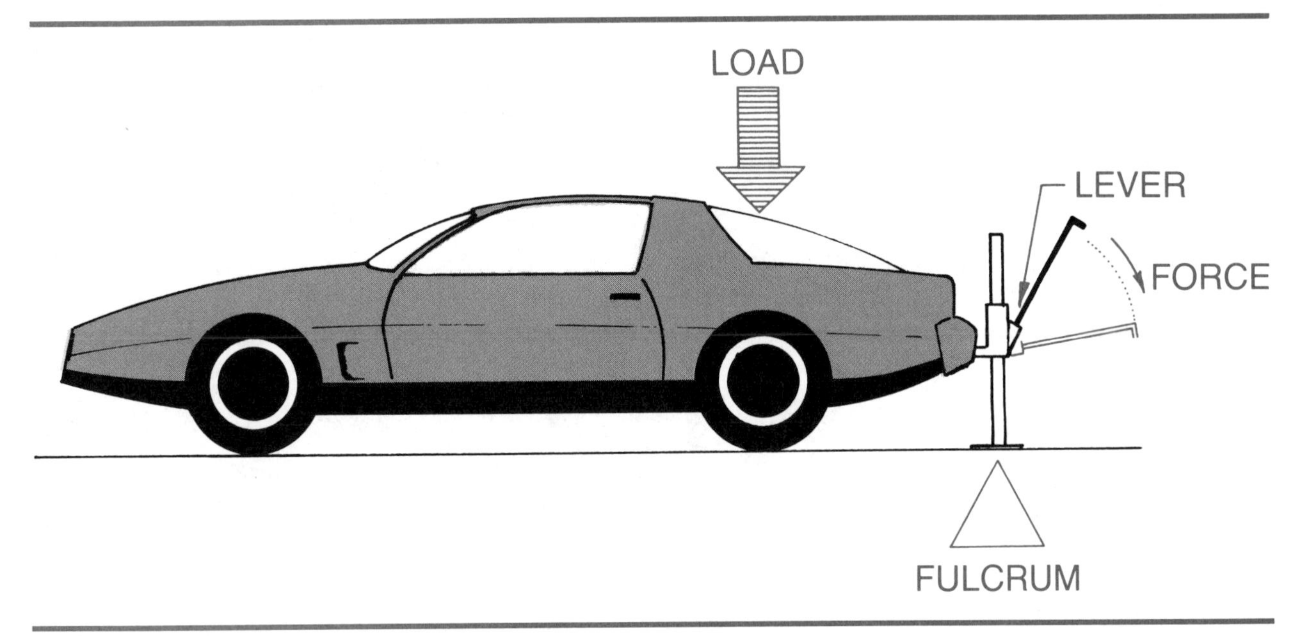

You use leverage when jacking up a car.

The principal of rotation.

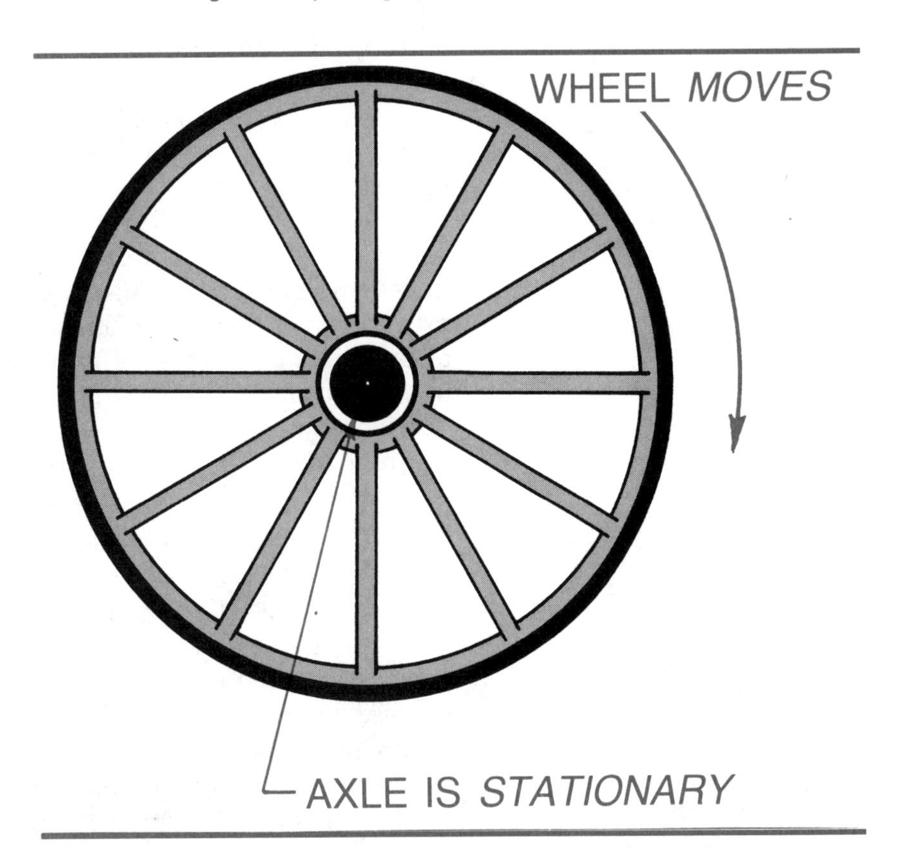

The principal of inclination (wedge).

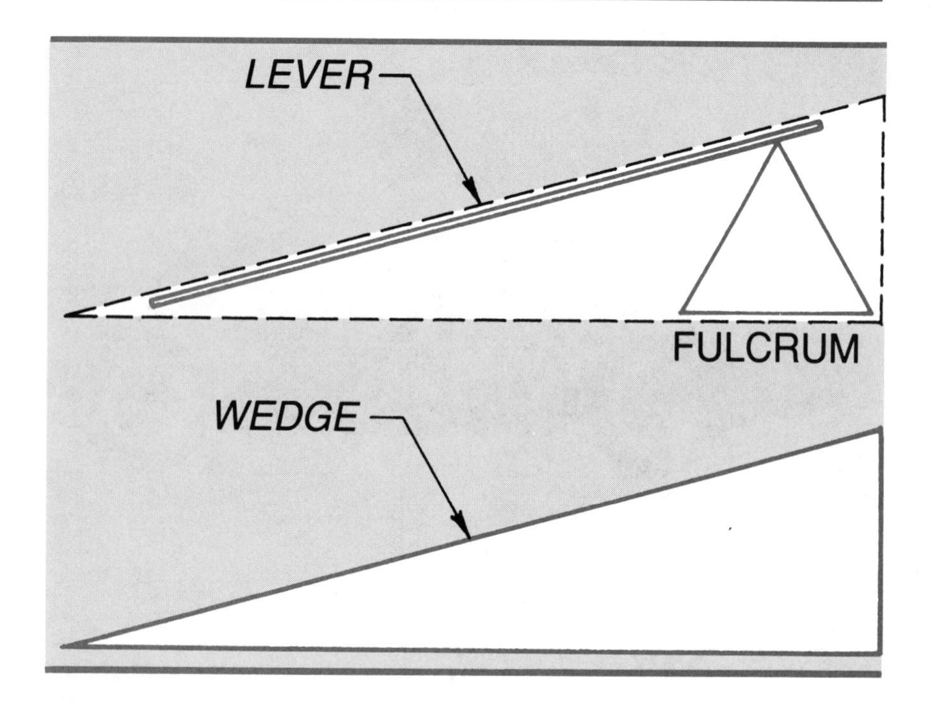

People use hundreds of different tools. Although these tools look different and are used for different jobs, they all have something in common. They are all based on the principles of leverage, rotation, or inclination. Now that you know how tools operate, let's look at how they have developed.

How Tools Have Developed Since humans first used a stone or a stick to kill a wild animal for food, they have been making and using tools to meet their needs. Many tools have since been developed to do many different things. You already know one thing that all of these tools have in common—the principles they are based on. They also have something else in common—the factors that controlled their development.

Three factors have always controlled the development of tools:

1. The availability of materials to make the tools
2. The availability of other tools to use in making the new tools
3. The needs of the people

To show how these factors have affected the development of tools, let's use the example of the hammer.

The first hammer was nothing more than a stone that could be held in one hand and used for pounding things. When people learned to cut dried animal skins into leather strips, they were able to tie stones

The hammer shows how tools have developed over the years.

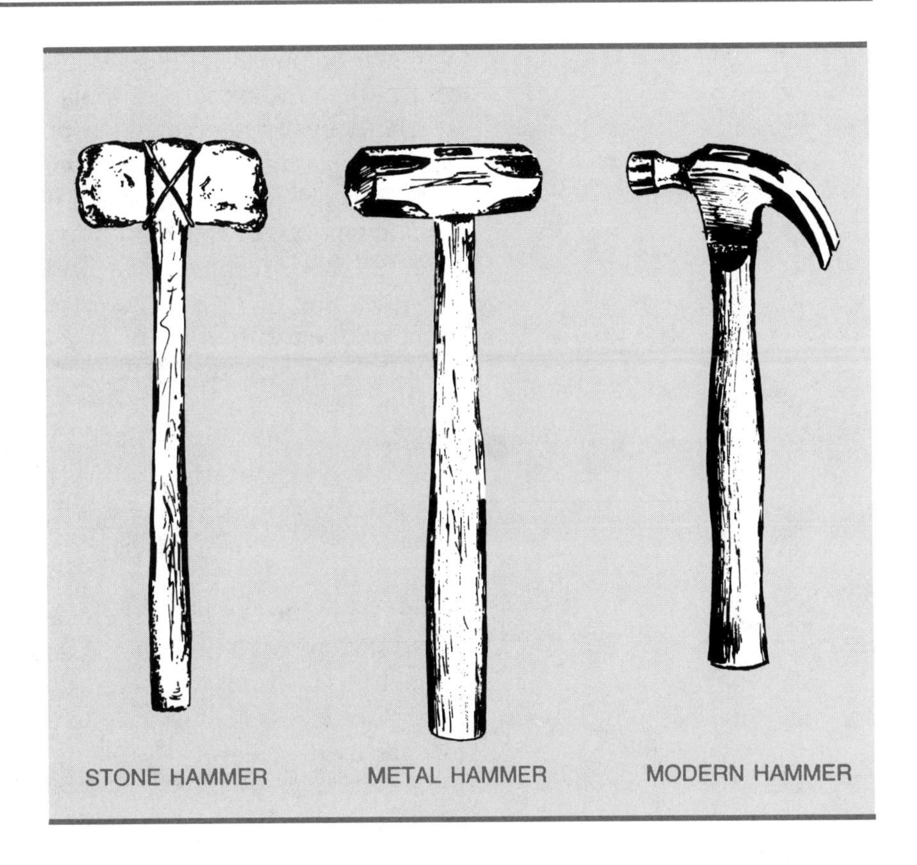

to wooden handles. This gave them more leverage and made better hammers.

Then people learned about metals: first bronze, then iron, then steel. They now made the head of the hammer from metal. The shape of the head also changed. When nails came to be used, the "claw" for pulling bent nails was added.

Now let's use this example to see how **materials**, other tools, and the needs of people affected the development of the hammer. When only stone was available as a material, hammers were made of stone. When leather strips became available, hammers were made of stone, leather strips, and wood for the handles. When metal became available, hammer heads were made of metal. So you can see how materials control the development of tools.

Other tools also contributed to the development of the hammer. The leather strips used to tie the stone hammer to a wooden handle were made with a stone knife. Without the knife, the stone hammer would still be just a stone. Fire was used as a tool to melt metals for the heads of metal hammers. Molds were used as tools to shape the heads of metal hammers. So you can see how other tools have helped control the development of tools.

The needs of people also contributed to the development of the hammer. Pounding on things with just a stone probably hurt the hand. It is also harder to pound things without the leverage you get from a handle. For this reason the wooden handle was added to the hammer. Stone hammers also broke a lot, especially when pounding on other stones or hard objects. The need for harder hammers eventually led to the use of metal. Then when nails were invented, some way was need to pull out old or bent nails. So you can see that the needs of people have helped control the development of tools.

You now know what a tool is, how it operates, and how it has developed over time. There is still one more thing you should know about tools: how they are used.

How We Use Tools There are so many different tools in the world that it would be impossible to name them all. But you should know how to group tools according to their use. When you can do that, there is less confusion in trying to keep up with all of the tools that are used.

Tools help us perform work in many different fields. For example:

- Construction
- Manufacturing
- Transportation
- Energy/power
- Communication
- Agriculture
- Medicine and health care
- National defense
- Entertainment
- Sports
- Leisure

Some tools used in one field might also be used in another. For example, we use wrenches in **manufacturing**, transportation, and agriculture. Can you name a tool that we use in all of these fields? Sure you can, it's easy. The wheel!

To illustrate how many tools we actually use for different jobs, let's take one of the fields from our list and see how many tools we can name that are used in that field. To make the task easier, let's limit our list to just hand tools that are used in **construction**.

Auger
Backsaw
Bits
Braces
Burnisher
Caliper rule
Chisels
Claw hatchet
Coping saw
Crosscut saw
Doweling jig
Drills
Files
Gauges
Gouge
Grinding wheel
Hammer
Jack planer
Jointer
Knives
Level
Mallet
Patterns
Planes
Plumb rule
Rules
Sandpaper
Saws
Scrapers
Screwdrivers
Squaring board
Tapes
Wrenches

This cordless electric screwdriver is an example of a modern hand tool. *Courtesy of Sears.*

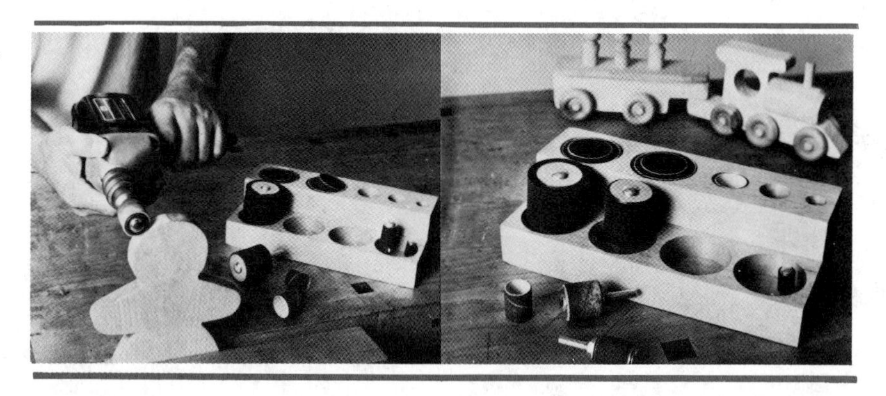

This sanding drug kit is an example of a modern tool. *Courtesy of Sears.*

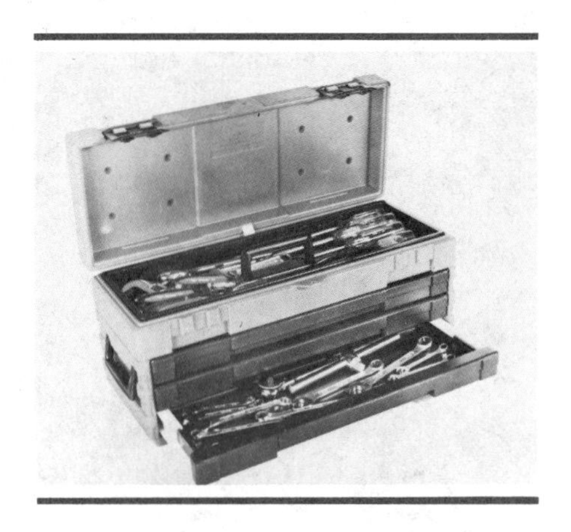

Many families have a tool box full of hand tools like this three-drawer plastic box. *Courtesy of Sears.*

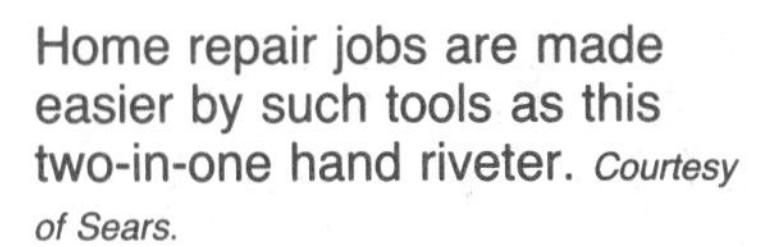

Home repair jobs are made easier by such tools as this two-in-one hand riveter. *Courtesy of Sears.*

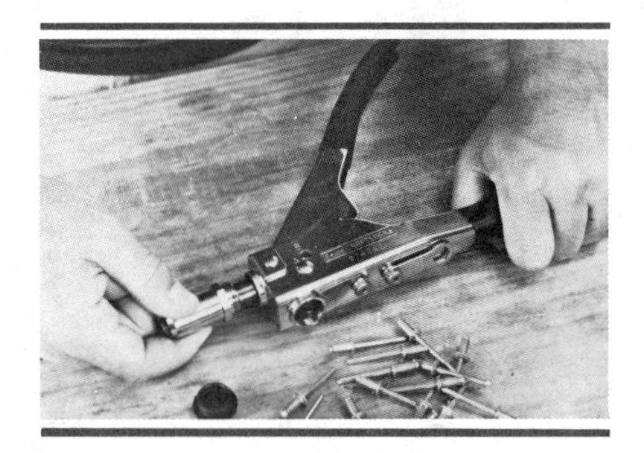

The list is already long, and it is not nearly complete. Can you add any other tools to the list? Remember, it's easier to put the tools we use into groups according to the field they are used in. Try an activity. See how many hand tools you can name that are used in the transportation industry. (*Hint*: Think of the tools that auto mechanics use.) Now that you know about tools, let's look at mechanisms.

Mechanisms

A tool is a device that helps people do things they couldn't do with their bare hands, such as hammering a nail or tightening a nut on a bolt. **Mechanisms** are more advanced than tools. They are devices that give people more power than they really have.

The most common mechanisms are the wheel, the pulley, and the gear. Let's look at each of these mechanisms and see how they give people more power.

The Wheel The **wheel** is the most widely used of all mechanisms. When we talk about the wheel as a mechanism, we are talking about

Common mechanisms.

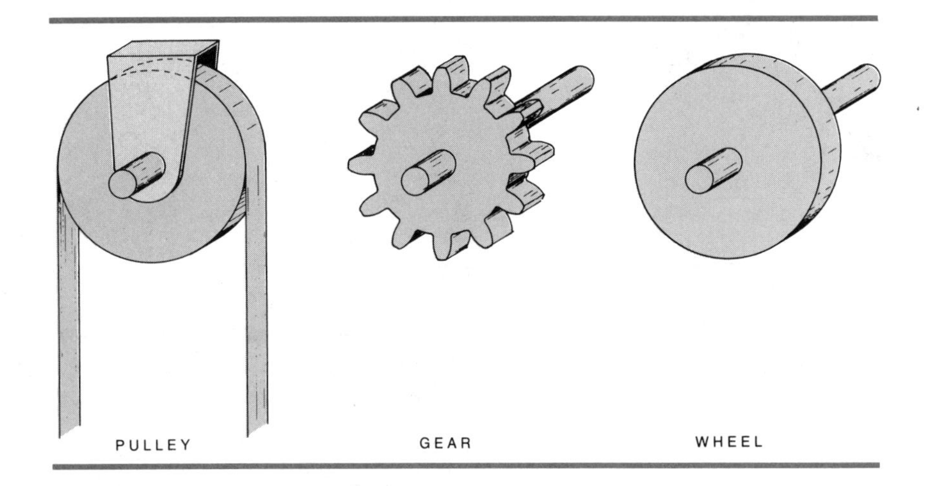

the wheel and the axle. As you know, the wheel is the round part, and the axle is the shaft that it rotates around.

You know what wheels are. You see them all the time. There are wheels on cars, trucks, buses, vans, bicycles, motorcycles, ATVs, wagons, carts, and many other things. We see wheels so much that we just take them for granted. We don't appreciate what an important mechanism the wheel is, but we should.

Without the wheel, the activities of transportation, construction, national defense, entertainment, sports, agriculture, and every other field would stop. The reason we depend on the wheel so much is that it is easier to roll heavy things than to pull or drag them. Think about this for a minute.

If you had to carry two heavy boxes home from the store, would you rather lift them up and carry them in your arms or put them in a wagon and pull them? Would it be easier to drag a tree you cut down or to put it on a cart and pull it? Would it be easier to carry a friend piggyback or to put him on a bicycle and push him? Of course, in each example it would easier to use wheels.

Another mechanism you should know about is the pulley.

The Pulley A **pulley** is a mechanism that we use to help lift heavy loads straight up (vertically). Have you ever seen a pulley? When heavy crates and other types of cargo are lifted off the dock and put in ships, pulleys are used. When a crane lifts heavy steel beams on a construction job, it uses pulleys. When an engine is lifted out of an automobile, a pulley is used. The weight machines that are popular in health clubs and gyms use pulleys.

There are two types of pulleys that you should know about. The first is the *fixed pulley*. A fixed pulley is attached to something such as a roof beam or truss. The rope is passed over it and attached to the load.

The pulley is an important part of this heavy construction shovel.

The two types of pulley systems.

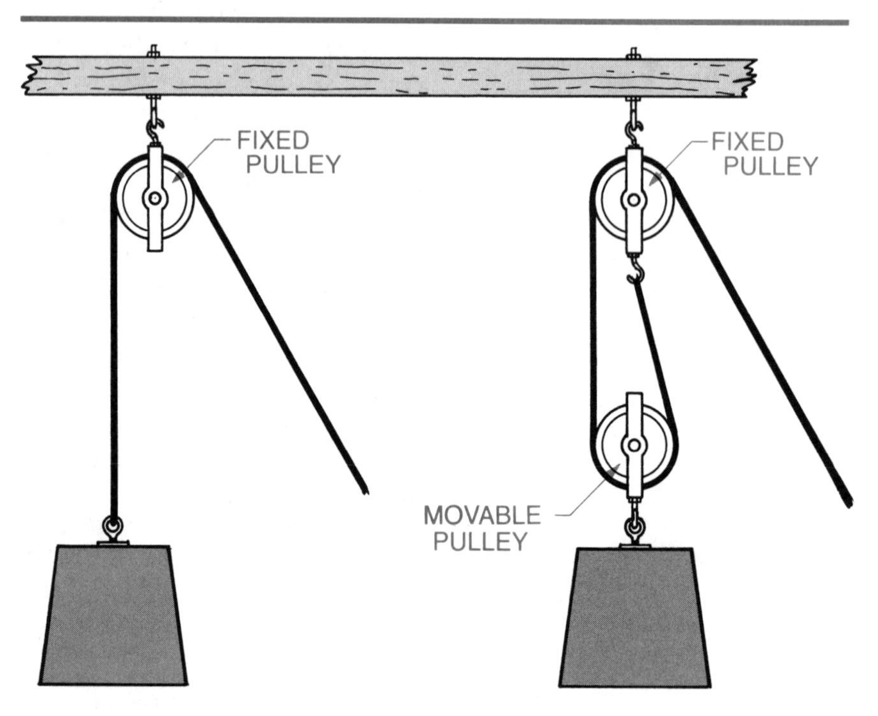

The second type of pulley is the *movable pulley*. A movable pulley is not attached to anything. Instead it is hung below a fixed pulley. Every time you add a pulley it makes lifting the load easier. The more pulleys you use, the easier it is to lift the load.

The last type of mechanism you need to know about is the gear.

The Gear A **gear** looks like a wheel with teeth. Like a wheel, a gear rotates on a shaft. Have you ever seen a gear? Maybe you have and don't know it. The sprocket on a bicycle is a gear. If you have a 10-speed bicycle, the 10 speeds come from using gears.

If you have ever ridden a 10-speed bicycle, you know that by changing gears you can make it easier to pedal. The reason is that the gears are doing what all mechanisms do—giving you more power by doing some of the work for you.

Gears are also used in cars. Have you ever heard a car referred to as an "automatic" or a "standard"? If it is an automatic, that means the gears change automatically. If it is standard, that means the driver must change the gears. Does your family car have an automatic transmission or a standard transmission?

Gears are used in many other ways. They are used in manufacturing machines such as lathes, drills, mills, and boring machines. They are used in kitchen appliances such as mixers, blenders, and food processors. Cement mixers on construction jobs have gears too.

If you will look at these examples closely enough, you might be able to see what they have in common. Then you will understand

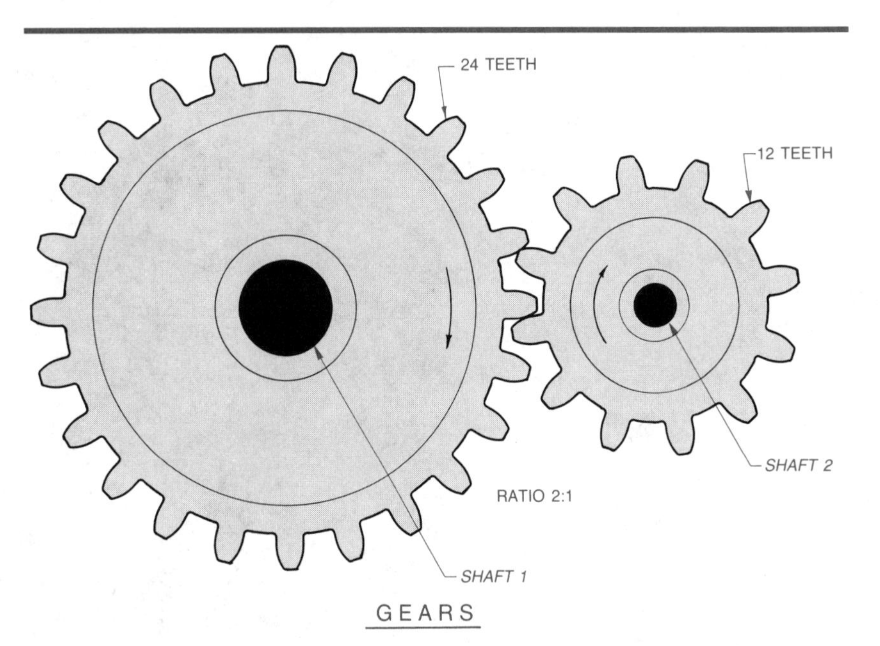

Gears.

when and why gears are used. What do a bicycle, a car, a drill, a blender, and a cement mixer have in common? Each is based on the principle of rotation for one thing. In addition, each requires some type of power to rotate. You power the bicycle. An engine powers the car. Electric motors power the blender and drill. A small gas engine powers the cement mixer. In each case, the gears give the power sources more power.

Now that you are familiar with mechanisms, we can go to the highest level of tool: the machine.

Machines

You hear the word "machine" all the time. In the house we have a washing machine. In the shop we have a milling machine or a welding machine. You hear the word all of the time, but do you really know what a machine is?

A **machine** is a tool that is made up of moving parts that all work together to do a certain job. A machine is usually more complicated than a mechanism. In fact, most machines contain mechanisms such as gears, pulleys, and wheels and combine the work of several tools.

A robot such as this G-36 is a machine. *Courtesy of PRAB Robots, Inc.*

AIR TRAFFIC CONTROL SYSTEMS

Technological systems need to be carefully controlled for reasons of safety and efficiency. For instance, aircraft are closely monitored to prevent accidents and delays as they land and take off at busy airports. The increasing number of flights has led to the use of computers and other electronic equipment. However, people are still the decision-makers in this important control system. (1) The airport control tower. (2) These controllers are working in the radar room. (3) Computers are an important part of the modern air traffic control system.

THE CHANGING WORLD OF CONSTRUCTION

When the Erie Canal was constructed in the early 1800s, construction technology was still very basic. Workers dug the canal by hand, and used horses and mules for moving heavy loads. Progress was slow and difficult. Modern construction technology has changed all that. (1) Architects, engineers and project managers carefully plan construction projects. (2) Modern machinery can move tons of earth efficiently. (3) Projects once thought to be impossible—like this undersea railway tunnel in Japan—are now being built.

Machines come in different sizes, shapes, and uses. A large drill press used in a manufacturing plant is a machine, but so is a small blender used in the kitchen. No matter how different they may look, all machines have three common elements.

The Basic Elements of a Machine All machines have three basic elements. If a tool does not have all three of these, it is not a machine. It might have more elements, but it must have at least these three:

1. Structural support and cover system
2. Power transmission system
3. Control system

An example of a car is a good way to explain these elements. A car is a transportation machine. The cover of a car is the body (the fenders, doors, roof, trunk, hood, and so on). The cover gives the car its shape and protects the other elements from dust, dirt, and the weather.

The structural support system of a car is the frame. The frame is the "skeleton" of the car. The frame holds the car up just as your skeleton holds you up. What would happen to your body if it had no skeleton?

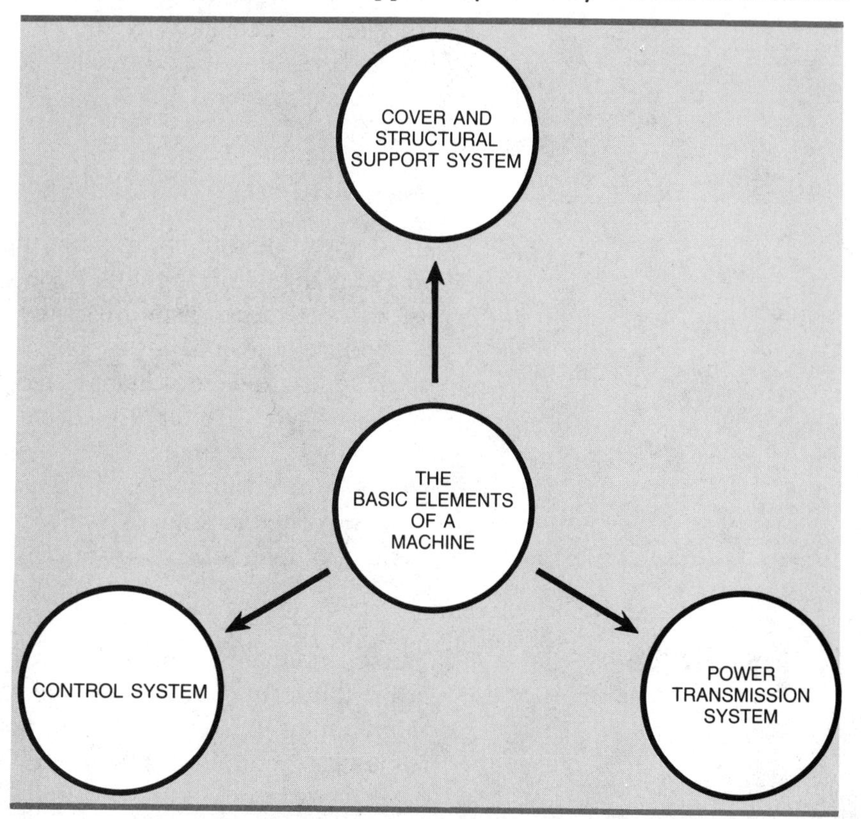

The basic elements of a machine.

It would collapse because it would have no support. The same would be true of a car without its frame.

The power transmission in a car is called the *transmission*. The transmission moves the power from where it is to where it needs to be to make the machine work. In a car, the power is in the engine, but it needs to be moved to the wheels to make them turn. The transmission does this.

The control system of a car lets you speed up, slow down, turn, and stop. In a car, the control system is made up of the steering wheel, the gas pedal, and the brakes.

Let's test your knowledge of machines. Is a bicycle a machine? Before answering, look back at the definition and the three elements of a machine. The definition says that a machine has moving parts that work together. Does a bicycle have moving parts that work together? Yes, the wheels, sprockets, pedals, and chain all move, and they work together. A bicycle passes the test of the definition. Now let's test it against the three elements of a machine.

Does the bicycle have a cover or support system? Yes, the frame is the support system. Does the bicycle have an energy transmission system? Yes, your energy is transmitted to the wheels by the pedals, sprockets, and chain. Does the bicycle have a control system? Yes, the handle bars, brakes, and pedals let you control it. It passes the test. A bicycle is a machine.

Systems

A system is one step above a machine. A **system** is a group of related elements that work together for a specific outcome. A system makes use of tools, mechanisms, machines, processes, and various resources to produce a desired result that will improve the quality of life.

Systems are so important to technology that they must be covered in great depth. Chapter 3 is devoted to a comprehensive treatment of technological systems.

Tools, mechanisms, machines, and systems make up the first main element of technology. Now it is time to look at the second main element of technology—resources.

RESOURCES

The second main element of technology is resources. Maybe you have heard the term "natural resources." A natural resource is one provided for us by nature. Water, air, land, and trees are all natural resources. Natural resources are only one type. There are several other types of resources.

Types of resources.

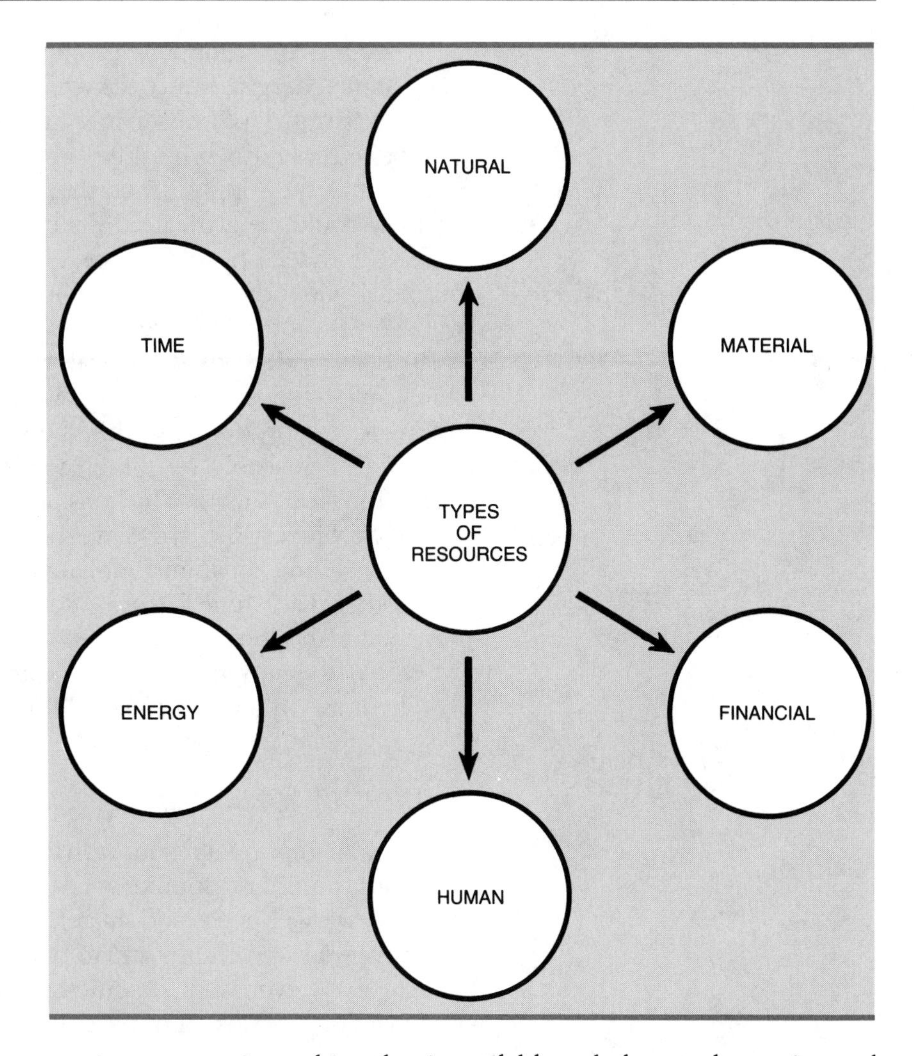

A **resource** is anything that is available to help people survive and improve the quality of life. Water is a resource that we need to survive; so is air. The tools, mechanisms, and machines you just read about are also resources. There are several groups of resources: natural, financial, human, material, energy, information, and time. Let's look at each of these groups of resources.

Natural Resources

Natural resources are the resources provided by nature. There are two types of natural resources:

1. Those that can be replenished
2. Those that cannot be replenished

Resources that cannot be replenished are such things as water and land. Some energy resources such as oil, coal, and natural gas also fall into the first group, as do many metals and minerals. Resources that cannot be replenished must be used wisely and sparingly because we can exhaust the supply. When they are gone, they are gone forever.

How could we grow food if all of the farm land was used up? On the other hand, how can we build houses, schools, businesses, hospitals, and industrial firms without using land? This is the dilemma we face with the natural resources that cannot be replenished.

The second type of natural resource is the replenishable or renewable resource. *Replenishable* means that a resource can be regrown or reproduced so that the supply is always available. Examples of replenishable resources are trees, plants, crops, and animals raised for food, such as chickens and cattle.

As we harvest the crops we need for food, we can grow more crops. Each time an animal is slaughtered for food, it can be replaced by another. Each time a tree is cut down to make lumber, it can be replaced with a seedling.

One of the important things to learn in studying technology is how to rely more on the replenishable natural resources.

Financial Resources

Money is an important resource in the modern world. It is often needed to pay for the other resources we need. For example, if we want to build a house, we will need land, material, people, and energy. To buy the land, material, and energy and to pay the people, we will need money.

Have you ever wanted something badly, but you didn't have the money to buy it? Most of us have. It is an object lesson in how valuable a resource money can be.

Human Resources

Human resources are people. However, not just anybody can be a resource. When there is a job to be done, the only people who are resources are people who know how to do the job. To be a resource, a person must be physically and mentally able to contribute to getting the job done.

If the job that needs to be done involves reading blueprints and you don't know how to read blueprints, you cannot be a resource. If the job involves carpentry and you have no woodworking skills, you cannot be a resource. For people to be resources, they must have the knowledge, skills, attitudes, and physical capabilities required.

Therefore school and learning are important. By going to school and learning, we become resources to each other in making the world a better place. Understanding why it is important to be a resource and how to become one is one reason for studying technology. Intelligence is our most important resource.

Material Resources

Some material resources are natural, and some are manufactured. For example, wood is a natural material, and plastic is a manufactured material. Some materials are a combination of both. How many different material resources can you name? Let's narrow the question down to building materials. What are some of the materials we would use to build a house? We would use glass, steel, wood, concrete, plastic, aluminum, copper, stone, brick, gypsum, and several other materials. For making clothing, we use natural materials like cotton and wool and synthetic materials like nylon and rayon. A *synthetic* material is a manufactured material.

Some materials are made of other materials. Concrete is made of cement, water, rocks, and sand. Glass is made of sand, lime, and soda. Some glues are made by boiling the bones, hooves, and hides of animals.

Energy Resources

We need energy to perform any type of work. We also need energy to power machines, light homes, and heat buildings. There are many kinds of energy. One of the most important is human energy. If people are to be resources, they must have the energy to do the work. People get energy from the food they eat; so food is an energy resource.

We need energy to power our various forms of transportation. Cars, boats, and airplanes all need energy as a resource to power their engines. The most common source of the fuel used in transportation is crude oil, or petroleum. The gas you put in a car is made from crude oil.

Electricity is one of the most widely used energy resources. Think of all the different ways you use electricity. The lights in your house and school run on electricity as do the blender, mixer, microwave oven, and dishwasher.

Natural gas is another energy resource used for cooking and heating. The sun is also an energy resource for solar hot water and heating systems. Every type of work requires energy.

Information Resources

Information is an important resource. Have you ever heard the saying "information is power"? Maybe you have heard that ours is becoming an information society. But what is information? It is the second element in a group of three very important elements:

- Data
- Information
- Knowledge

The word "data" is plural. So when using it, remember to say "data are" instead of "data is." *Data* are raw facts. They are random and unrelated. If you could take a lot of names, places, and other facts and throw them in a box, you would have a box full of data.

Data becomes *information* after it has been processed. Data processing involves organizing and composing related data. Data that have been processed in this way are called information.

In the example of the box full of data, which consisted of dates, names, and places, if this data were processed, you could have such information as the names, addresses, and places of birth of all of your classmates.

Information that is learned and used by people is *knowledge*. For example, this book is full of information about technology. As you learn this information, it becomes knowledge.

Time Resources

We measure time in seconds, minutes, hours, days, weeks, months, and years. Time is an important resource. We cannot do a job or task unless we have the time to do it.

Have you ever had so many things to do that you didn't have time to do them all? Have you ever run out of time before finishing something? If so, you have had an object lesson in how valuable a resource time is.

When planning a project, it is easy to get so wrapped up in thinking about all of the other resources you will need that you forget about time. Be sure it doesn't happen to you. Nothing gets done unless there is time to do it.

Now that you know the seven resource groups, let's look at an example of how we use resources and why they are so important. Let's say we want to build a house, and we need to determine what resources will be needed. The easiest way to approach this problem is to ask what resources will be needed in each of the seven groups.

What natural resources will be needed? We will need land to build the house on. We will need water for mixing cement and for the workers to drink. We will need clean air for the workers to breathe.

What material resources will we need? We will need concrete for the footing and concrete blocks for the foundation walls. Lumber will be needed for the floor, walls, ceiling, and roof. Plastic and metal pipes will be needed for the plumbing system. Metal wires will be needed for the electrical system. Paint, carpet, nails, shingles, and many other materials will also be needed.

What financial resources will be needed? Money will be needed to buy the land, hire someone to **design** the house and draw the plans, buy the materials, and hire the workers.

What human resources will be needed? Bankers will be needed to loan the money. Real estate agents will be needed to sell the land. Designers and drafters will be needed to make the plans. Carpenters, plumbers, electricians, painters, and several other types of workers will also be needed.

What energy resources will be needed? Human energy will be needed by all of the human resources. Fuel will be needed for all of the trucks and other vehicles used. Electrical energy will be needed to power tools and machines.

Last, but very important, time and information will be needed.

From this example, you can see how important resources are to technology. You already know how important tools are. In the example of the house, many different tools and resources were needed. But more than just tools and resources will be needed. The third element of technology is also needed. That element is processes.

PROCESSES

The third main element of technology is processes. If we have the tools and materials, the only other things needed to solve a technological problem are processes. A **process** is a set of steps that leads to a specific result. There is a process used to make a cake. It is explained on the cake mix box under "directions." If you go through the process exactly as explained, you will make a good cake.

There are many different processes, and the steps in them vary. For example, the process used to wash a car is not the same as the process used to pull a tooth. Two of the most important technological processes are the design process and the manufacturing/construction process. The design process is the process we use for designing the products we need in our daily lives. The manufacturing/construction process is the process we use for making those products. These two processes make good examples. If you understand them, you can understand any process.

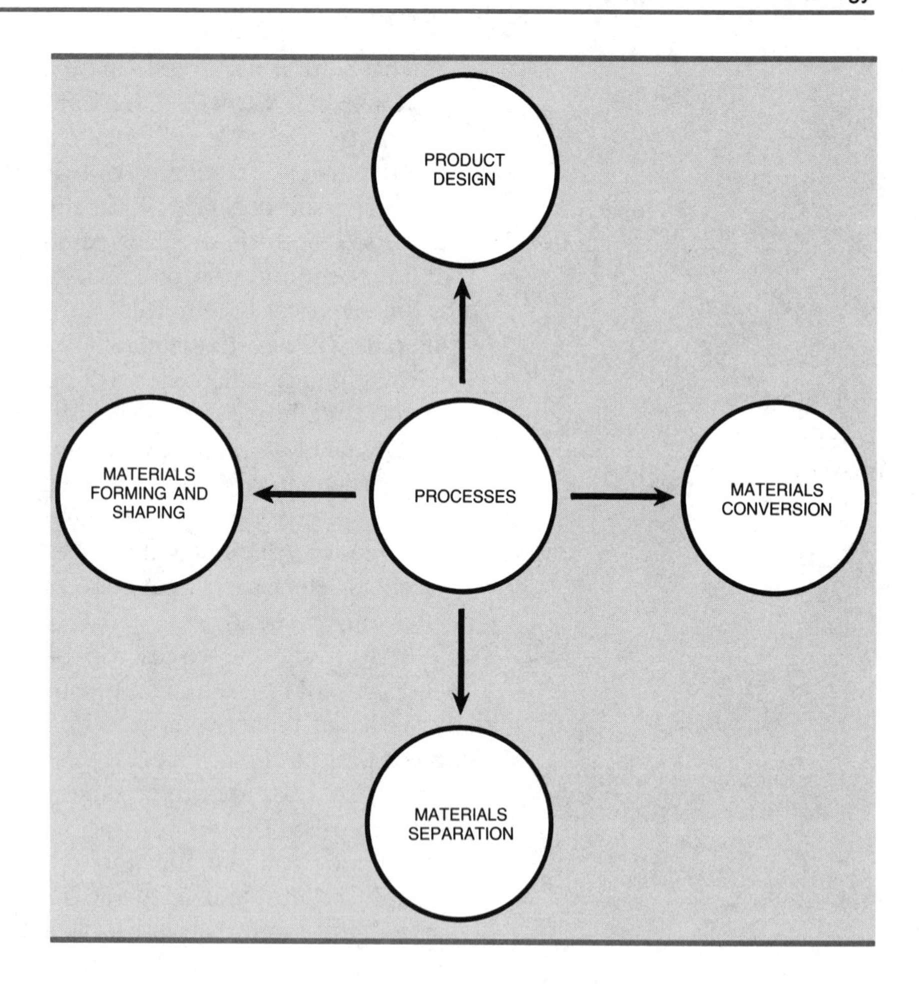

Types of processes.

The Design Process

Think of all the products you use in a day. Your bicycle, television, watch, VCR, clothes, shoes, baseball glove, and anything else you use is a product that was manufactured or constructed. But before it could be manufactured or constructed, it had to be designed.

The design process is a five-step process used in designing the products we need. Before making a tool to solve a problem, we have to design it first. The five steps used in designing a product are the same no matter what the product is. The same five steps used to design a house would be used to design a pair of running shoes or a machine part. These five steps are

1. Define the problem.
2. Try out ideas.
3. Select the best idea.

The design process.

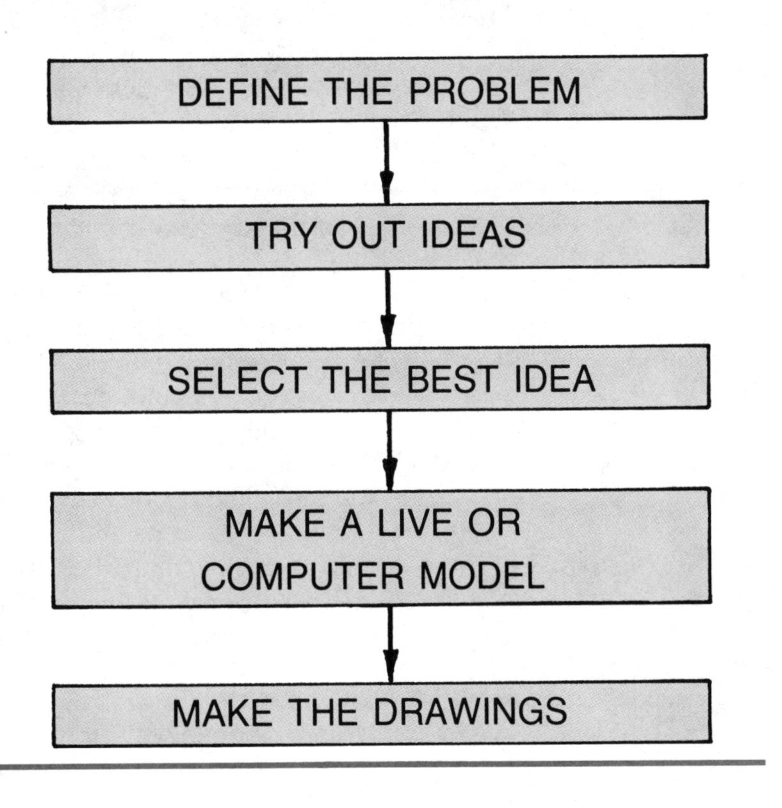

4. Make a model.
5. Make the drawings.

After all five steps of the design process have been completed, the product is ready to be manufactured or constructed. Then the manufacturing/construction process takes over. Let's look at each step in the design process before moving on to the manufacturing/construction process. The best way to do this is with an example. Let's use the example of a house. Suppose you want to build yourself a house. First you must design the house. To do this, you would use the design process.

DESIGNING A HOUSE

STEP 1 Define the Problem. In this step, you have to decide on the kind and size of house you want. Do you want a large house or a small one? Do you want a brick house or a wooden one? Do you want two or three bedrooms? Do you want a carport or a garage? Do you want carpeted or wooden floors? Do you want a fireplace?

By asking such questions, you define the problem. A well-defined house problem might be as follows: The problem is to design an 1800-square-foot house with three bedrooms, two bathrooms, a large kitchen, a living room with a fireplace, and a two-car garage. The house should be off-grade with wooden studs and brick veneer and cost less than $50,000. Now you know where you are going. The problem has been defined.

STEP 2 Try Out Ideas. In Step 1, the problem was defined. However, even with this definition there are hundreds of different designs that would work. Many houses could have 1800 square feet, three bedrooms, and two bathrooms. And these houses would satisfy the specifications in the definition even though no two of them would look alike. In this step, you start trying out some of the ideas you have for houses that will satisfy the definition.

You know you need three bedrooms, but there are many ways to arrange them and many sizes they could be. For every idea you have, sketches or preliminary drawings are made so that other people can see your ideas on paper. Some people call Step 2 "brainstorming."

STEP 3 Select The Best Idea. In Step 2, you might have had 10 or more good ideas for the house. In this step, you have to choose the one you like best. Sometimes it's one of the ideas from Step 2, and sometimes you might combine several of the ideas into a new one. If you combine ideas, a new sketch or preliminary drawing has to be made.

STEP 4 Make a Model. Before building the house, it is a good idea to build a scale model of it so that you can see exactly what it will look like and how it will fit together. A model will help you find problems that might not show up on sketches or drawings. This step is very important.

You want to find all problems and solve them before you start building. It is very expensive to make corrections to the design after the house has already been built. Three-dimensional computer models are sometimes used instead of live models.

STEP 5 Make the Drawings. After all the problems found with the model have been worked out, the working drawings are made. Unlike the sketches or preliminary drawings made earlier, these must have complete details. They must contain all of the information the workers will need to build the house. Nothing can be left off, because the people who do the building are not the same people who did the design.

The Manufacturing/Construction Process

When something has been designed, it is ready to be made. If it is a product such as a watch, a radio, a television, a machine part, or a car, we say that we manufacture it. If it is a structure such as a house, a school, or a football stadium, we say that we construct it or build it. Enough similarities exist between manufacturing and construction so that they can be combined into one process and looked at together.

The manufacturing/construction process has six steps:

1. Examine the design.
2. Select the materials.
3. Obtain the materials.
4. Convert the materials.
5. Assemble the product or structure.
6. Finish the product or structure.

MANUFACTURING/CONSTRUCTING A HOUSE

STEP 1 Examine the Design. The design is communicated to the people who will manufacture or construct it by the working drawings made in Step 5 of the design process. Using the drawings, manufacturing and construction workers can plan all of the different phases of their work.

STEP 2 Select the Materials. In this step, workers decide what materials will be needed. Then they make up a bill of materials (BOM), which is a complete list of all materials needed. If 2 × 4 lumber is needed, the BOM will list 2 × 4 and how many feet of this type lumber will be needed. If nails are needed, it will list the type of nails and how many boxes. The BOM is like a shopping list for the job.

STEP 3 Obtain the Materials. In this step, the BOM is used to actually obtain the materials. If Step 2 is making the shopping list, Step 3 is doing the shopping. It involves ordering each item on the list from the different places that sell them. Everything isn't ordered at once. Instead it is ordered as it is needed. For example, when buiding a house you would order cement, lumber, and nails right away. But you wouldn't order carpet until the house is almost completed.

STEP 4 Convert the Materials. This step is what most people think of when they think about manufacturing or construction. It is the step in which the product or structure begins taking shape. In manufacturing, it is the step in which drilling, milling, shaping, planing, sawing, and turning take place. In construction, it is the step in which concrete is made and lumber is sawed.

STEP 5 Assemble the Product or Structure. In this step, the materials that have been worked on in Step 4 are put together to make the product or structure. In manufacturing, this step might involve putting in screws, tightening nuts on bolts, welding metal, or soldering wires. In construction, assembly involves nailing, bolting, and gluing. When this step is complete, a rough version of the product or structure is ready.

STEP 6 Finish the Product or Structure. Before a product is released to the market, it must be finished. Finishing gives it the "brand new" look we are used to seeing when we buy something. In manufacturing, finishing involves cleaning, grinding, polishing, painting, and so on. In construction, it involves cleaning, sanding, painting, carpeting, and trimming.

These are a few of the many processes of technology. Many more—some large, some small—could be listed. But the important thing to remember about all processes is that they are step-by-step and lead to a specific outcome.

SUMMARY

The most important elements of technology are tools, resources, and processes. When talking about technology, we sometimes use the word "tools" in a general way to mean tools, mechanisms, machines, and systems. A tool is a device or object that allows us to perform tasks we couldn't perform with our bare hands.

All tools operate according to the basic principles of leverage (the most basic), rotation, and inclination. Tools have developed as our need for them has developed. Three things control the development of tools: materials, other tools, and human needs.

Mechanisms are more complex than tools. A mechanism is a device that helps us generate more power than we could without it. The most common mechanisms are gears, wheels, and pulleys.

Machines are even more complex than mechanisms. In fact, machines usually contain mechanisms. A machine is a tool made up of moving parts that work together to do a job.

A system is a group of related elements that work together for a specific outcome.

The second main element of technology is resources. A resource is anything that is available to help us survive and improve the quality of our lives.

The final main element of technology is processes. A process is a set of step-by-step procedures that leads to a desired outcome.

KEY WORDS

Tool
Machine
Process
Construction
Inclination
Principle
Pulley
Wheel
Mechanism
Resource
Manufacturing
Leverage
Rotation
Material
Gear
Design
System
Information

SOME QUESTIONS TO TRY

1. Define tool.
2. Define mechanism.
3. Define machine.
4. What things control the development of tools?
5. Name the three elements of a machine.
6. Define system.
7. What is a resource?
8. Name the seven groups of resources.
9. What are replenishable natural resources?
10. Define a process.
11. Name the five steps in the design process.
12. Name the six steps in the manufacturing/construction process.

TECHNOLOGY LEARNING ACTIVITIES

1. Keep a list of all the tools you use in one day. (*Hint*: A spoon or a fork is a tool.)
2. Locate a sturdy board at least 6 inches wide and 6 feet long and two concrete blocks. Using the board as the lever and the concrete blocks as a fulcrum, try an experiment. Find a friend who is heavier than you. Position the board so that the fulcrum is in the middle. Have your friend sit on one end of the lever. You climb on the other. What happens? Now move the board so that the fulcrum is very close to your friend's end and try again. What happens this time?
3. Locate a 10-speed bicycle. Ride it around the block, changing gears until you have used all 10 gears. Notice the effect on pedaling each time you change gears.
4. Your job is to develop a plan for building a birdhouse. Make a list of all tools and resources needed for the job.
5. Locate a second-hand lawnmower that can be used for a class project. Separate this machine into its three basic elements: structural support system, power transmission system, and control system.

CHAPTER 3

TECHNOLOGICAL SYSTEMS

OBJECTIVES

After completing this chapter, you should be able to

- define a system.
- list the elements of a system and explain each one.
- explain how systems and subsystems are related.
- define open- and closed-loop systems.
- explain how people use systems to control their lives.
- explain how systems are used in solving problems.
- list the main elements of the public education system.
- list the main elements of the health care system.
- list the main elements of the transportation system.
- list the main elements of the communications system.
- list the main elements of the production system.

In Chapter 2, you learned about the three main elements of technology: tools, resources, and processes. When people use these things, what sometimes results is a technological system. As we said in Chapter 2, a **system** is a group of related elements that work together for a specific outcome. Technological systems are created by people to benefit people.

SYSTEMS AND SUBSYSTEMS

One of the most important concepts in a study of technology is the system. To understand technology, you must understand systems and subsystems. The only difference between the two is that a subsystem is smaller; it is a part of a system.

A good example of a system and a series of subsystems is the automobile. A car is a system. It consists of a group of related elements that work together for a specific outcome. What are the related elements of a car? Let's see how many we can name.

The wheels, engine, transmission, brakes, steering column, frame, and seats are all related elements that, along with many others, make up the car. What is the specific outcome that these related elements work together to accomplish? You can answer with just one word. Do you know what it is? Sure you do. It's **transportation**!

If an automobile is a system, what are some of its subsystems? A car has a power system (the engine), a brake system, an electrical system, a cooling system, and a frame system. Each of these systems contains a number of subsystems. The power system contains such subsystems as the fuel injection system, the lubrication system, and the cooling system (air or water). Can you name any other subsystem of the power system?

The electrical system also has a number of subsystems: for example, the ignition system, the exterior lighting system, and the interior lighting system (dashboard lights, mirror light, and overhead light).

The two basic types of systems are open-loop systems and closed-loop systems. There are important differences between the two with which you should be familiar.

Open-Loop Systems

An open-loop system is the simplest form of system. An open-loop system has four elements: *purpose*, *input*, *processes*, and *output*. With only these four elements, an open-loop system is the least effective, least efficient type of system.

The problem with an open-loop system is that it has no built-in feedback mechanism to tell if it is satisfying its purpose. Let's look at an example that will illustrate this weakness.

People used to heat their houses with space heaters. A basic space-heating system is an open-loop system. The purpose of this space heater is to heat the house to a desired temperature. The input is fuel oil piped into the heater. Processing involves burning the fuel oil. The output of this processing is heat. But is the heater producing

An open-loop system.

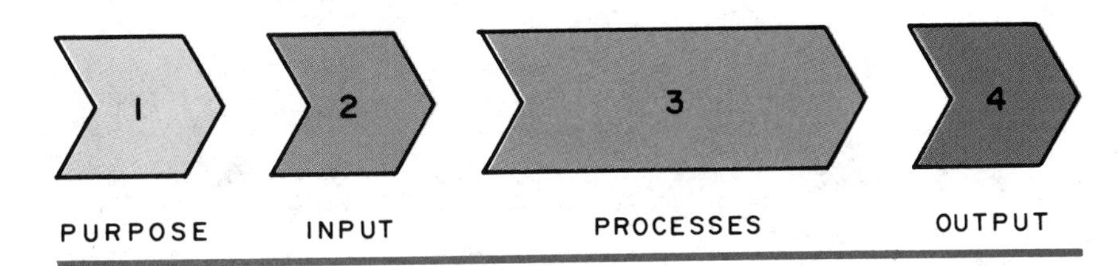

enough heat? Is it producing too much? Is it heating evenly? The system is not capable of answering such questions because it has no feedback mechanism, such as a thermostat. By adding a feedback mechanism to an open-loop system, you can have a closed-loop system.

Closed-Loop Systems

A closed-loop system is a more complex form of system. Such a system has five principal elements: *purpose*, *input*, *processing*, *output*, and *feedback*. This last element is what distinguishes a closed-loop system. It is an important element because it is a self-correcting mechanism that tells if a system is accomplishing its purpose and allows it to respond accordingly.

Let's go back to the example of the space heater and replace it with a heating system that is a closed-loop system. The new heating system consists of several subsystems, including the heater for burning the fuel oil to produce heat, a blower unit for pushing the heat through the distribution system, the distribution system (ductwork and registers) for carrying heated air to various parts of the house, and the thermostat for providing continuous feedback.

This feedback allows the system to be self-correcting. If the house gets too hot, the system can shut itself off. If it is too cool, the system will run until the house reaches the desired temperature.

A closed-loop system has another advantage over an open-loop system. It allows people to set their own guidelines for the system. The purpose of the closed-loop system we talked about was to heat a house.

But how much heat is enough? What is the desired criteria? (That is, what are the guidelines for the system?) With a closed-loop system you can decide. You set the criteria. If you want the temperature to be maintained at 72 degrees Fahrenheit, you set the thermostat to 72 degrees. Then, all feedback relates to this criterion. When the

A closed-loop system.

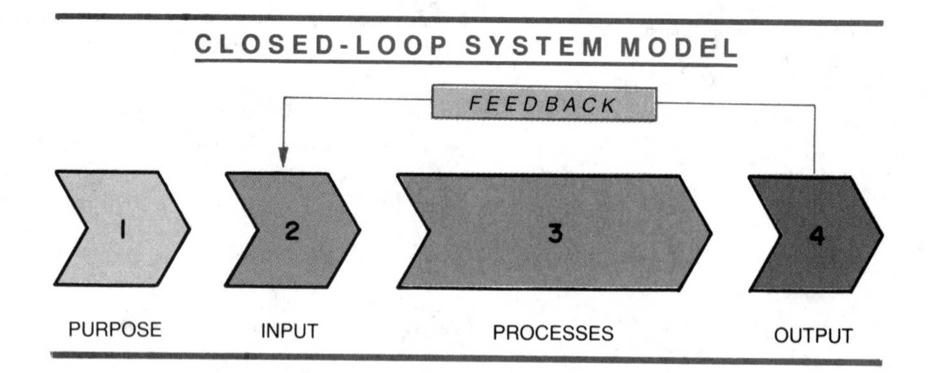

temperature is 72 degrees, the feedback mechanism shuts off the system. When it is less than 72 degrees, it causes the system to run.

Such a system allows people to control their environment. And it allows them to decide what they want their environment to be. Allowing people to control their lives is one of the most important purposes of systems.

CONTROLLING OUR LIVES THROUGH SYSTEMS

It is important for people to exert as much control over their lives as possible. Through the continuous development of technological systems, people have gained more and more control over their lives.

The earliest people had very little control over their lives. They often found themselves at the mercy of nature. When it rained, they got wet. If they failed to catch an animal, they did not eat. If they were not careful, they became the meal for some larger animal. If the temperature dropped, they froze. If the temperature rose, they sweltered. With no means of travel other than walking, their range was limited. If they got sick, chances were they would die. Over the years, people began to learn how to take control of their lives by developing systems to solve their problems.

SYSTEMS AND PROBLEM SOLVING

The problems confronting the earliest people were very basic. Most of them revolved around providing food, clothing, and shelter for themselves and for their families. The systems they developed for solving these problems were simple at first. They have since become increasingly sophisticated.

The need for food eventually led to the development of a system of agriculture, which included special subsystems for planting, growing, harvesting, storing, processing, and distributing food. The need for clothing led to a simple system, at first, which involved skinning animals and processing the hides into clothing. This simple system eventually became the complex, international system that is today's textile industry.

The need for shelter led to the development of a system of constructing simple shelters. This system has grown into one of the most sophisticated of all systems. Modern shelters are complex and have many subsystems. Today's houses, apartments, townhouses, and condominiums have floor, wall, roof, electrical, plumbing, and heating-ventilation-air conditioning subsystems. Let's take a look at some of today's most important technological systems.

There are several major systems that have been created by people to benefit people. Some of the most important of these are the

1. Public education system
2. Health care system
3. Transportation system
4. Communications system
5. Production system

There are other systems, some large and some small, but these are the most important. Let's look at each of these systems. You need to know the purpose of each, how they benefit people, and what elements they contain.

The Public Education System

People have created the public education system to make sure that the maximum number of people in this country have an opportunity for at least 12 years of formal education. Most people believe that education is the best way to help people become self-sufficient, contributing citizens.

What are the elements of the education system? How is education made available to the public? The main elements of the public school system are

1. Elementary school
2. Junior high or middle school
3. High school
4. College

Each of these elements satisfies a part of the overall purpose of the system. The ages of people served by each element vary from state to state. But generally, elementary schools serve students from 5 to 12 years of age. Junior high or middle schools serve 13- to 15-year-olds.

Elements of the public school delivery system.

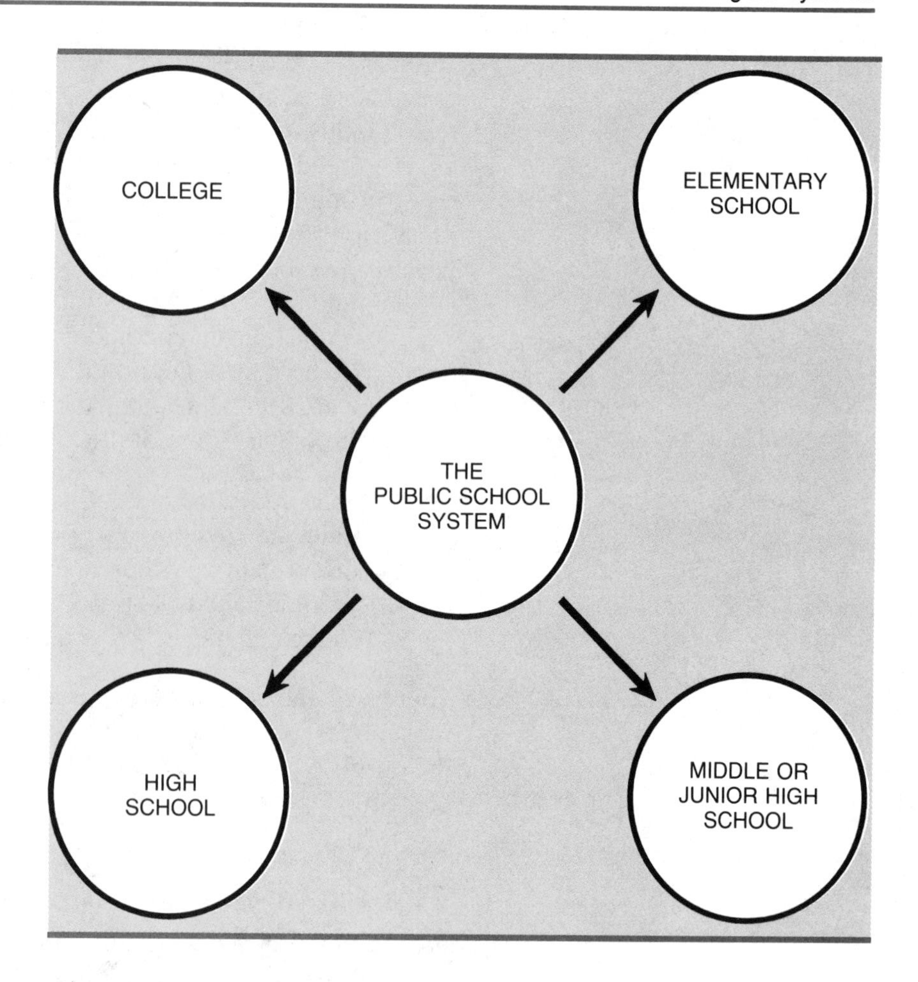

High schools serve 16- to 18-year olds. And colleges serve people who are out of high school, regardless of their age.

You are in the public school system right now. The purpose of the classes you take is to help you learn. What are some of the tools used to help you learn? There are books, workbooks, pencils, chalkboards, marker boards, chairs and desks, film projectors, wall charts, globes, video monitors, and computers.

What are some of the resources needed by the public education system? Land is needed to build the school buildings on. Many different material resources are needed to build the buildings. Human resources are needed for teachers, administrators, counselors, janitors, maintenance workers, and clerical workers.

What are some of the processes used in public education? They include traditional, teacher-directed instruction; individual, self-paced instruction; and computer-assisted instruction.

The Health Care System

People have created the **health care** system to ensure that health care and medical services are available when needed. Advances in health care technology are helping people live longer. The average life expectancy of people in America is over 70 years. Health care technology has become so effective at keeping even critically ill or injured people alive that it has spawned a new and controversial social issue—a person's right to die.

What are the elements of the health care system? How is health care made available to the public? The main elements of health care are

1. Doctor's offices
2. Hospitals
3. Clinics
4. Immediate-care centers
5. Wellness centers

Have you ever been to a doctor's office or a hospital? When you don't feel well or when you are injured, you go to a doctor's office. If your problem is minor, it can be treated there. For more serious problems, people go to hospitals. Have you ever had to go to the emergency room of a hospital?

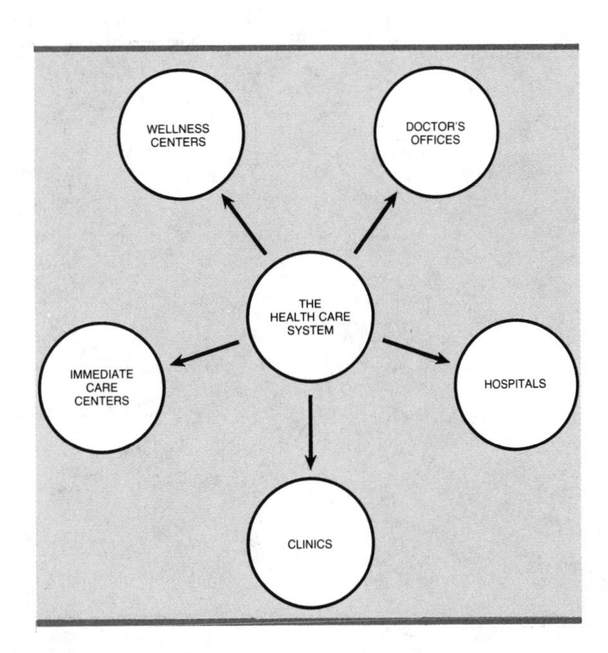

Elements of the health care delivery system.

When people need shots and vaccinations, they can go to clinics. For emergencies such as broken bones or bad cuts, people can go to immediate-care centers or the emergency room of a hospital.

Wellness centers are to help people prevent health problems. They offer exercise classes, nutritional information, checkups, and self-help advice and counseling.

What elements of the health care system are available in your community? Is there a hospital? Is there a clinic? Is there an immediate-care center? Is there a wellness center? Have you been to any of these?

The Transportation System

People have created the transportation system for two reasons: (1) to move people from place to place; (2) to move goods from place to place. The transportation system includes land, air, and water transportation.

What are the elements of the transportation system? How is transportation made available to the public? The main elements of the transportation system are

1. Land **vehicles**, **routes**, and **facilities**
2. Water vessels, routes, and facilities
3. Aircraft, routes, and facilities

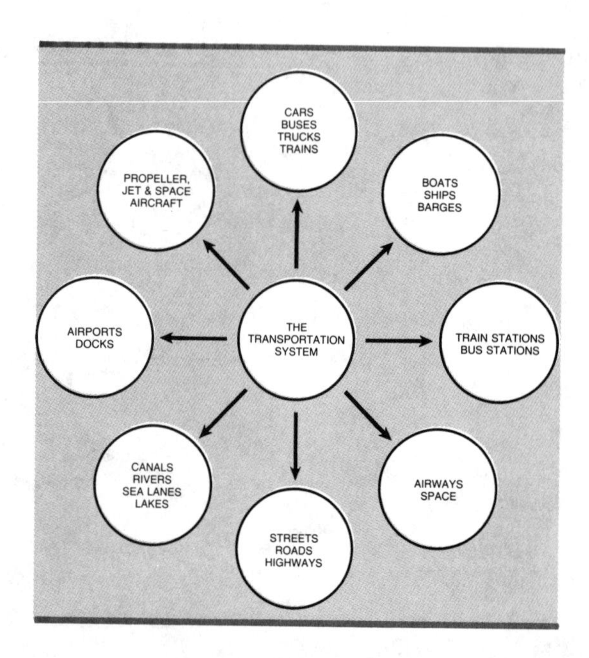

Elements of the transportation delivery system.

To provide transportation for themselves, people have had to create a number of different types of vehicles, vessels, craft, routes, and facilities. How many land vehicles can you name? People and goods are transported over land by cars, trucks, buses, trains, motorcycles, and many other types of vehicles. Land routes include roads, streets, highways, and railroad tracks. Land transportation facilities are bus stations, subway stations, train stations, loading/unloading docks for trucks, and maintenance/repair shops.

The vessels used for water travel range from canoes and rowboats to huge, ocean-going ships. How many different types of vessels can you name? There are canoes, rowboats, motorboats, sailboats, tugboats, cabin cruisers, merchant ships, ocean liners, military ships, submarines, and many more.

The routes used for water transportation are rivers, canals, lakes, and sea lanes. Did you know that the oceans are marked off into sea lanes and that ships must use these lanes? Water transportation facilities include loading docks, warehouses, and dry docks.

Aircraft range from the smallest, ultralight craft to NASA's space shuttle. Have you ever flown in an airplane? Have you ever watched as the space shuttle was launched? The air above the earth is marked off into airways. Air traffic controllers are responsible for ensuring that all aircraft stay in their assigned airways. Air transportation facilities include airport terminals, hangars, runways, air traffic control towers, and warehouses.

If you wanted to visit a friend in another state, how many different ways might you get there? It's too far to ride your bicycle or ATV. You could take a bus. You could fly in a commercial airliner. You could take a train. You could ride in a car. Transportation technology has come a long way since the days of walking and riding animals.

Commercial trucks such as this "18 wheeler" depend on good roads. *Courtesy of Fruehauf Division of Fruehauf Corporation.*

Chrysler's Liberty 1 prototype is representative of cars of the near future. *Courtesy of Chrysler Corporation.*

Sailing is one of the oldest forms of water travel, and it is still popular as a leisure activity.

Modern ocean liners can transport hundreds of passengers in luxurious accommodations.

The Boeing 767 is a popular commercial airliner.

The Communications System

People created the **communications** system for two reasons: (1) to be able to communicate with other individuals no matter where they are; and (2) to be able to communicate with a large number of people at one time. Communications technology has become so advanced that we can talk with people traveling through space, place a telephone call while flying in a commercial airliner, and, in a matter of seconds, talk with a friend in another country. In fact, you could even talk with someone on the moon!

What are the elements of the overall communications system? The main elements are

1. Telephone
2. Radio
3. The mail system
4. Television
5. Newspapers
6. Magazines

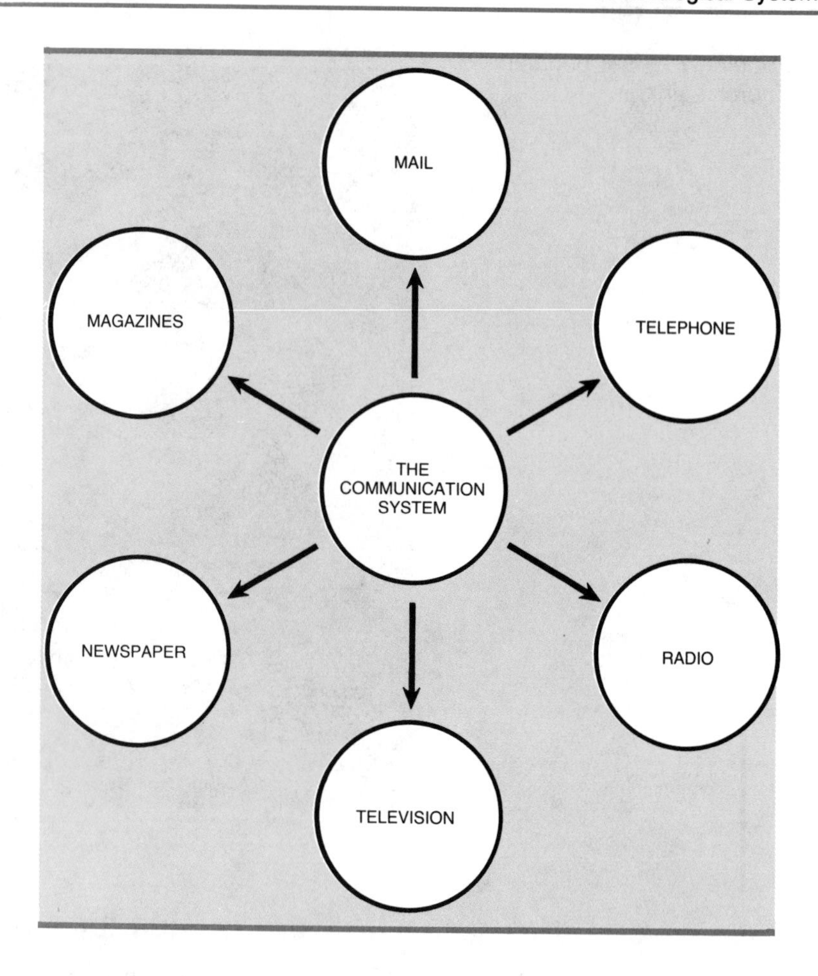

Elements of the communications delivery system.

A communications system has three essential ingredients: (1) a sender, (2) a carrier or **transmitter**, and (3) a receiver. Let's look at some examples. The telephone system has all three ingredients. When you call someone and talk on the phone, you are both a **sender** and a **receiver**. The part of the telephone that you talk into is a sending device. The part you listen to is a receiving device. Your voice is transmitted either over telephone lines or through the air by radio waves.

Television is also part of a communications system. The company that broadcasts a television program is the sender. The program is transmitted by radio waves, and you are the receiver. The television set in your house is a receiving device.

The telephone is an example of communication between two individuals. Television is an example of mass communication. **Mass communication** means that a large number of people receive the

The AT & T personal terminal is a modern communications device.

MESSAGE SENDER	+	MESSAGE CARRIER	+	MESSAGE RECEIVER	=	COMMUNICATION SYSTEM

A communications system.

message at the same time. When certain programs are broadcast on television, thousands and sometimes millions of people watch them at the same time. For example, each year millions of people watch the Super Bowl to determine who will be the champion of the National Football League.

Radio is also a form of mass communication. Radio and television are particularly effective at keeping large numbers of people up-to-date as to what is going on in the world. If there is an important event taking place, you can watch it or hear about it as it happens because of television and radio.

When the president has an important message he wants to communicate, he goes on television or radio and millions hear him. When politicians are running for office, they buy time on television so that the greatest number of people possible can see and hear them. Communications technology is one of the most advanced of the technology areas.

The Production System

People have created the **production** system to provide the food, clothing, shelter, and consumer goods they need. What are the elements of the production system? How do we produce the things we need? The main elements of the production system are

1. Raw materials processing
2. Construction
3. Manufacturing
4. Agriculture and food processing
5. Textile production
6. Energy production
7. Chemical production

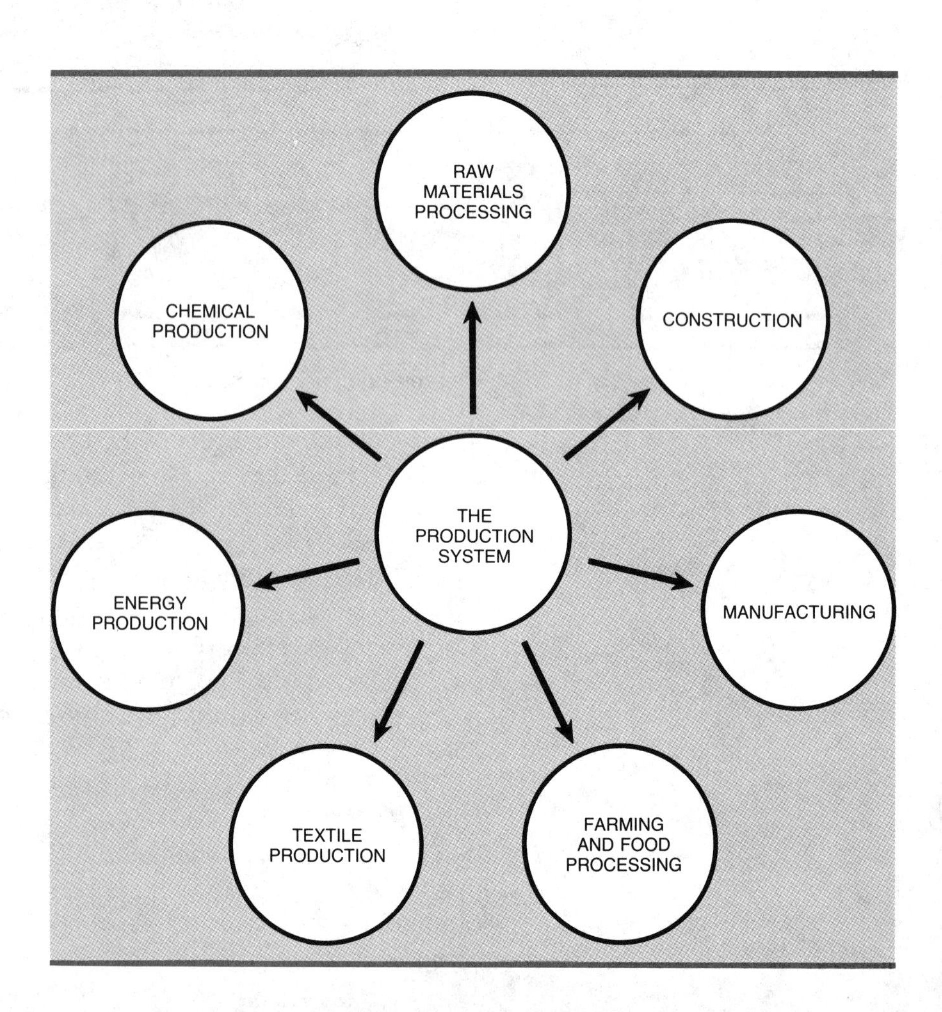

Elements of the production system.

Raw materials **Raw materials** are natural materials that we can collect and convert into other useful materials. A tree is a raw material that can be processed into other useful materials such as lumber and paper. Trees are processed into lumber in sawmills. The processing involves cutting, trimming, and planing.

Iron ore is a raw material that can be processed into steel. Animals are raw materials that can be processed into food and clothing. For example, the processing of a cow into food and clothing involves (1) slaughtering, (2) skinning and cleaning, (3) butchering, (4) drying the skin, and (5) using the dried skin to make leather products such as belts, shoes, boots, and garments.

Construction **Construction** is the way we produce the structures we need: for example, houses, churches, stadiums, apartments, schools, hospitals, office buildings, bridges, roads, dams, seawalls, and many other things. How many other types of structures can you name?

Manufacturing **Manufacturing** is the way we produce the products we need. Manufacturing involves converting raw materials into usable materials and converting these materials into finished products. The list of manufactured products is so long it would fill a whole stack of books. We manufacture cars, furniture, books, televisions, shoes, radios, paper, computers, cooking utensils, tools, and many other things.

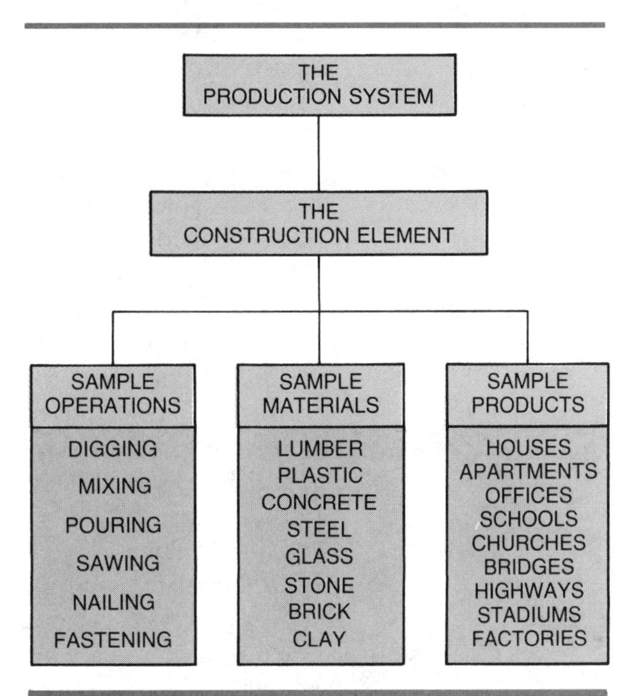

The construction element of the production system.

The manufacturing element of the production system.

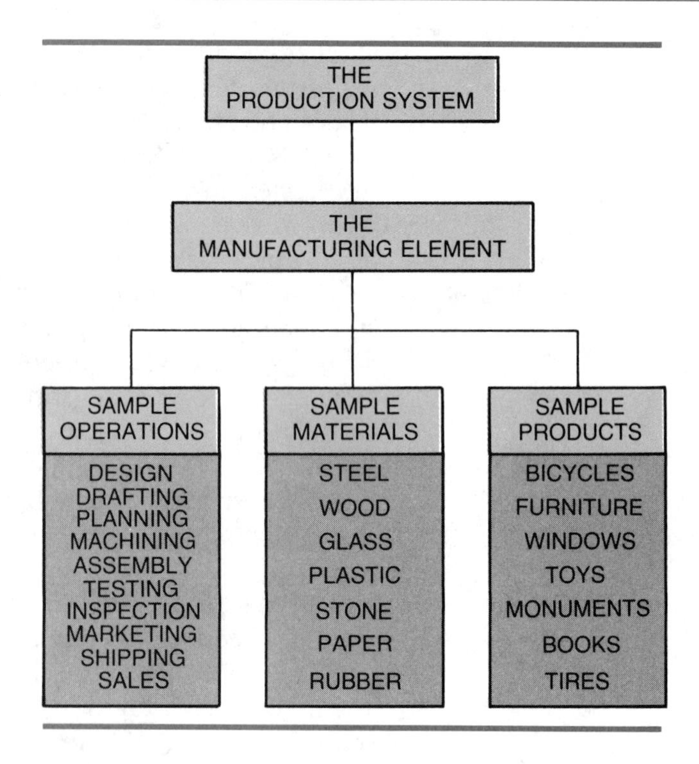

Agriculture and Food Processing **Agriculture** and **food processing** are the ways we produce most of the food we need. Agriculture involves planting, growing, and harvesting crops as well as breeding, reproducing, and growing animals such as chickens and cattle.

Food processing is what happens after the crops are harvested and the animals are slaughtered. Processing of food means converting it from rough form into the form it's in when we see it in a grocery store. Frozen, canned, boxed, and bottled food have all been processed.

One of the most frequently used processing steps is adding chemical preservatives to food as it is processed. Preservatives are chemicals that keep food from spoiling while it is waiting to be consumed. Preservatives are an important part of food processing. Even with refrigeration some food products will spoil quickly. Preservatives lengthen the time food can be stored or sit on a shelf before it must be used.

The containers of many food products are stamped with an expiration date. You should use the product by this date or else it will spoil. Have you ever noticed the expiration date on a food package? The next time you are in a grocery store, look at some cans, boxes, and bottles and see if you can find the expiration dates on them.

Textile Production **Textile** production is the way we produce the clothing we need. It involves converting both natural and synthetic materials into clothing. *Natural* materials are such things as wool, cotton, silk, and leather. *Synthetic* materials are machine-made materials such as nylon, rayon, and dacron. Some materials are a combination of natural and synthetic materials. These materials are called *blends*.

Natural, synthetic, and blended materials are converted into shirts, pants, hats, socks, dresses, skirts, sweaters, overcoats, and many other items of clothing. The conversion process involves cutting, sewing, and trimming.

Most garments have a tag that tells what they are made of and where they are made. Look in several of your shirts or blouses and find this tag. What is the clothing made of? Where was the clothing made?

Energy Production **Energy** production is the way we produce the energy we need for working, cooking, powering machines, heating and cooling, transportation, and many other uses. We produce several different types of energy for meeting our needs. We use **gasoline**, natural gas, **electricity**, steam, coal, water, **solar energy**, nuclear fuel, **geothermal energy**, and even garbage as forms of energy.

Gasoline is used in cars, trucks, boats, and aircraft. Natural gas is used for heating and cooking. Coal, nuclear fuel, water, geothermal energy, and burning garbage are all used to generate the steam that turns the turbines that make electricity.

Have you ever seen a geyser such as Old Faithful erupt? Each time a geyser erupts, it gives off geothermal energy in the form of steam. Geothermal energy is energy that escapes or is captured from the center of the earth.

Is your home heated with natural gas or electricity? Is the cooking in your home done with natural gas or electricity?

Chemical Production **Chemical** production is the way we produce the chemicals used in fertilizers, medicine, **preservatives**, and many other applications. We use chemicals all the time. In fact, we depend on them. We use chemicals to make plants and trees grow faster and bigger. We use chemicals to preserve food during processing. We use chemicals in making the various medicines we need. We use chemicals to make paint, dye, paper, synthetic fabrics, building materials, and many other products.

Most processed food products contain chemicals. Have you ever read the label on a food product such as a can of soup, a loaf of bread, or a box of cereal? Most containers have labels that list the chemicals contained in the food. Look at a box of cereal or a loaf of bread. How many different chemicals are listed on the label? What are the names of the chemicals?

So far we have looked at large systems: education, health care, transportation, communications, and production systems. Systems do not have to be large. There are also small systems. Remember, a system is a group of related elements that work together for a specific purpose.

A car contains several systems. There is the steering system, braking system, and electrical system. A house contains several systems. There is the heating and cooling system, the electrical system, and the plumbing system. Let's see if you can name the elements of the heating system in your house.

There is the heater to generate warm air. There is the ductwork to take the warm air to the various rooms. There are the registers or vents to let the air out and distribute it evenly throughout the rooms. And there is the thermostat to let you control the temperature. All of these things are elements of the heating system of your house. From this you can see that systems do not have to be large. There are many technological systems both large and small.

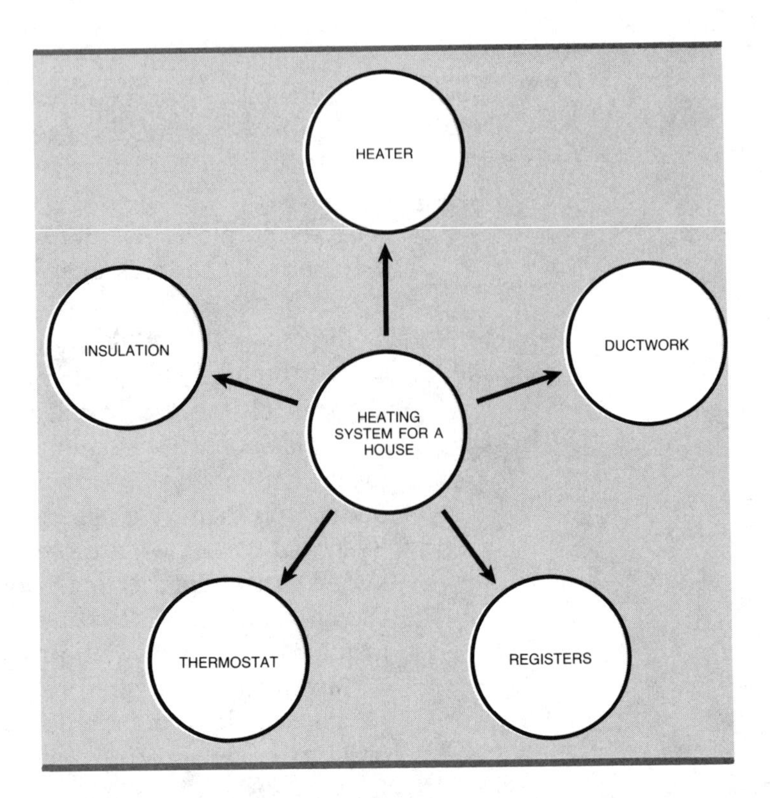

Elements of the heating system of a house.

MATH AND PROBLEM SOLVING

You have seen how the design process is used to solve problems and how systems are used to solve problems. One of the most important skills with regard to problem solving is math. You can solve very few technological problems without using math.

In the design process, math is used in every phase. In developing systems to solve problems, we use math extensively. Let's look at an example of the various ways math is used solving a technological problem. The problem is to design a heating and cooling system for a residential dwelling. Math would be required to

1. Calculate the square footage of the house
2. Calculate the heat loss and heat gain through the floor, walls, ceiling, windows, and doors
3. Calculate the heating and cooling capacity required of the heater and air conditioner
4. Calculate the size of ductwork and registers

There are almost no occupations in the world today that do not require math. Some require more than others, but almost all require some math. Scientists and engineers use analytical geometry, calculus and higher level math every day. Technicians frequently use intermediate levels of math such as algebra, geometry, and trigonometry. Mechanics and other craftsmen often use the basic levels such as arithmetic. Therefore math should be considered an essential skill area for living and working in an age of technology. A person cannot study too much math. It is that important.

SUMMARY

A system is a group of related elements that work together for a specific outcome. There are open- and closed-loop systems. Some of the most important technological systems are public education, health care, transportation, communications, and production systems. All systems are made up of individual elements. For example, elements of the transportation system are land vehicles, routes, and facilities; water vessels, routes, and facilities; and aircraft, routes, and facilities. Elements of the communications system are the telephone, radio, mail service, television, newspapers, and magazines. Smaller systems within systems are called subsystems.

There are many different technological systems, large and small. The braking, electrical, and steering systems in a car are examples of small systems. The heating and cooling system in your house is another example of a system.

KEY WORDS

System
Transportation
Health care
Communication
Production
Vehicle
Route
Facility
Sender
Food processing
Textiles
Energy
Chemicals
Receiver
Transmitter
Gasoline
Electricity
Mass communication
Raw materials
Construction
Manufacturing
Agriculture
Preservatives
Solar energy
Geothermal energy

SOME QUESTIONS TO TRY

1. Define a system.
2. Name three large systems.
3. Name three small systems.
4. What are the elements of the health care system?
5. What is the purpose of the production system?

TECHNOLOGY LEARNING ACTIVITIES

1. Build a model or construct a three-dimensional wall chart that illustrates the main elements of
   - The health care system
   - The transportation system
   - The communications system
   - The production system
2. Design a system for bringing about a specific outcome such as
   - Calmly and safely evacuating the classroom for a fire drill
   - Loading and unloading the buses at your school
   - Collecting and returning papers in your class

LASER DISC AUDIO
NAVSTAR
LASER ATLAS

PART TWO

THE PHYSICAL TECHNOLOGIES

CHAPTER 4

CONSTRUCTION TECHNOLOGY

OBJECTIVES

After completing this chapter, you should be able to

- explain the need for construction.
- list and explain the various things people construct.
- list and explain the most frequently used construction tools, materials, and processes.
- list some of the careers associated with construction.

Construction is one of the oldest technology areas. There has been construction technology since the first thatched hut was built to shelter people. The products of construction technology are the shelters and other structures we need. In this chapter, you will learn why construction technology is needed, what types of things people construct, some of the most important tools, materials, and processes used in construction, and careers in construction technology.

NEED FOR CONSTRUCTION TECHNOLOGY

The three most basic needs people have are food, clothing, and **shelter**. Construction technology is the way we make the shelters and other structures we need. Although we tend to take it for granted, the need for shelter is a critical need. The house you live in is a shelter. It provides you with protection, comfort, and convenience. These are the three needs we have with regard to shelter: protection, comfort, and convenience.

Protection

Protection was the first reason for building shelters. The earliest people needed protection from rain, snow, sun, wind, cold, predatory animals, and other people who might harm them or take their belongings. Even the most modern shelters still serve the same purposes.

Think of the house you live in. The floor, walls, and roof keep out the sun, wind, snow, and rain. People can lock their doors and windows to keep out thieves. Predatory animals are not as much of a problem as they used to be, but houses still keep out the neighborhood animals.

Comfort

If all people wanted from a shelter was protection, they could live in a cave. However, people also want comfort. Over the years, people have continually added features to their shelters to make them more comfortable. Even the earliest people built fires in their caves for warmth and put leaves and straw on the floor so that they could have a softer, dryer place to sleep.

The basic human needs.

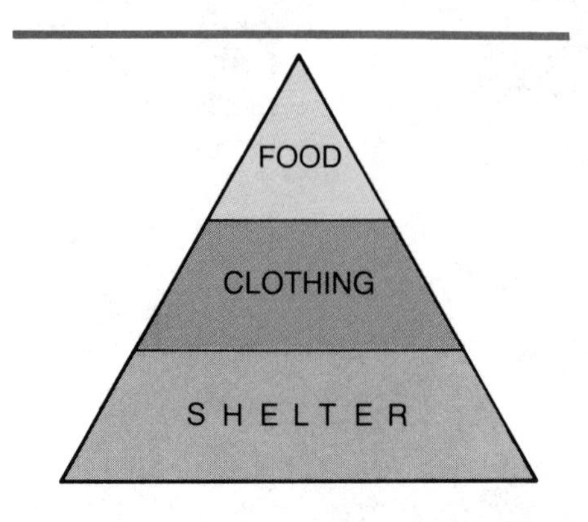

Modern shelters provide protection from the elements.

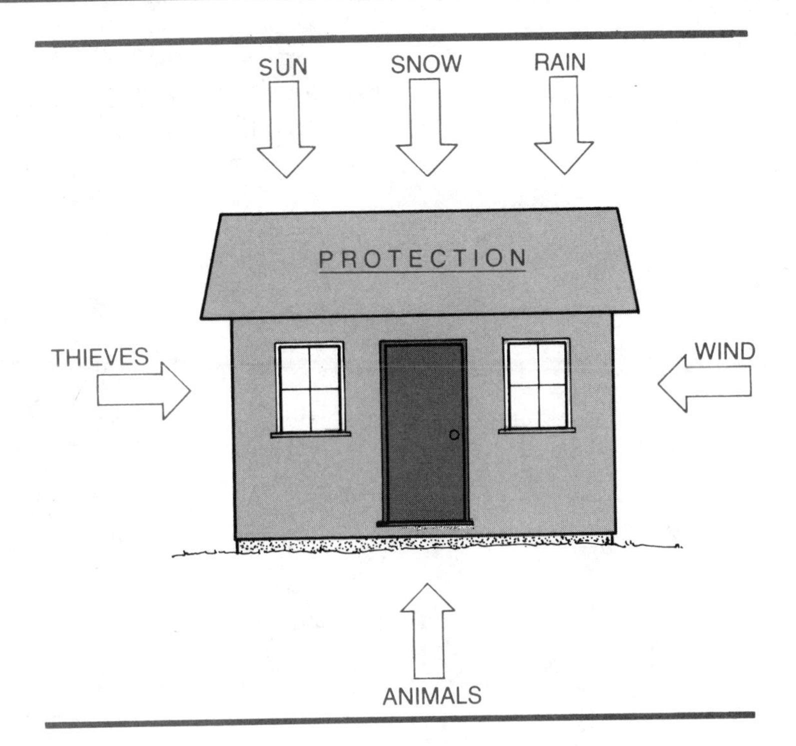

Comfort features provided by modern shelters.

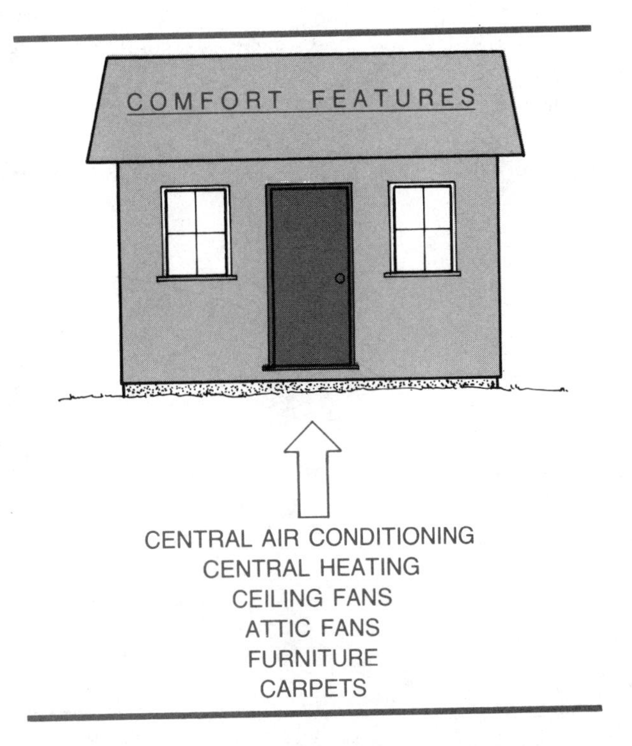

Modern shelters are designed for maximum comfort. Think of the many comfort features you can find in a new house: central heating and air conditioning, carpet, attic fans, furniture, and hot running water. All of these things make the occupants more comfortable.

Convenience

As time passed, people began to want more than just protection and comfort from their shelters. They also wanted convenience. In early times, people had to walk outside in the rain, snow, or cold to draw water from the well. They had to go outside to bring in food stored in root cellars or barns. You are lucky. Think of how it would be to have to go outside whenever you wanted a drink of water or something from the refrigerator. Suppose you had no electricity for your television, radio, stereo, or VCR?

Modern shelters are designed for convenience. They have plumbing systems, hot and cold running water, electricity, stoves, sinks, bathtubs, and refrigerators. You don't have to go outside to get water or food. You don't have to go outside to bathe or wash clothes. You don't have to build a fire to heat water or cook.

So far we have talked only about shelters. Shelter is the most important need satisfied by construction technology. But we also need other types of structures. In the next section, we will look at all of the various structures people make to improve the quality of their lives.

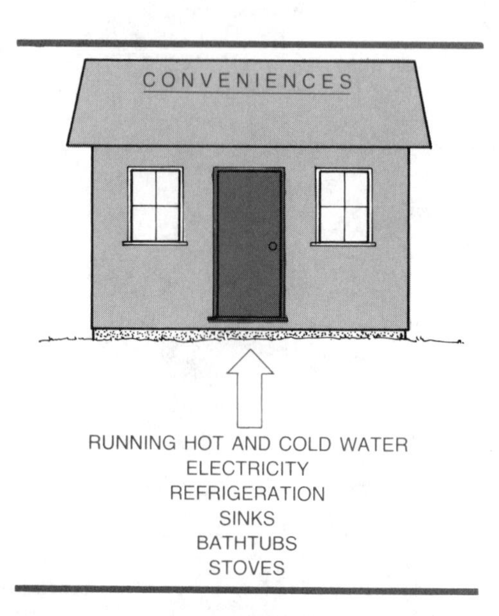

Conveniences provided by modern shelters.

THINGS WE CONSTRUCT

The things we construct are the products of construction technology. These products can be divided into three groups: shelters, structures, and pathways or routes. Let's look at some of the products of construction that fall into each of these groups.

Shelters

What do you think of when you hear the word "shelter"? Most people think of where they live. Homes or dwellings are shelters. A **dwelling** is a **structure** designed for people to live in. There are many kinds of dwellings. How many can you name? There are houses, apartments, condominiums, cottages, cabins, and townhouses. Can you name any more?

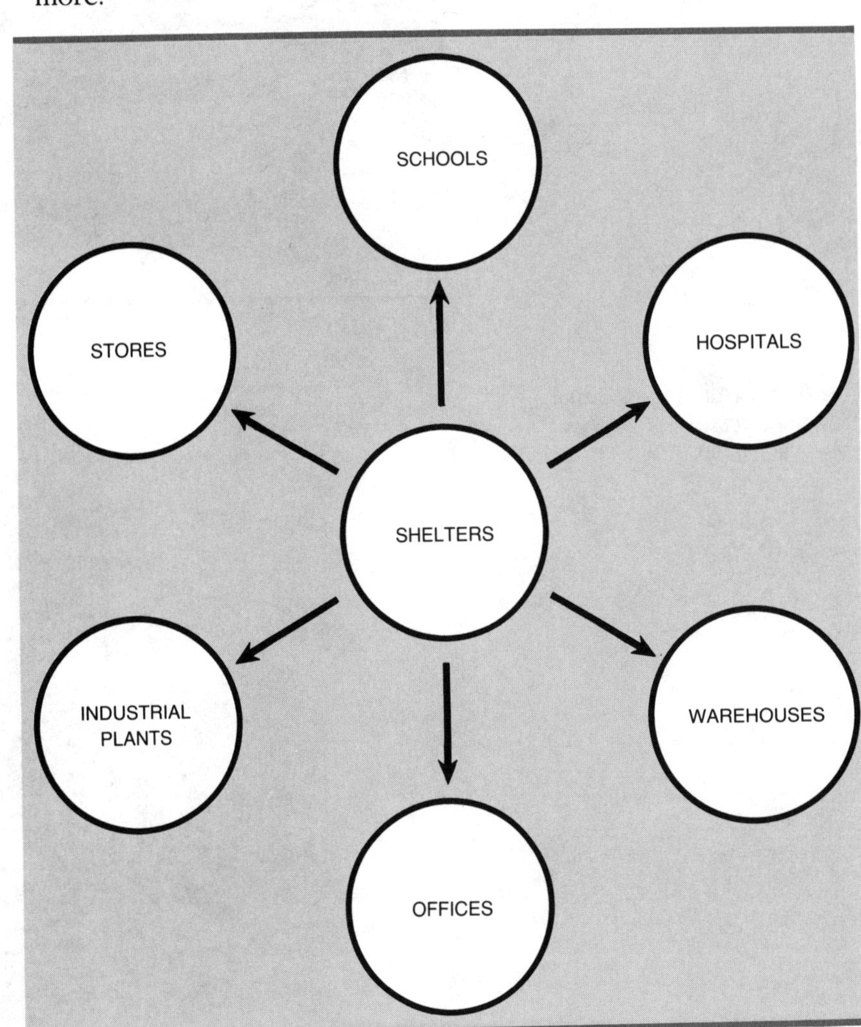

Types of shelters.

The Calgary International Airport is a prestressed concrete shelter. *Courtesy of Prestressed Concrete Institute.*

The Alcoa Building in San Francisco. *Courtesy of Bethlehem Steel Corporation.*

Dwellings are the shelters we live in, but there are other types of shelters too. Schools, hospitals, office buildings, industrial plants, warehouses, and airport terminals are shelters. Any structure that is designed to provide people with **protection**, **comfort**, and **convenience** while performing some type of activity is a shelter.

Schools give students protection, comfort, and convenience while they learn. Office buildings provide these things for the people who work in them. Stores provide these things while people shop. Can you name any other types of shelters?

The Federal Office Building and Post Office in Santa Rosa, California. *Courtesy of Prestressed Concrete Institute.*

The Guam Reef Hotel in the Mariana Islands. *Courtesy of Prestressed Concrete Institute.*

The administrative headquarters for the Citizens First National Bank of New Jersey. *Courtesy of Bethlehem Steel Corporation.*

The Woonsocket Institution for Savings in Providence, Rhode Island. *Courtesy of Bethlehem Steel Corporation.*

The Shelby Mutual Insurance Company's headquarters building in Columbus, Ohio. *Courtesy of Bethlehem Steel Corporation.*

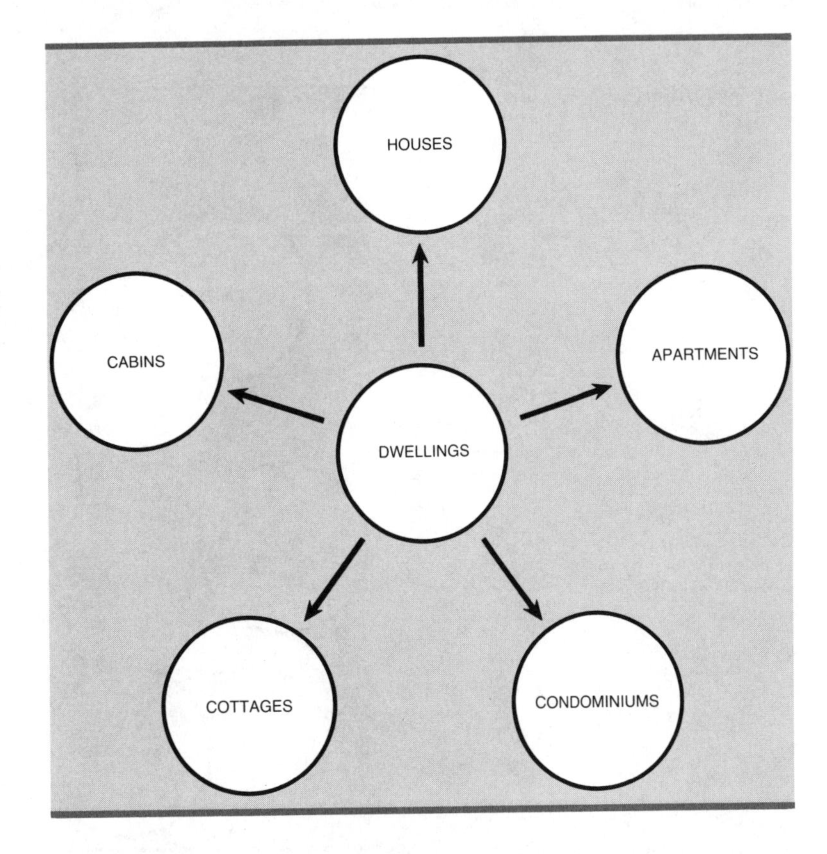

Types of dwellings.

Structures

Not all of the products of construction technology are shelters. How many structures can you name that are not shelters? There are bridges, overpasses, dams, stadiums, walkways, water towers, and monuments. Can you name any others?

Bridges allow us to be transported across rivers, canals, bays, lakes, and other bodies of water. An overpass is a bridge over dry land. Overpasses allow one road to pass over another. Dams are constructed to hold back water.

Stadiums are constructed to house sporting events such as football, soccer, baseball, and track and field games. Walkways give people safe passage over heavily traveled roads, streets, and highways. Water towers are constructed to store and distribute the water needed by communities.

Monuments are built to honor people, ideas, and events. The Statue of Liberty is a monument. Have you ever seen the Statue of Liberty? What other monuments have you seen? Can you name any other types of structures?

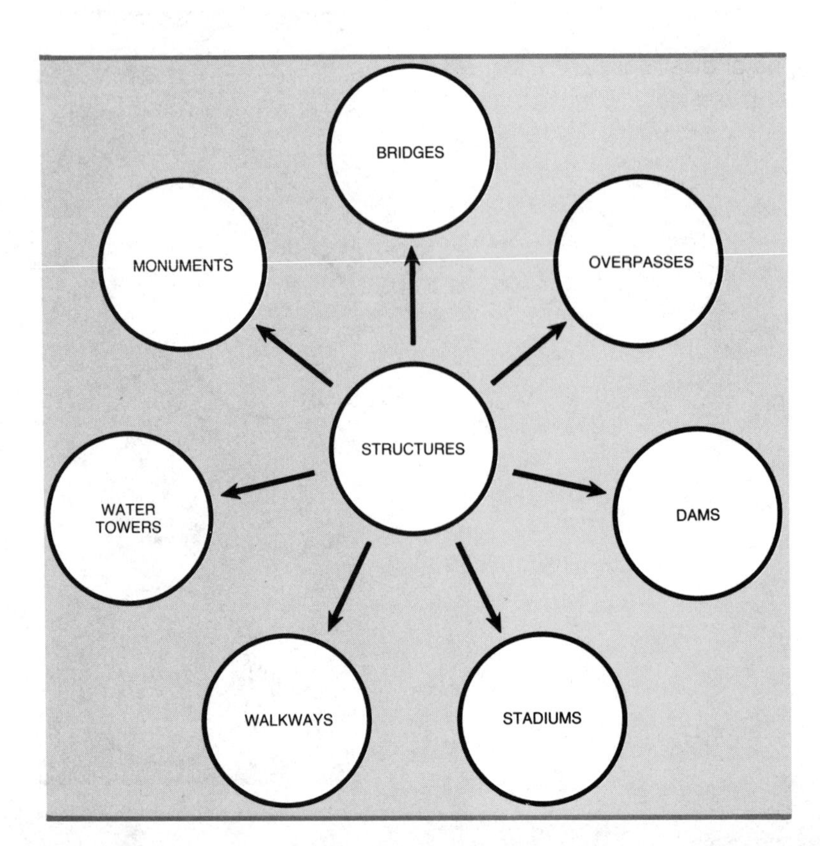

Types of structures.

The longest steel arch bridge span in the world over the New River Gorge near Charleston, West Virginia. *Courtesy of United States Steel.*

Erecting the New River Gorge Bridge. *Courtesy of United States Steel.*

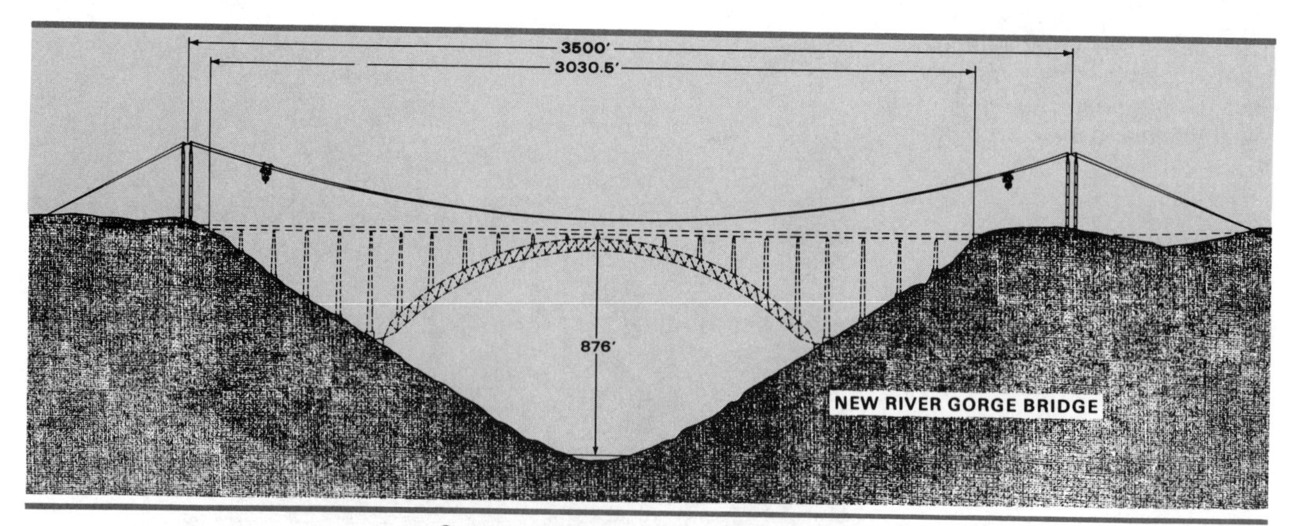

Sectional drawing of the New River Gorge bridge showing the total span and height. *Courtesy of United States Steel.*

The Big Sandy Bridge in Kenova, West Virginia. *Courtesy of United States Steel.*

The Cedar River combination bridge and dam in Cedar Rapids, Iowa. *Courtesy of Untied States Steel.*

The Twelfth Street viaduct in St. Louis, Missouri. *Courtesy of Bethlehem Steel Corporation.*

The bridge over the Cuyahoga River on the Ohio Turnpike. *Courtesy of Bethlehem Steel Corporation.*

The Napa River Bridge in Napa County, California. *Courtesy of the Prestressed Concrete Institute.*

Pathways

Pathways are the things we construct to make transportation easier. How many different kinds of pathways can you name? People build roads, streets, highways, and tracks to make land transportation easier. People build runways so that aircraft can take off and land. Sidewalks and bicycle trails are also pathways. Can you name any other types of pathways?

CONSTRUCTION TOOLS, PROCESSES, AND MATERIALS

It takes a lot of different tools, processes, and materials to construct the houses, office buildings, schools, hospitals, bridges, dams, and stores we need. You should be familiar with some of these tools, materials, and processes.

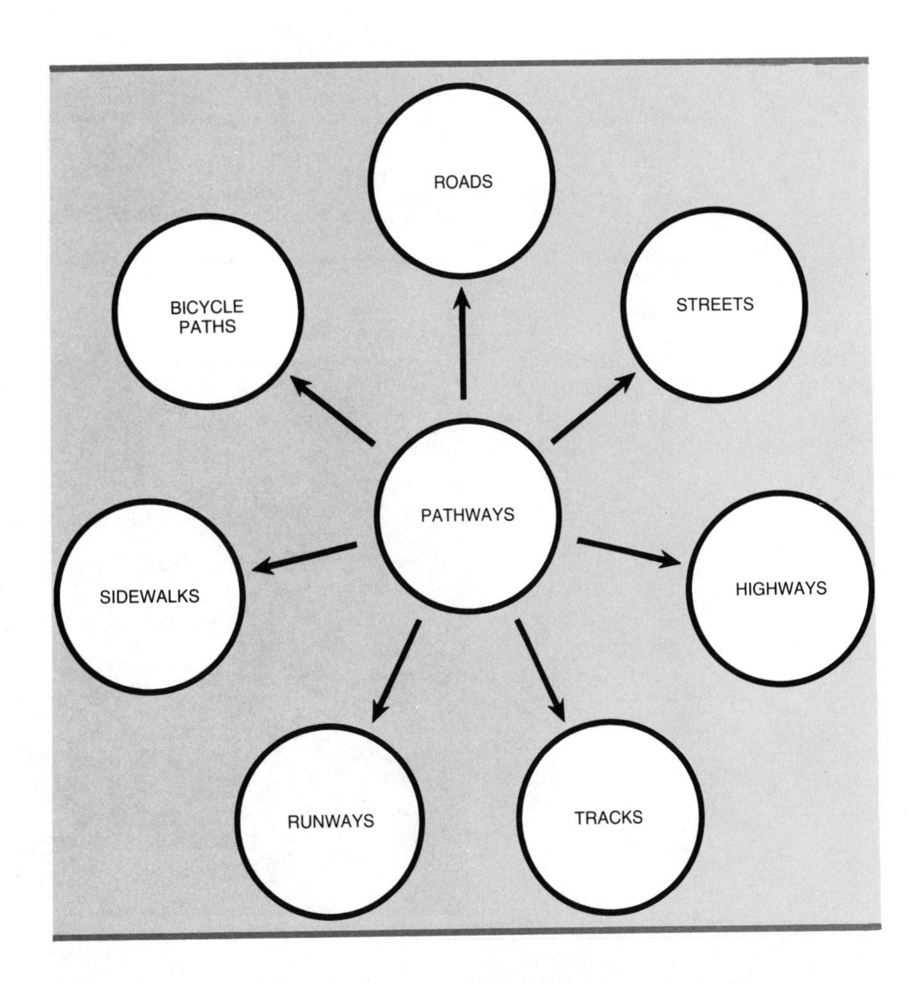

Types of pathways.

Construction Tools And Processes

In Chapter 2, you learned that the word "tool" is sometimes used in a general way to mean tools, mechanisms, and machines. Such is the case in this section. There are too many tools and processes in construction to look at them all, but the ones presented in this section are good examples.

The front loader is a tool used in construction projects.

The bulldozer is a tool used to clear and level large construction sites.

Design and Drafting Before something can be constructed it must be designed. The design process was presented in Chapter 2. It is a five-step process:

1. Define the problem.
2. Try out ideas.
3. Select the best idea.
4. Make a model.
5. Make the drawing.

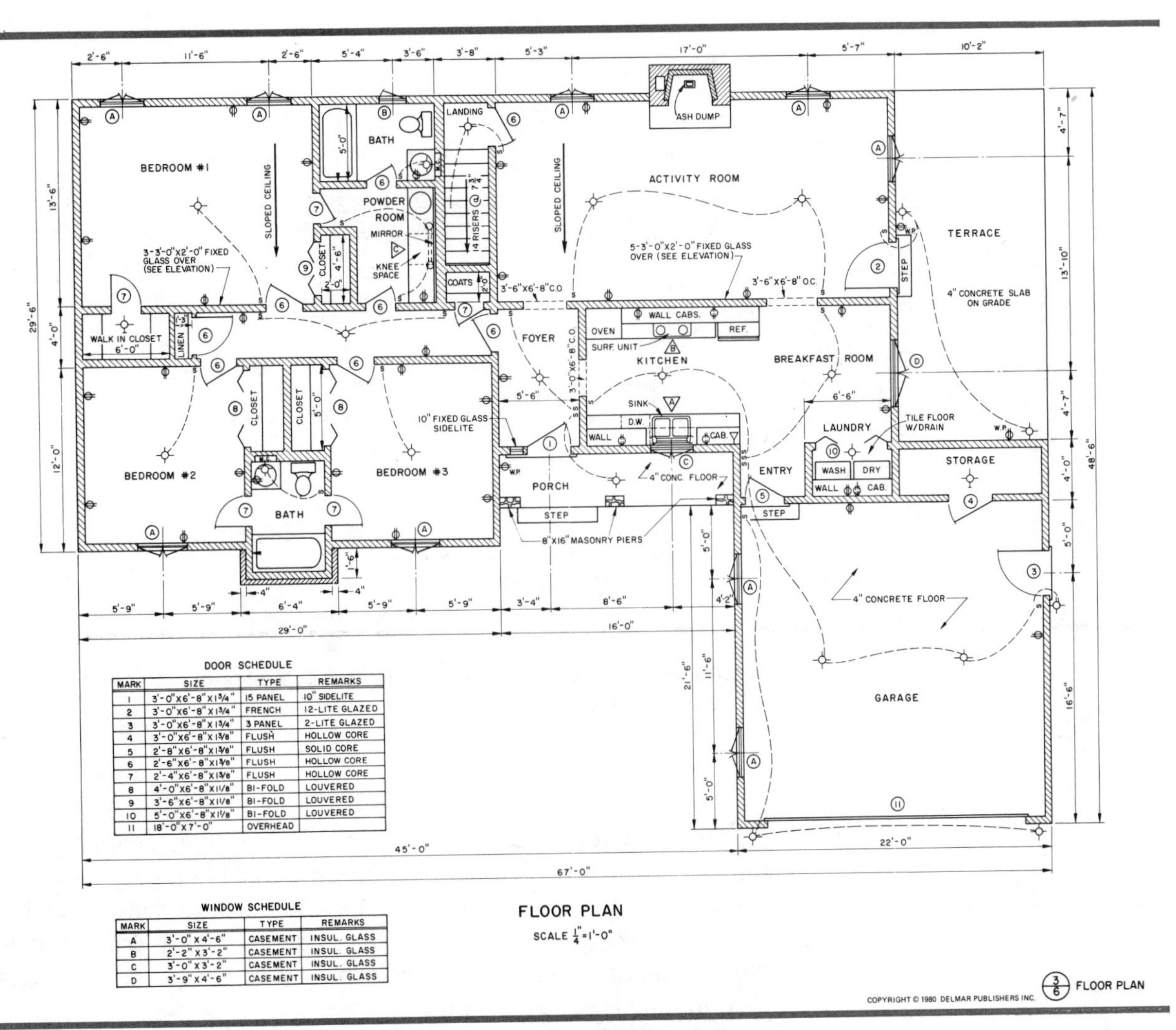

DOOR SCHEDULE

MARK	SIZE	TYPE	REMARKS
1	3'-0"X6'-8"X1 3/4"	15 PANEL	10" SIDELITE
2	3'-0"X6'-8"X1 3/4"	FRENCH	12-LITE GLAZED
3	3'-0"X6'-8"X1 3/4"	3 PANEL	2-LITE GLAZED
4	3'-0"X6'-8"X1 3/8"	FLUSH	HOLLOW CORE
5	2'-8"X6'-8"X1 3/8"	FLUSH	SOLID CORE
6	2'-6"X6'-8"X1 3/8"	FLUSH	HOLLOW CORE
7	2'-4"X6'-8"X1 3/8"	FLUSH	HOLLOW CORE
8	4'-0"X6'-8"X1 1/8"	BI-FOLD	LOUVERED
9	3'-6"X6'-8"X1 1/8"	BI-FOLD	LOUVERED
10	5'-0"X6'-8"X1 1/8"	BI-FOLD	LOUVERED
11	18'-0"X7'-0"	OVERHEAD	

WINDOW SCHEDULE

MARK	SIZE	TYPE	REMARKS
A	3'-0"X4'-6"	CASEMENT	INSUL. GLASS
B	2'-2"X3'-2"	CASEMENT	INSUL. GLASS
C	3'-0"X3'-2"	CASEMENT	INSUL. GLASS
D	3'-9"X4'-6"	CASEMENT	INSUL. GLASS

Floor plan for a residential dwelling. *Courtesy, from Goetsch, CAD Applications: Architectural,* © 1986 by Delmar Publishers Inc.

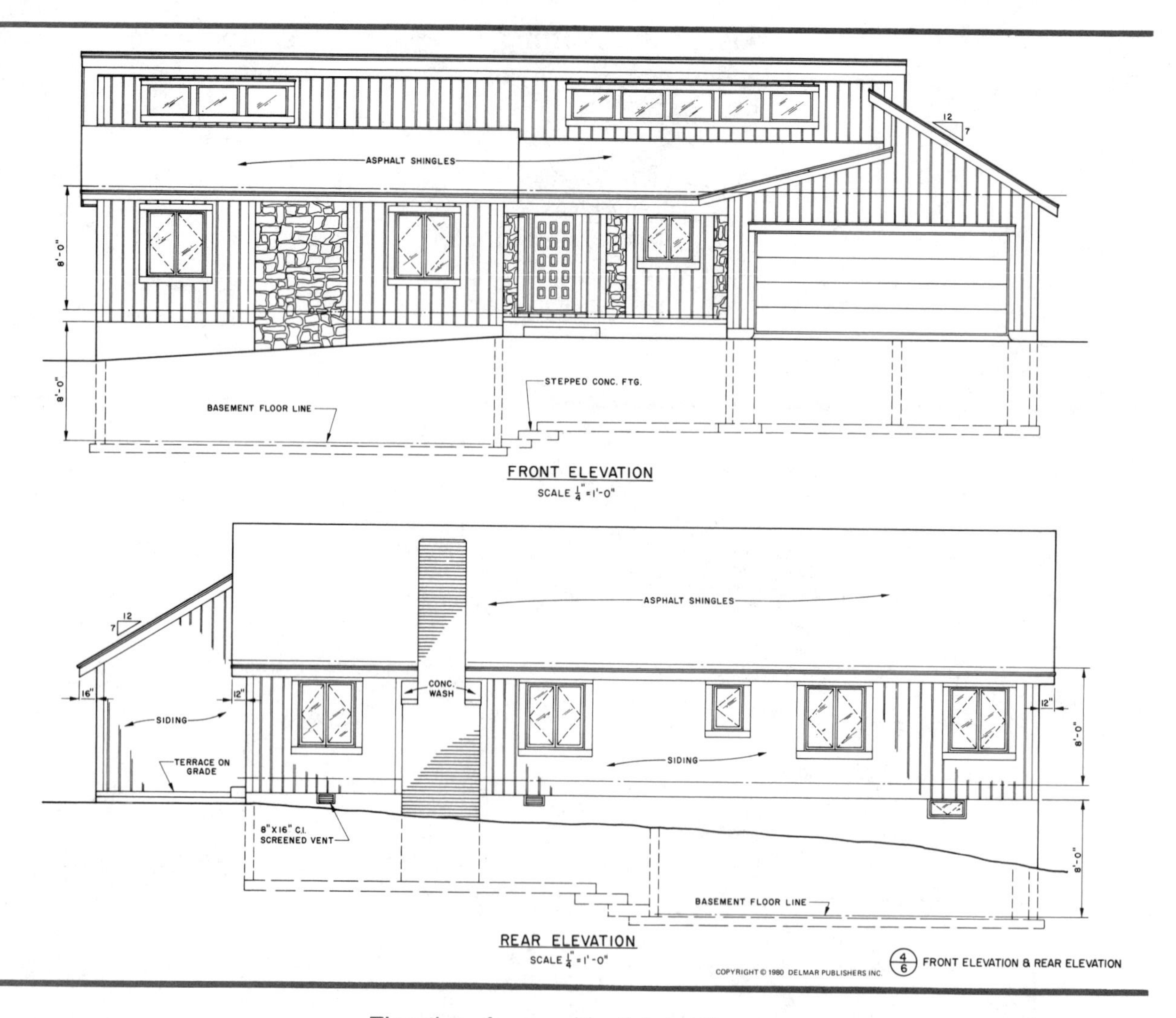

Elevations for a residential dwelling. *Courtesy, from Goetsch, CAD Applications: Architectural,* © 1986 by Delmar Publishers Inc.

There are tools used in each of these five steps. Traditionally, the tools used in design and drafting have included pencils, dividers, scales, templates, parallel bars, drafting machines, triangles, technical pens, and inking sets. In more modern design and drafting situations, these traditional tools have been replaced by computer-aided design and drafting (CADD) systems.

A **CADD** system consists of a processor, disk drives, a keyboard, display terminal, digitizer, plotter, and some type of interactive device

Students learn CADD. *Courtesy of Robert C. Atwood.*

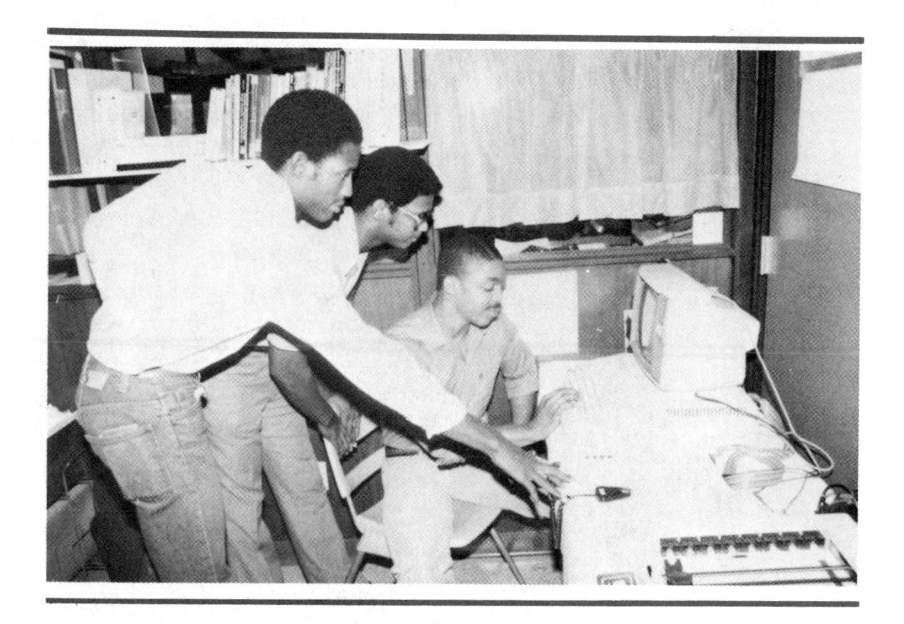

such as a light pen, mouse, or puck. The process of design and drafting is not changing very much, but the tools used are.

When a design has been completed and plans have been produced, the construction processes can begin. These processes include measurement, layout, testing, sawing, boring, smoothing, and fastening.

Measurement Many things have to be measured when constructing a shelter, structure, or pathway. To make **measurements**, people use rulers, scales, meter sticks, tapes, folding rules, and calipers. Land must be measured before construction begins. Lumber must be measured before it is cut. Holes must be measured before they are drilled. Footing trenches must be measured before they are dug. The ingredients for concrete must be measured carefully before they are mixed.

Layout **Layout** is an important construction process. Before building begins on any type of structure, the footings must be laid out. Floor beams and joists must be laid out. Wall studs, doors, and windows must be laid out. Ceiling joists and beams must be laid out. Roof trusses or rafters must be laid out.

Some of the most frequently used layout tools are the framing square, marking gauge, panel gauge, butt gauge, and sliding T bevel. Can you name any other layout tools? Of course, tapes, rules, chalk, string, pencils, and shovels are also used in laying out a structure.

Testing Each operation performed in a construction project should be tested. Is it straight? Is it level? Is it square? Is it strong enough?

These are important questions that are answered through various types of tests. A variety of tools are used in **testing**.

Some of the most frequently used tools are the level, 6-8-10 triangle, and the builder's level. The level tests for flatness and straightness. The 6-8-10 triangle tests for squareness in laying out a building. The builder's level tests the level of land. Can you name any other testing tools?

Sawing **Sawing** is an important construction process. People saw wood, concrete, and even steel in construction projects. There are a number of different tools used for sawing. The hand saw is the one most people think of. It is used for making straight cuts and angular cuts. The keyhole saw is used for cutting along curves. Miter saws are used along with miter boxes for making square and angular cuts. The coping saw is used for making fine irregular cuts.

There are also a number of power tools used for sawing. The circular saw is used for straight cutting, crosscutting, ripping, and bevel cutting. The radial arm saw is one of the most versatile of construction tools. It can do crosscutting, bevel cutting, ripping, mitering, dado cutting, shaping, and routing. Saber saws and reciprocating saws are two other types of power saws used often. Can you name any other power tools used for sawing?

Boring **Boring** is the process used for making holes in construction materials. People use a variety of drills and bits for boring holes. There are hand drills and power drills. Both can be used for boring holes in wood, concrete, and even steel. Have you ever seen a set of drill bits?

Smoothing Edges, corners, and surfaces have to be smoothed before they can be painted or finished. Different tools are used for **smoothing**. Concrete can be smoothed with a trowel or a broom. Wood can be smoothed with scrapers, planers, and files. Can you name any other tools used for smoothing?

Fastening **Fastening** is one of the most frequently used construction processes. Fastening is the way people join construction materials and make them stay together. Materials can be fastened by nails, bolts, screws, welds, and other means.

The tools used for fastening materials are hammers and nails, screwdrivers and screws, wrenches, nuts, and bolts, and welding rods and machines. Can you name any other fastening tools?

Construction Materials

Different materials are used in construction technology. You need to be familiar with the most frequently used materials: wood, stone, brick, steel, aluminum, and concrete.

Radial arm saw. *Courtesy of Sears.*

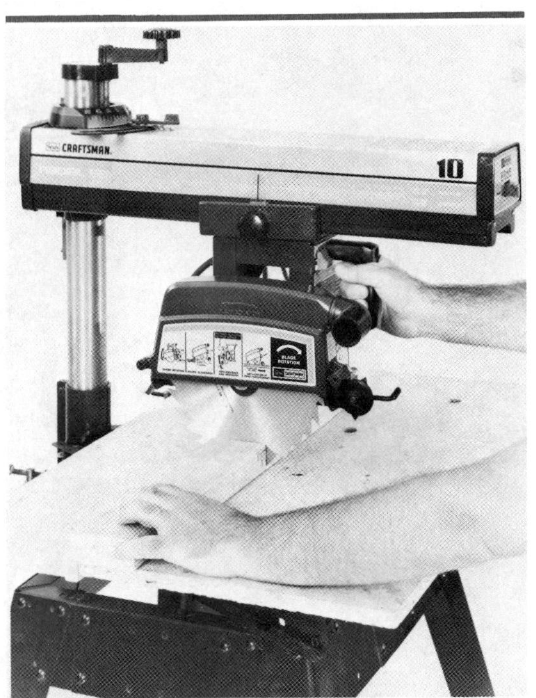

Cutting a 2 × 4 on a radial arm saw. *Courtesy of Sears.*

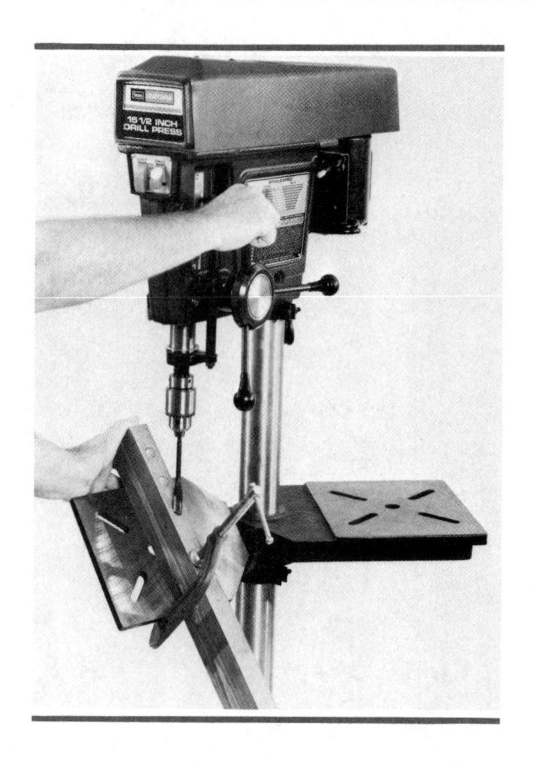

15 1/2-inch drill press. *Courtesy of Sears.*

12-inch bandsaw-sander. *Courtesy of Sears.*

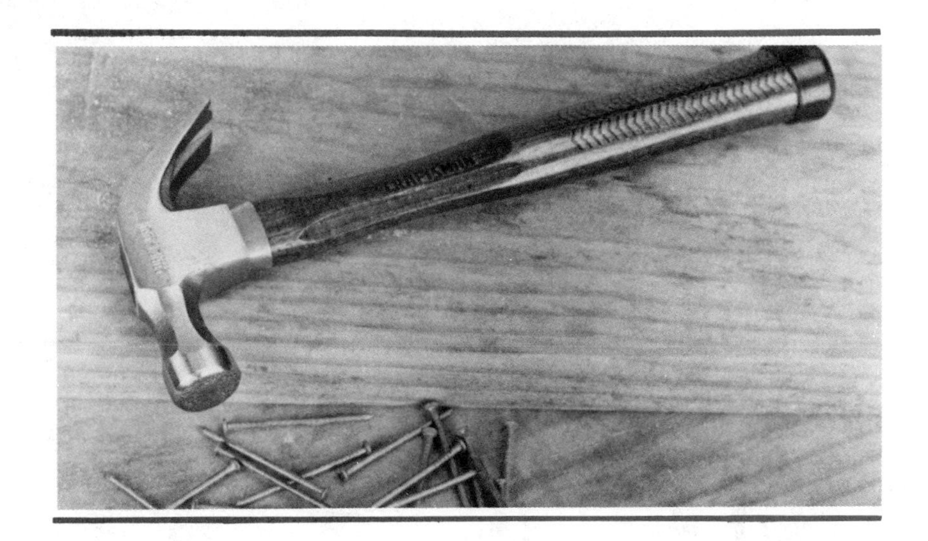

Three-claw hammer. *Courtesy of Sears.*

Wood **Wood** is the most frequently used material in residential construction. Residential construction involves building dwellings such as houses, apartments, condominiums, and townhouses. It is also used a great deal in commercial construction. Wood is typically used in making floors, walls, ceilings, and roofs.

Wood is used in several forms in construction. Lumber is the most common form, but it is also the most expensive. To save money and get as much wood as possible out of each tree, processes have been developed to make wood products out of wood chips and glue. These products include plywood, paneling, and particle board. Wood is also used in making other construction products such as floor and roof trusses.

Stone and Brick **Stone** and **brick** are used mostly for decoration in modern construction projects. Stone footings have been replaced by the concrete variety for the most part, and brick walls today are more likely to be brick veneer walls.

Brick veneer is a thin layer of brick that is attached to the outside of a wood or metal wall to make an attractive appearance. Brick veneer walls are not "structural." In other words, they are not involved in carrying the load of the building.

Stone is still popular for decorative wall facings and in making fireplaces. Bricks are made by baking different types of clay. Stone is cut in different shapes in quarries.

Steel and Aluminum **Steel** and **aluminum** are used in all types of construction projects, but particularly in commercial projects. A commercial project is an office building or a store. Steel and aluminum are converted into several different standard shapes in

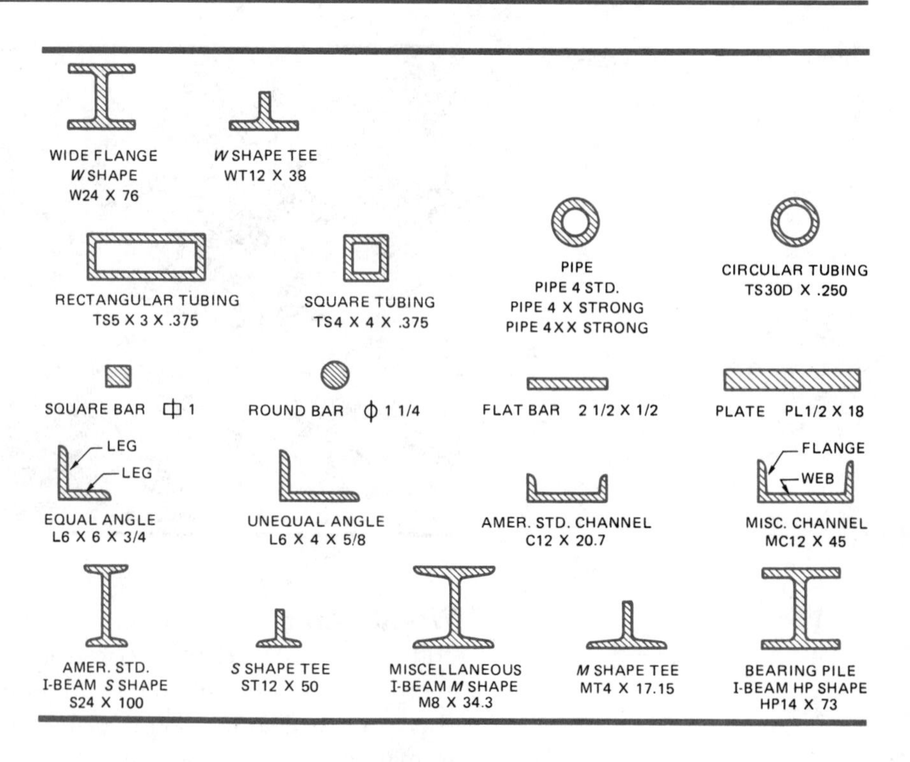

Structural steel shapes. *Courtesy, from Taylor, Blueprint Reading for Machinists, 4th Edition, © 1985 by Delmar Publishers Inc.*

mills. Then they are cut, bent, drilled, punched, bolted, and welded as needed.

Concrete **Concrete** is made by mixing cement, sand, water, and rocks. Steel bars, strands, or mesh can be added to give the concrete more strength. Concrete is used in several different forms in construction.

Poured-in-place concrete is mixed on the job site and poured into metal or wooden forms or footing trenches. Poured-in-place concrete is usually reinforced with metal bars and welded wire mesh. Precast concrete is like poured-in-place except that it is poured into forms at a concrete plant, allowed to dry, and then shipped to the job site and erected.

Prestressed concrete is similar to precast except that high-strength steel strands are pulled through the forms before the concrete is poured. When the concrete has dried long enough, the strands are cut and the tension on them is transferred to the concrete giving it more strength. Prestressed concrete products are shipped to the job site and erected.

The three phases of construction.

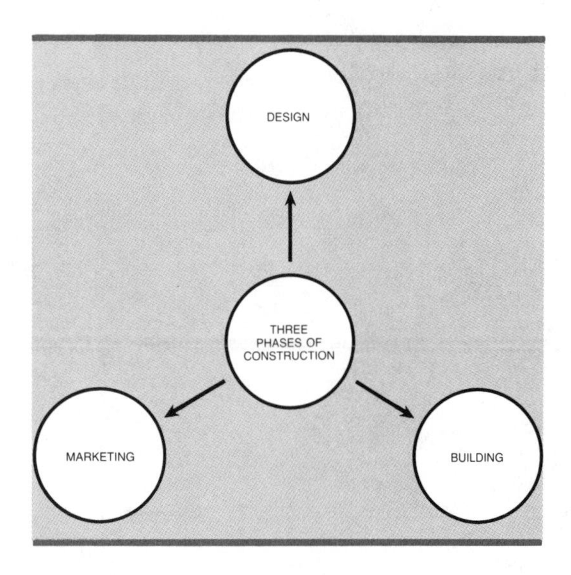

A SAMPLE CONSTRUCTION PROJECT

The best way to understand how the tools, processes, and materials of construction technology produce shelters or structures is by studying an example. Let's look at how a house is constructed after it has been designed and the plans have been drawn.

CONSTRUCTING A HOUSE

STEP 1 Site Work. Using the plot plan as a guide, workers get the site ready for construction. They clear trees, stumps, roots, underbrush, and anything else that might be in the way. They fill in low spots and cut away high spots so there is a level area on which to build.

STEP 2 Footings and Foundations. In this step, the footing trenches are dug, and reinforcing bars are placed in the bottom of the trenches. Concrete is mixed and poured into the trenches. When the footing is set, the foundation walls are built on top of it. Foundation walls are usually made out of concrete or concrete block.

STEP 3 The Floor System. In this step, the floor is constructed. If it is an off-grade floor system, it consists of floor beams, joists, and plywood subflooring. If it is an on-grade system, fill dirt is poured inside the foundation walls and tamped down until it is solid and firm. Then the soil is treated for termite control, covered with a thin layer of sand, and covered with a sheet of plastic for moisture control. Welded wire mesh is placed on top of the plastic, and the concrete is poured. Most concrete floors are 4 inches thick.

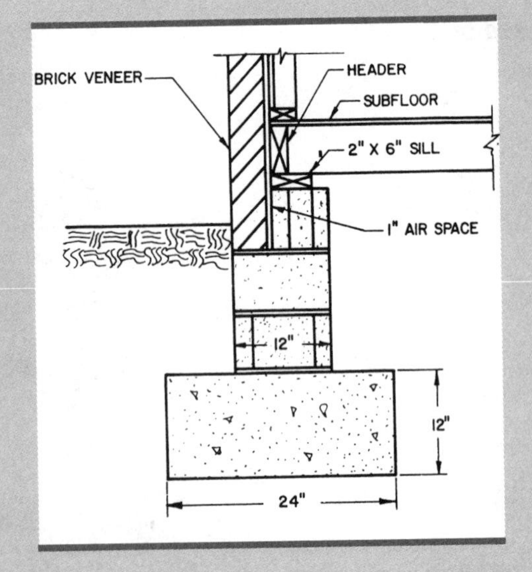

12-inch block foundation detail.
Courtesy, from Goetsch, CAD Applications: Architectural, © 1986 by Delmar Publishers Inc.

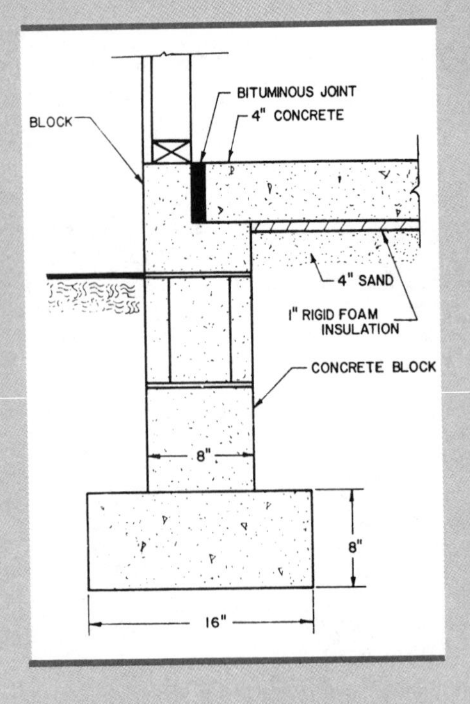

8-inch foundation wall detail.
Courtesy, from Goetsch, CAD Applications: Architectural, © 1986 by Delmar Publishers Inc.

STEP 4 The Walls. In this step, the walls are constructed. The wall members are usually made of wood but can be made of aluminum. They consist of the sole plate, regular studs, cripple studs, trimmer studs, headers, and a double top plate. The main vertical members, the studs, are usually placed 16 inches apart from center to center.

TAPPING ALTERNATE SOURCES OF ENERGY

Scientists and technologists are developing new ways to produce electricity. (1) This geothermal plant in California uses steam from deep in the ground. The steam powers large turbines, which generate electricity. (2) This solar energy plant harnesses energy of the sun. (3) Modern windmills are now used in many parts of the world. The "eggbeater" shape of this type, called the Darrieus windmill, allows it to move no matter what the wind direction may be. These three alternate energy sources are unlimited—that is, they cannot be used up, like oil or coal. Also, they do not pollute our environment the way fossil fuels do.

INDUSTRY: THE NEW LOOK

100 years ago manufacturing was done in noisy, crowded factories; hundreds of workers put in long, tiring days in harsh working conditions. Now machines do much of that work, and fewer people are needed to produce manufactured goods. (1) Industrial robots are used in this auto factory. (2) This is a flexible manufacturing system. Automatic machinery can be moved around the factory as needed. Operators in the control booth move the machines with special computers. (3) A student using CNC (Computer Numerical Control) equipment in school. The control panel can be used to program the machine to drill holes exactly where needed on the workpiece.

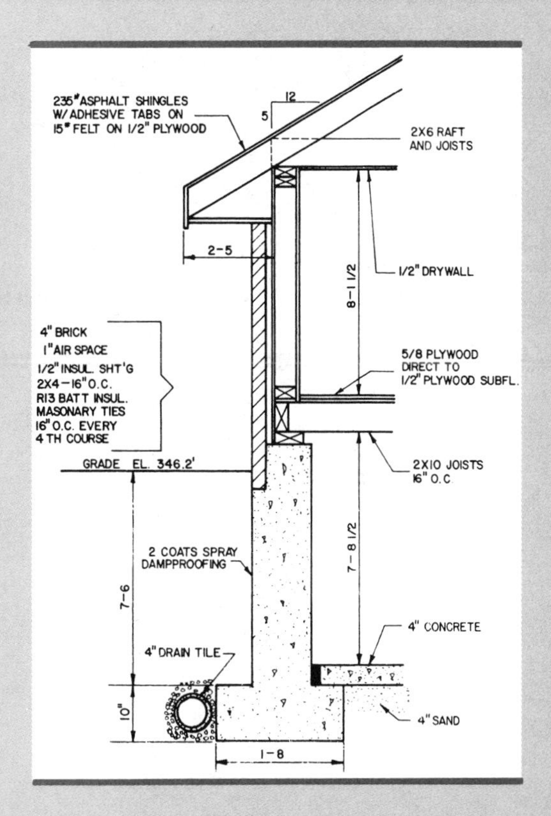

Typical wall section for a residential dwelling. *Courtesy, from Goetsch, CAD Applications: Architectural, © 1986 by Delmar Publishers Inc.*

STEP 5 The Ceiling and Roof. In this step, the ceiling is constructed along with the roof. One method of construction involves ceiling joists and rafters. Another uses trusses. A truss is a prefabricated product that contains the joists and rafters in one piece. The ceiling is constructed by attaching either plywood or gypsum board to the bottom of the ceiling joists or trusses. The roof is constructed by attaching plywood, a moisture barrier, and roofing materials such as shingles to the top of the rafters or trusses.

STEP 6 Enclosing the Structure. In this step, the house is shut off from the outside elements by adding the cornice to the eaves of the house, by installing the windows and doors, and by applying exterior wall coverings such as wood siding, vinyl siding, or brick veneer.

STEP 7 The Plumbing System. In this step, all fittings, pipes, and plumbing fixtures are installed. Most modern dwellings have both copper and PVC (plastic) pipe. The fixtures include bathtubs, sinks, toilets, shower stalls, and water heaters.

STEP 8 The Electrical System. In this step, the house is wired for electricity. The electrical system in a house includes the distribution panel, circuits, switches, outlets, junction boxes, and lighting fixtures.

STEP 9 The Climate Control System. In this step, the heating and cooling system is installed. Most heating-ventilation-air conditioning systems consist of the furnace, air conditioning unit, ductwork, registers, and thermostat.

STEP 10 Finishing. The last step is to put the finishing touches on the house. In the house, this involves installing carpeting, wall coverings, trim work, and painting. Outside the house, it involves cleaning, painting, staining, and preserving. After the finishing step has been completed and the house has been inspected, the house is ready to be sold and occupied.

CAREERS IN CONSTRUCTION

Many different careers are associated with construction technology. Most of them fall into one of four categories: professional, technical, skilled trades, and services. Several career fields are associated with each category. This section describes some of the most common careers.

Professional and Technical Careers

Professionals in construction are the architects and engineers, who design what is to be built, and the top-level managers who run construction companies. Technical personnel work with architects and engineers in communicating their designs and have titles such as drafter, designer, checker, and CADD technician.

Architects **Architects** decide how a dwelling or structure will be shaped, how it will look, how it will fit into its surroundings, what materials it will be made of, and how available space will be used. Architects must complete a college education ranging from four to six years and pass a licensing test in order to practice.

Engineers Different types of **engineers** work on construction projects. Civil engineers oversee the calculations and design relating to site preparation. Structural engineers design the structural components of the structure called for in the architect's overall design. These components include the columns, beams, joists, wall members, and roof members. A structural component is one that carries some load.

Mechanical engineers in construction design the heating, cooling, and plumbing systems for the structure. Electrical engineers design the electrical system. Engineers must complete a college education that specializes in their chosen field (structural, civil, electrical, mechanical) and pass a licensing test in order to practice.

Drafting Technicians **Drafting technicians** are the people who communicate the design ideas of architects and engineers. They might be called drafters, draftsmen, designers, checkers, CADD technicians, or an assortment of other localized titles. Regardless of what they are called, drafting technicians are responsible for producing working drawings and plans needed to communicate the design for a dwelling or structure. Drafting technicians must undergo specialized training in a high school, area vocational school, trade school, or community college. A technical certificate or associate degree in drafting is usually needed for entry into drafting technician careers.

Skilled Trade Careers

Skilled tradespeople are the workers who take the architect's and engineer's plans and actually build the structure. Skilled tradespeople include carpenters, masons, electricians, plumbers, heating and cooling mechanics.

Carpenters **Carpenters** build the wooden parts of a structure and special forms for the concrete parts. There are "rough" carpenters and "finish" carpenters. A structure might have wooden floors, walls, ceiling joists, rafters, or trusses. Such components are built by rough carpenters. Finish carpenters apply the trim work and molding and make shelves and cabinets.

Masons **Masons** do the brick, stone, and block work for construction projects. This includes building the foundation wall, laying block, stone, or brick walls, veneering walls, and building fireplaces.

Electricians **Electricians** install the electrical systems in construction projects. This involves mounting the distribution panel, running wires for the electrical circuits, installing outlets and switches, and hanging fixtures.

Plumbers **Plumbers** install the plumbing systems for construction projects. They run lines for hot and cold water and drains. They install the water heater, bathtubs, sinks, toilets, dishwasher, and tie-ins for the washing machine.

Heating and Cooling Mechanics **Heating** and **cooling mechanics** install the heating and cooling systems for construction projects. They install the heater and air conditioner, ductwork, return air vent, registers, and thermostat.

These are only some of the many skilled trade careers associated with construction technology. There are still others. All of those listed require specialized training that can be obtained through formal study in a high school, area vocational school, trade school, community college, apprenticeship program, or through on-the-job-training.

Service Careers

A number of service careers are associated with construction technology. People in these fields are involved in the marketing of building products and materials or marketing the structures that are built. The most prominent of these fields are real estate agent, wholesale sales representative, and retail sales representative.

Real Estate Agent **Real estate agents** handle the selling of land and the products of construction such as houses, apartments, and buildings of all types. Many homes are built on "speculation." This means the builder constructs them without having definite buyers for them. He is speculating that the house will sell.

Builders are helped in this regard by real estate agents. Real estate agents undertake all of the things necessary to sell the houses, such as advertising, putting up "For Sale" signs, and showing houses to prospective buyers. Real estate agents must have a formal course in real estate and pass a licensing test in order to practice.

Wholesale Sales Representatives **Wholesale sales representatives** work for companies that make and sell construction-related products and fixtures such as windows, doors, cabinets, electrical fixtures, plumbing fixtures, and appliances. Wholesale sales representatives sell these products to construction companies or retail outlets, which then sell them to the public. Wholesale sales representatives must have specialized sales training, but it can be obtained in a wide variety of ways, ranging from on-the-job-training to a full college degree.

Retail Sales Representatives Stores and retail outlets that sell construction materials, products, and fixtures employ **retail sales representatives** to meet the public and assist them in finding what they need. For example, the person who waits on you at a hardware store that sells lumber is a retail sales representative. Retail sales personnel also require specialized training, but the amount of training required and the way it can be obtained vary widely.

SUMMARY

Construction is the way people make the shelters, structures, and pathways they need. Shelters provide people with protection, comfort, and convenience, and include houses, apartments, condominiums, townhouses, schools, offices, industrial plants, and hospitals.

Commonly built structures include dams, bridges, parking decks, stadiums, water towers, and walkways. Pathways are the avenues of transportation. They include streets, roads, highways, airways, and shipping lanes.

Some frequently used processes in construction are design and drafting, measurement, layout, testing, sawing, boring, smoothing, and fastening. Many different tools are used in these processes.

Tools used in the construction processes include such things as tapes, rules, calipers, gauges, levels, saws, drills, bits, scrapers, planers, hammers, wrenches, screw drivers, and many others. The principal materials of construction are wood, stone, brick, steel, aluminum, and concrete.

The main career fields related to construction are architect, engineer, drafting technician, carpenter, mason, electrician, plumber, and heating and cooling technician.

KEY WORDS

Shelter
Structure
Pathway
Protection
Comfort
Convenience
Dwelling
CADD
Measurement
Layout
Sawing
Testing
Boring
Smoothing
Fastening
Wood
Stone
Brick
Steel
Aluminum
Concrete
Architect
Engineer
Drafting technician
Carpenter
Mason
Electrician
Plumber
Heating and cooling mechanic
Real estate agent
Wholesale sales representative
Retail sales representative

SOME QUESTIONS TO TRY

1. What do shelters provide for people?
2. Name the three categories of things that people construct.
3. Name three construction processes and the tools used in them.
4. What are the principal materials of construction?
5. Name three construction careers.

TECHNOLOGY LEARNING ACTIVITIES

Check out or refer to a book on construction techniques and processes to assist you in completing the following technology learning activities:

1. Design a three-bedroom, two-bathroom, one-story house, and sketch the floor plan on 1/4-inch graph paper.
2. Using plaster of paris for concrete, balsa wood for wooden members, and glue as a fastener, construct a model of the frame skeleton for your house. It should include
   a. An actual dirt site
   b. A concrete footing
   c. A concrete foundation wall
   d. A concrete floor
   e. Wooden stud walls
   f. Wooden ceiling joists
   g. Wooden rafters
3. Using actual materials, build a scaled-down corner of your house. The completed replica of a corner should be approximately 4 feet tall when finished.
4. Make a timeline of important technological developments in construction.

CHAPTER 5

ENERGY AND POWER TECHNOLOGY

OBJECTIVES

After completing this chapter, you should be able to

- explain the need for energy and power.
- define energy and power and distinguish between the two.
- list and explain the main sources of energy.
- explain the ways in which energy is converted into power.
- explain how power is transmitted.
- explain how people use energy and power.
- list some of the careers associated with energy and power.

All technology areas depend on energy and power. Technology, in general, cannot progress without these things. In the earliest times there was only people power. How much power people had depended on many things, but the most important was the food they ate. Food is the source of energy for people power. What people eat is very important. When people power was the only type of power used, they were very limited in what could be done.

Early cave dwellers could only trap or kill small animals because larger ones would be difficult to drag back to the cave. They could only use small trees for building other shelters and fires because large ones were too heavy to carry.

Because people power was so limited, they began to seek different and better types of power. The first major improvement in this direction was to domesticate animals and harness their power. One sturdy animal such as a bull, ox, horse, or mule had the power of several men.

Over the years people have continually developed better and better ways of obtaining the power they need to improve the quality of their lives. Early forms of water transportation were powered by oars. Then came sails, then the steam engine, then the diesel engine, and now nuclear-powered engines.

In this chapter, you will learn about energy and power and the contributions they make to technology. You will learn the definitions of energy and power, some of the many sources of energy available to people, how energy is converted into power, how power is transmitted from where it is generated to where it is needed, and some of the many careers in energy and power technology.

ENERGY DEFINED

Energy is the ability to do work. You already know something about energy, even if you have never thought about it. Do you ever feel worn out after such activities as running, dancing, or playing? If so, it is

Important steps in the development of energy and power technology.

NUCLEAR POWERED SHIP 1954
ELECTRICITY FROM ATOMIC ENERGY 1951
JET ENGINE 1937
LIQUID FUEL ROCKET 1926
DIESEL ENGINE 1892
ELECTRIC MOTOR 1888
ELECTROSTATIC GENERATOR 1880
INTERNAL COMBUSTION ENGINE 1860
ELECTROMAGNETISM 1819
STEAM TRAIN 1801
FIRST BATTERY 1800
STEAM BOAT 1787
STEAM ENGINE 1775
WATER TURBINE 1775
POWER ENGINE 1673
WINDMILL 200 B.C.
WATERWHEEL 600 B.C

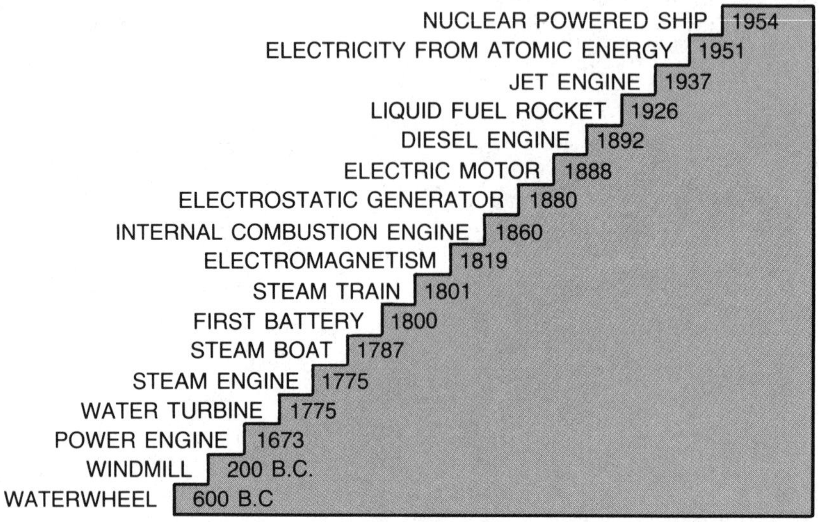

because you have run out of energy. Then if you eat a good meal and rest for a while, you will feel completely restored and full of energy.

Gasoline provides the energy to run a car. Electricity provides the energy for your television, radio, stereo, VCR, and tape player. The sun provides the energy to heat water in a solar hot water system. Everything we do every day requires some type of energy—even sleeping!

POWER DEFINED

You just learned that energy is the ability to do work. **Power** is the amount of work done in a given amount of time. Power can also be defined as the rate of energy conversion. The ability of something to convert energy is a measure of how powerful it is.

If you have ever been in a weight room at the gym, you know that some people are more powerful than others. Some people can easily pick up weights that others can barely move. They are more powerful. They are better able to convert energy into power. They can do this because they have more muscle tissue, which is what converts the energy into power.

Have you ever heard an automobile's engine rated according to **horsepower**? A 100-horsepower engine is supposed to have the power of 100 horses: the greater the horsepower, the more powerful the engine. If two cars each have to transport four people up a steep hill, the one with the greater horsepower will get to the top first.

Notice that energy and power go hand-in-hand. People need power in almost everything they do, and they get it by converting some form of energy. Later in this chapter, you will learn about some of the ways that energy is converted into power, but first you need to know some of the different sources of energy.

SOURCES OF ENERGY

There are several different sources of energy for the power people need every day. The most important energy sources are food, sun, wind, water, fossil fuels, nuclear materials, and the earth's core. With the exception of food, all of these energy sources are used in generating **electricity**, which is an important source of power in the modern world. They are also used in other ways, as you will see.

Food

Human muscle power and animal power are not as important as they were in the days of the cave dwellers, but they are still important. Although modern people don't use their muscles or animals as much as early people did, they still use them a lot. When you walk, run,

Sources of energy.

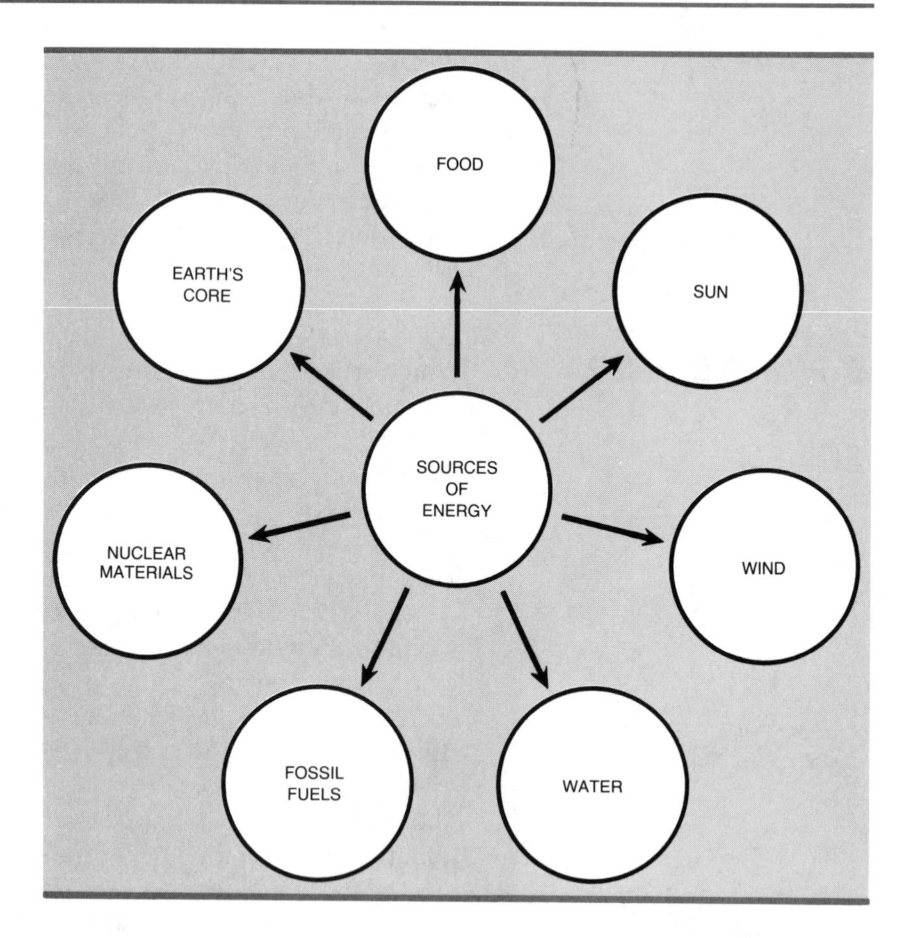

dance, or play you are using **muscle power**. When you use muscle power, the energy that is being converted comes from food.

People get energy from eating foods that contain the right amounts of vitamins, minerals, protein, and, especially, carbohydrates. Energy actually comes from the carbohydrates. Protein builds bigger, stronger muscles that enable us to convert energy better. Vitamins and minerals enable our bodies to absorb and use the carbohydrates. So it is important for you to eat well-balanced meals.

Do you know anybody who runs in long-distance races or marathons? The night before a marathon, some long-distance runners load up on carbohydrates by eating pizza, spaghetti, cake, and other foods that are high in carbohydrates. This is one reason why it is fun to be a long-distance runner! By eating large amounts of carbohydrates, they are able to store away enough energy to run long distances.

On the other hand, when we don't use all of the energy we take in from food, it is stored away as fat. People who eat a lot but don't

Energy from food is the fuel for muscle power. *Photo by Deborah M. Goetsch.*

participate in physical activity tend to gain weight. They are taking in more energy than they are using. Consequently, it is getting tucked away in the form of fat.

The Sun

Actually all sources of energy found on earth come at least indirectly from the sun. Food cannot grow without the sun. Fossil fuels resting hundreds of feet beneath the earth's surface would not have formed without the sun. Although the sun is involved in some way in producing all the energy sources found on earth, in this section we will discuss only the energy that comes directly from the sun.

The most common use of direct energy from the sun is heating. We call the sun's energy *solar energy*. Have you ever heard this term used? Solar energy can be used to heat water, so more expensive, less plentiful sources such as electricity do not have to be used. Have you ever seen a house with solar collectors on the roof?

In a solar hot water system, the sun's rays heat a special fluid contained in copper tubes inside of the solar collectors you might have seen on someone's roof. The special fluid is like the antifreeze people put in the radiators of their cars during the winter. Like antifreeze, this special fluid won't freeze when the temperature drops outside. In addition, it is easier to heat than water.

Solar hot water system.

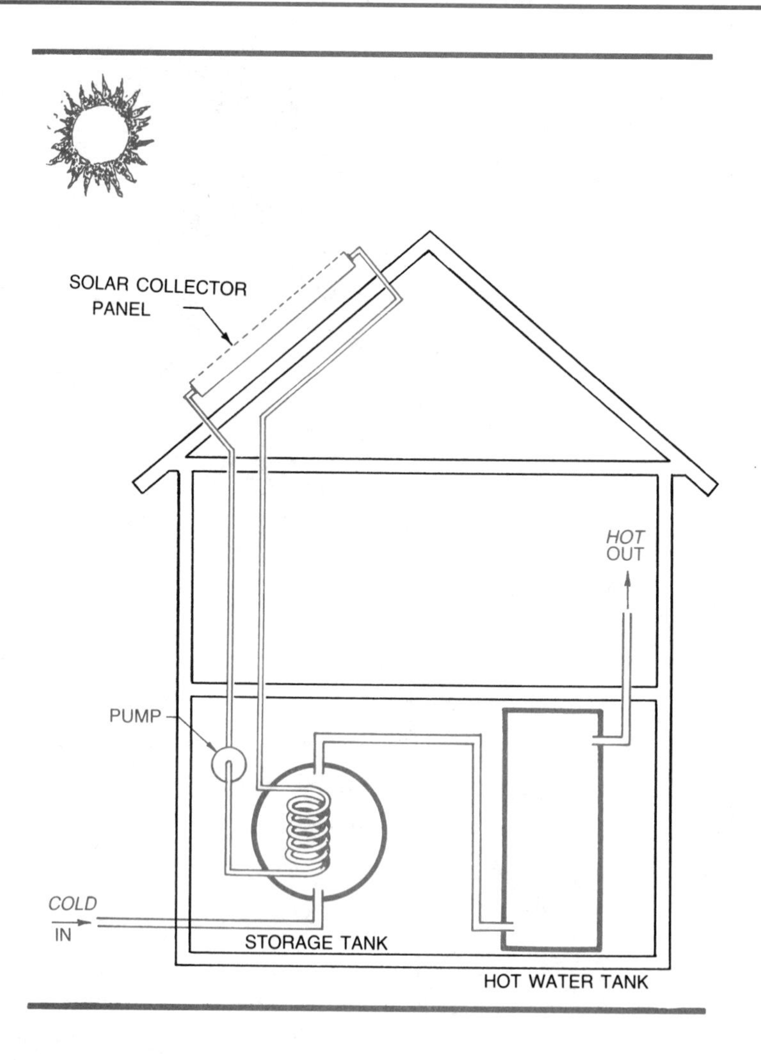

As sunshine strikes the collectors, the special fluid heats up, causing the copper tubes to also heat up. These copper tubes are surrounded by potable water. *Potable* water is water that is fit for drinking, cooking, and other human uses. The potable water heats up and passes down from the collectors through pipes and into a special hot water storage tank. When hot water is taken from the regular hot water tank, it is replaced by solar-heated water instead of cold water. Consequently, the regular hot water heater doesn't have to turn on to heat the water, and energy is saved.

Solar-heated water gets hot enough to be used for taking showers and baths, washing dishes or clothes, and any other normal household use of hot water. By saving energy used in homes, it can make more energy available for use in business and industry.

The Wind

The wind is one of the oldest sources of energy. Have you ever seen a picture of a windmill? As long ago as A.D. 600, windmills were used to harness the wind's energy and convert it to power. A **windmill** has large wind vanes that look something like giant propellers. When the wind blows, it causes the vanes to turn. Through the use of gears, the wind's energy is converted into power that can be used to grind grain or pump water.

Wind energy is also used to propel sailing ships and boats. Sailing represented an important improvement over rowing and paddling. It allowed ships to travel long distances, cross oceans, and even travel around the world.

Large ocean-going vessels no longer rely on sails for power or the wind for energy, but the wind is still a source of energy. Instead of windmills, people now use wind turbines. Wind turbines are much more efficient at harnessing the wind's energy, but instead of pumping water or grinding grain, the wind's energy is now used for generating electricity.

Although it still supplies a small amount of our energy needs, the wind has one serious limitation as an energy source. Do you know what it is? Of course, you do. It only works when the wind is blowing!

The energy of the wind is captured by the windmill.

Water

Like the sun and wind, water is one of the oldest sources of energy. It is still an important source. In earlier times, running water from streams or rivers was used to turn waterwheels. Have you ever seen a waterwheel? There are still some waterwheels in operation today.

The earliest waterwheels were turned by water that ran under the wheel. This version was called the *undershot* waterwheel. Water running under the wheel pushed against the blade, causing it to turn. The turning wheel could be used to power a water pump or turn a grinding wheel.

A new and better type of waterwheel was developed in the 1700s. This version was called the *overshot* waterwheel. In this version, water running over the top of the wheel turned it. We still use water as an important energy source. It is used primarily for turning turbines that generate electricity. By damming rivers, we learned to harness the energy of water. As the river flows through the dam, it turns large turbines that make electricity. The plants where this is done are called *hydroelectric* plants.

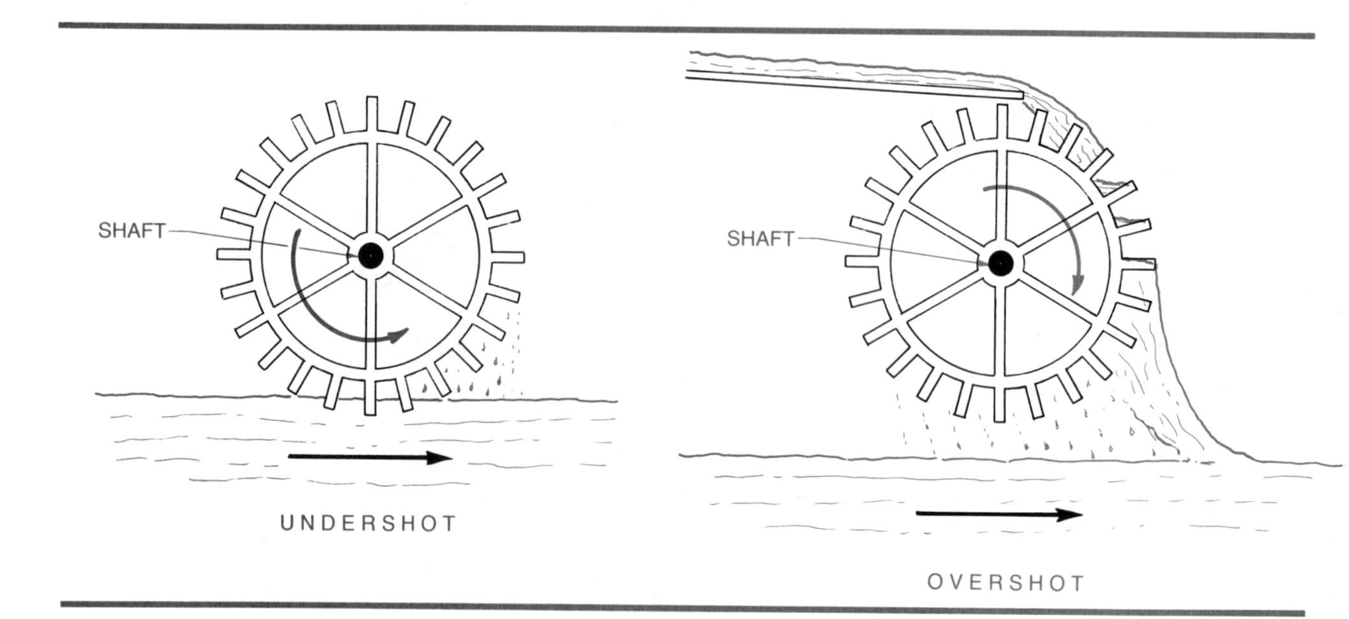

The energy of water is captured by the waterwheel.

Water flowing through a dam is used to generate electricity at the hydroelectric plant. *Courtesy of Philadelphia Electric Company.*

Fossil Fuels

Right now the most popular sources of energy are fossil fuels. **Fossil fuels** are such things as coal, oil, and natural gas. Fossil fuels account for over 90% of all the energy used in this country. Fossil fuels are those energy sources that come from animals and vegetation that died and rotted under the surface of the earth thousands of years ago.

We use coal for heating and making the fires that power steam turbines. Steam turbines are used for making electricity. We use natural gas for heating and cooking. But petroleum oil is the fossil fuel we use most.

Oil is refined as gasoline, diesel fuel, and motor oil. Cars, trucks, boats, trains, and airplanes depend on petroleum as an energy source.

There is an important difference between fossil fuels and other sources of energy. Fossil fuels are a limited source of energy. We can use them up very rapidly. Therefore it is important for people to (1)

Fossil fuels account for over 90% of the energy used in this country.

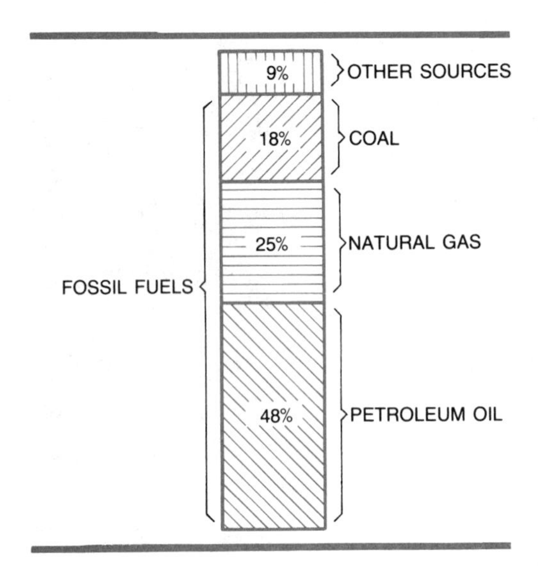

conserve fossil fuels by using them wisely and sparingly, and (2) constantly look for alternative sources of energy.

Nuclear Materials

Nuclear material, such as **uranium**, can be an important energy source. In fact, some day **nuclear energy** might replace fossil fuel. But nuclear energy has a bad side: It can be dangerous.

Nuclear energy comes from uranium or plutonium. Both of these materials are radioactive and dangerous to people. The uranium and plutonium are used to produce heat, and the heat is used to make steam for powering steam turbines and making electricity.

After the nuclear materials have been used, nuclear waste is left. This waste is radioactive and very dangerous. Consequently, it must be disposed of very carefully. At this time, there has still been no way found to safely dispose of nuclear waste. This is a serious problem that scientists are trying to solve. The problem with disposing of nuclear waste is that it remains in a dangerous state for hundreds of years. Consequently, it cannot be buried, burned, or cast into the ocean without contaminating the earth or the sea.

Another problem with nuclear energy is that it can be used to make bombs and nuclear warheads for missiles. Many people are concerned that as the use of nuclear energy spreads, so will the threat of nuclear war.

The Earth's Core

Heat that builds up at the **earth's core** turns underground water into steam. In certain places, cracks in the earth's surface let the steam escape. Wells can be drilled into the earth to let the steam escape where there are no natural cracks. Energy from the earth's core is called **geothermal energy**.

Geothermal energy can be harnessed to power steam turbines to make electricity. As the steam cools and turns into water, it can be pumped back into the earth and reheated. Although there are several places around the world that use geothermal energy to make electricity, it is one of the least-used sources of energy at this time. Unless the water approaches 130 degrees Fahrenheit, the process is too costly.

CONVERTING ENERGY INTO POWER

Energy cannot be used directly. It has to be converted into some type of power. Various devices convert energy into power. The most common are human and animal muscles, turbines, steam engines, and internal combustion engines. The following sections will examine how these various devices convert energy into power.

How energy is converted to power.

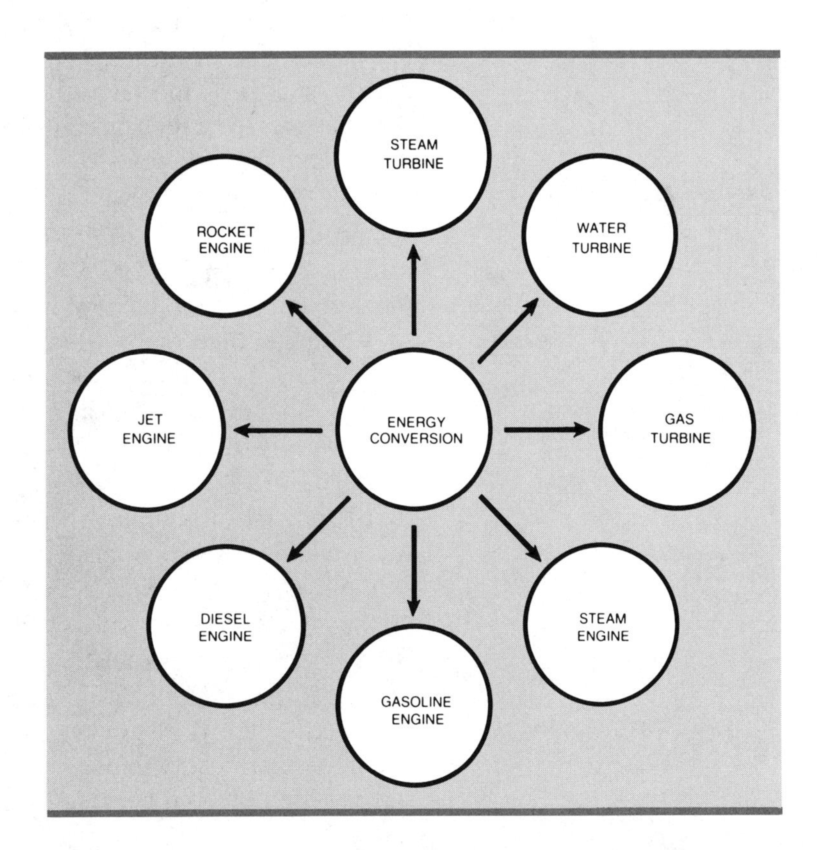

Human and Animal Muscles

People and animals get energy from the food they eat. Energy is converted into power by muscles. Therefore heavily muscled animals such as elephants, oxen, bulls, and horses are more powerful than smaller animals such as birds, dogs, and cats. It is why heavily muscled people such as weight lifters and football players are more powerful, usually, than people with smaller muscles. Their muscles enable them to convert energy faster and better than their small-muscled counterparts.

Turbines

You read earlier that several different sources of energy are used to turn turbines, which in turn power generators that make electricity. Wind, geothermal, water, nuclear, and fossil fuel energy are all used to turn turbines for making electricity. **Turbines** are very important energy conversion devices because they help make electricity, and electricity is our most important form of energy.

A turbine is a wheel with blades that can be turned by steam, water, air, or some other means. A turbine converts the energy of the water, air, or steam into mechanical energy for powering electrical generators. **Generators**, then, actually make the electricity. There are steam, water, and gas turbines.

Steam Turbines A **steam turbine** is one that is especially designed to allow steam to put pressure on the blades of the turbine, causing a turning motion. Water is heated until it turns to steam. It is then channeled through a pipe to the blades on the turbine wheel.

Water can be heated by burning oil, coal, and even garbage. It can also be heated by nuclear or geothermal energy. As the wheel turns, it provides the mechanical energy needed for running electrical generators, which in turn make the electricity.

Water Turbines The pressure against the blades of a **water turbine** comes from falling water. By damming a river, we can make water fall on the turbine blades from very high. The pressure of water falling from a great height is enough to turn the turbine. As it turns, it powers the generator, which makes electricity.

Gas Turbines **Gas turbines** can be used to power different modes of transportation as well as to power electric generators. A gas turbine is turned by the pressure of gas emitting from burning fuel. The jet engine is the most common use of the gas turbine. Trains and ships can also be powered by gas turbines.

Jet engines such as this PW2037 turbofan are a form of turbine. *Courtesy of Pratt and Whitney.*

Steam Engines

The **steam engine** is not used much any more. It has been replaced by gasoline, diesel, and nuclear-powered engines. However, at one time, it was an important means of converting energy into power. During the industrial revolution, the steam engine was the chief source of power for industry and transportation. Steam engines were used for powering industrial machines and water pumps. Sailing ships gave way to steam-powered ships that were faster and more efficient. The steam engine also allowed for the development of the locomotive. Steam for steam engines is made by burning some type of fuel, principally coal.

Internal Combustion Engines

The steam engine is an external combustion engine because the fuel is burned outside the engine. **Internal combustion engines** burn the fuel inside the engine. There are different types of internal

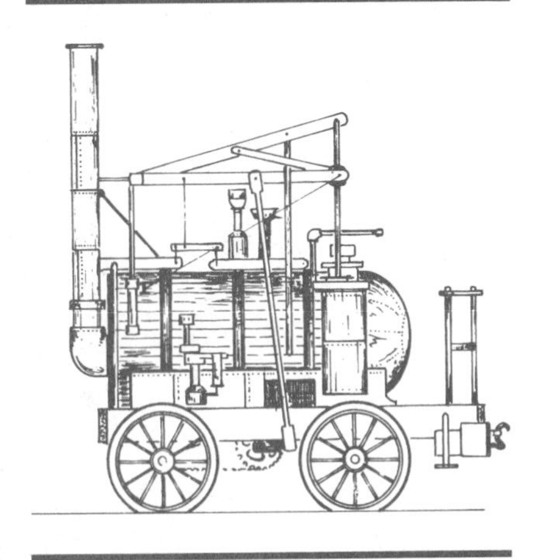

The steam engine was one of the earliest ways to convert energy into power.

Gasoline is one of the most popular forms of energy. *Photo by Deborah M. Goetsch.*

combustion engines: gasoline, diesel, jet, and rocket engines are internal combustion engines.

Gasoline Engines Gasoline is a major source of energy. It is made by refining petroleum oil. The **gasoline engine** is used for converting gasoline into power. Gasoline engines are used in cars, trucks, motorcycles, ATVs, lawnmowers, and chain saws. There are two-cycle and four-cycle gasoline engines. Most that we use are four cycle. A four-cycle gasoline engine goes through four phases of operation or cycles: intake, compression, combustion, and exhaust.

Intake is when fuel mixed with air is drawn into the cylinders. *Compression* is when the piston comes up, pushing this mixture into an increasingly smaller chamber. *Combustion* is when the spark plug ignites the compressed fuel, driving the piston back down. At this time,

The cycles of a four-cycle gasoline engine.

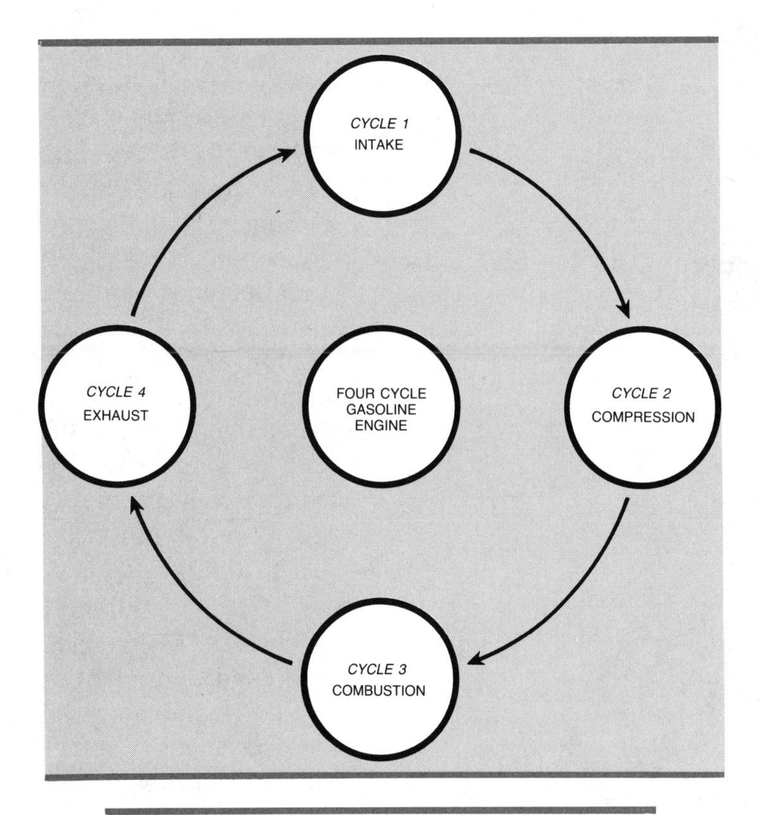

Operation of the gasoline engine.

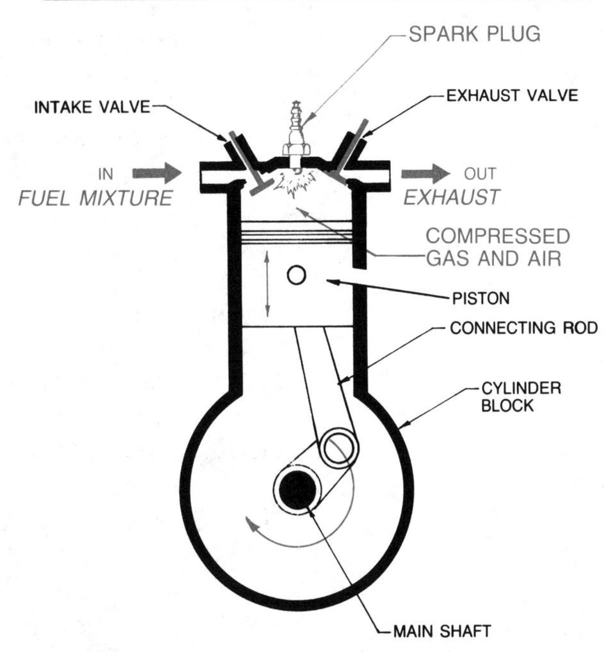

the energy of gasoline is actually converted into mechanical energy. *Exhaust* occurs when the piston goes back down and leftover gases from the ignited fuel escape through the exhaust valve.

Diesel Engines **Diesel engines** are used in large trucks, buses, tractors, heavy equipment, boats, ships, and industrial equipment. A diesel engine is similar to a gasoline engine, but there are some important differences.

Diesel engines require much higher levels of compression than gasoline engines. The reason is that the diesel engine creates the heat needed for igniting the fuel by compressing air. The piston comes up and compresses air in the cylinder. This causes the air to become very hot. When the piston is at the top of the stroke, fuel is injected into the cylinder. As it mixes with the hot air, it ignites and drives the piston down. This converts the energy of diesel fuel into the mechanical energy needed to power the engine.

Jet Engines Jet aircraft came into use at the end of World War II. The most advanced military and commercial aircraft are now powered by jet engines. A jet engine is much more powerful than the old gasoline engines used on propeller aircraft.

In a jet engine, a compressor pumps large amounts of air into a combustion chamber. A fuel injector sprays fuel into the chamber. As the air and fuel mix, they are ignited. The explosion turns a turbine and thereby converts the energy of the fuel into power. The leftover gases escape through an exhaust duct at the rear of the engine. There are three types of jet engines: **turbojet**, **turbofan**, and **turboprop**.

Rocket Engines Rocket technology was experimented with during World War II. But it really caught on after World War II when the United States launched its space program. In the early years of the space program, the most important problem scientists and engineers faced was developing an engine powerful enough to launch satellites, rockets, and capsules into space. This problem was solved by the development of the rocket engine.

A rocket engine is similar to a jet engine with the exception that it is self-contained. It does not take in air from the outside. Self-containment is necessary because there is little oxygen in the air in outer space, and oxygen is the ingredient that is needed from the air. This lack of oxygen is why astronauts must wear the special suits with air tanks that you have probably seen on television or in photographs. With this exception, a rocket engine operates much like a jet engine. The oxygen and fuel are mixed together in a combustion chamber and ignited. The explosion that results converts the energy of the fuel into power or *thrust*. Thrust boosts the rocket into space. The two types of rocket engines are liquid fuel and solid fuel.

The 4JH-DTE diesel engine.
Courtesy of Yanmar Diesel Engine Co., Ltd.

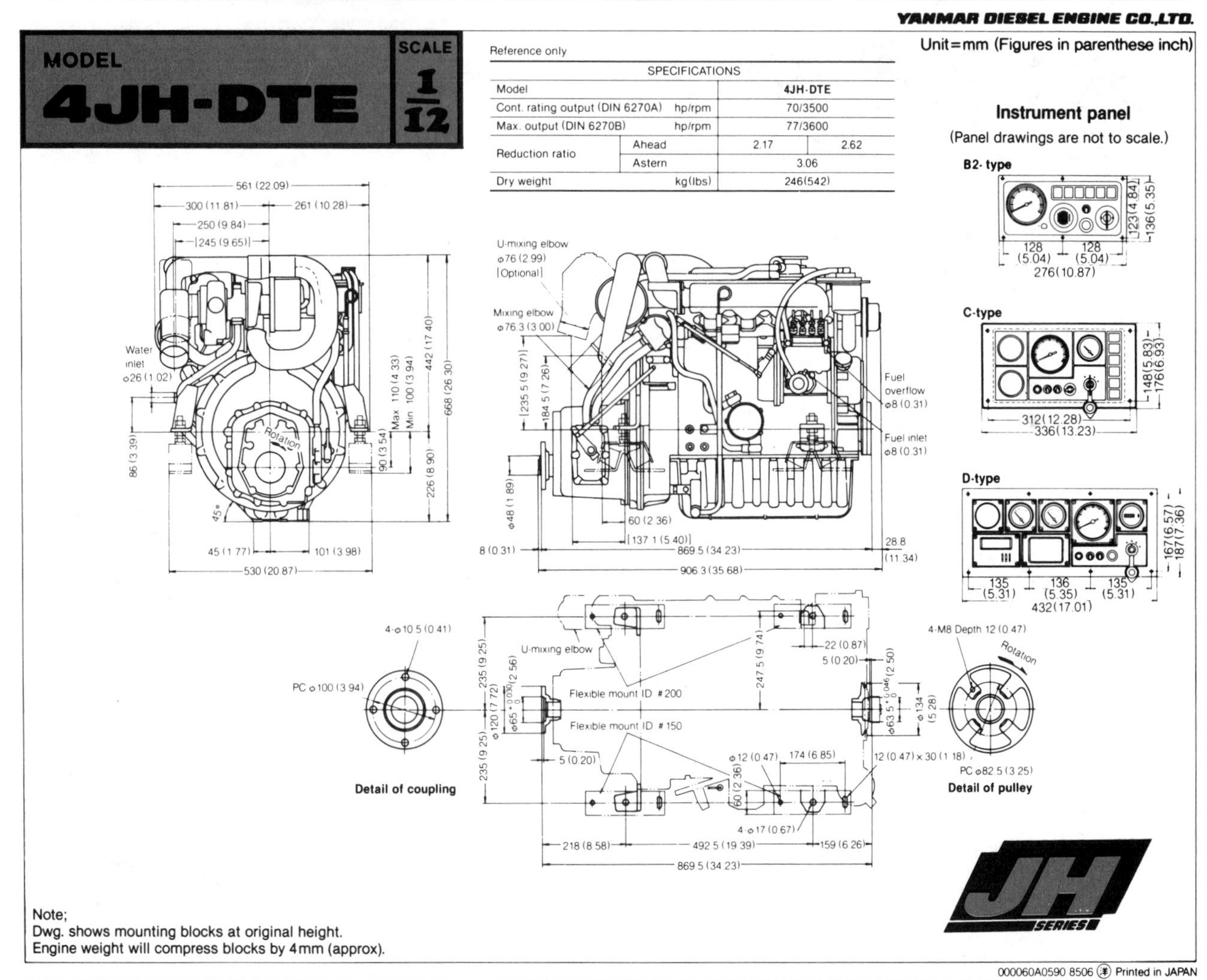

Reference only

SPECIFICATIONS

Model			4JH-DTE	
Cont. rating output (DIN 6270A)		hp/rpm	70/3500	
Max. output (DIN 6270B)		hp/rpm	77/3600	
Reduction ratio	Ahead		2.17	2.62
	Astern		3.06	
Dry weight		kg(lbs)	246(542)	

Drawing of the 4JH-DTE diesel engine. *Courtesy of Yanmar Diesel Engine Co., Ltd.*

The 4JH-HTE diesel engine.
Courtesy of Yanmar Diesel Engine Co., Ltd.

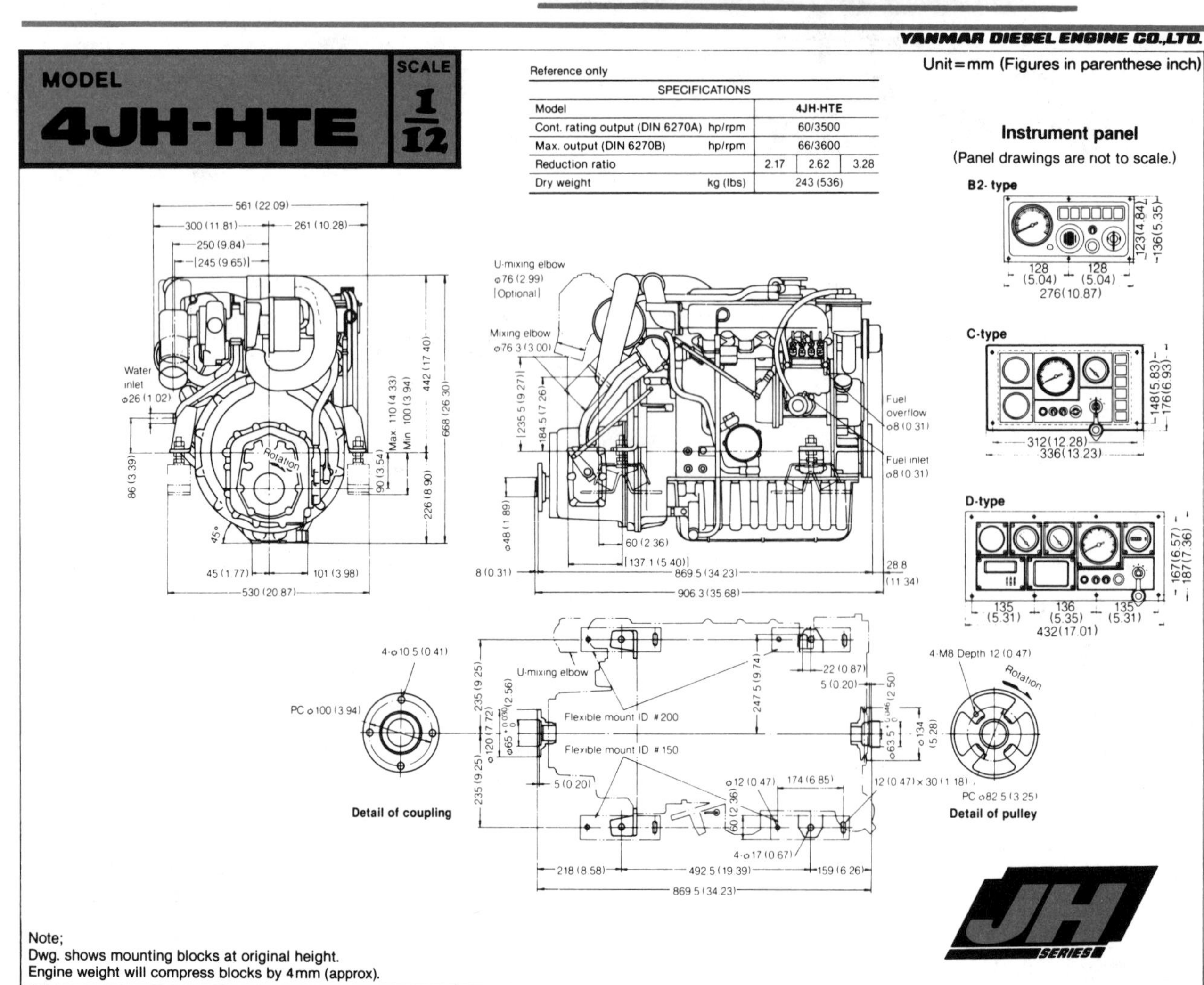

SPECIFICATIONS				
Model		4JH-HTE		
Cont. rating output (DIN 6270A)	hp/rpm	60/3500		
Max. output (DIN 6270B)	hp/rpm	66/3600		
Reduction ratio		2.17	2.62	3.28
Dry weight	kg (lbs)	243 (536)		

Drawing of the 4JH-HTE diesel engine. *Courtesy of Yanmar Diesel Engine Co., Ltd.*

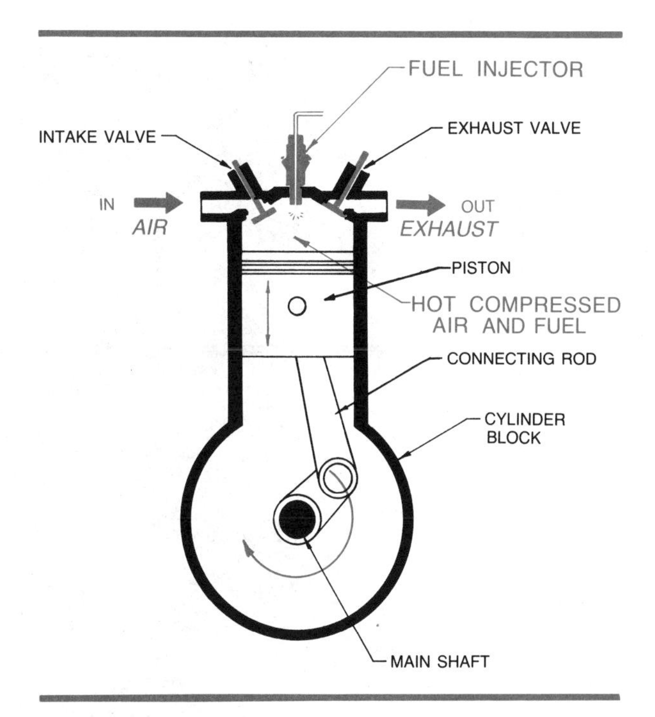

Operation of the diesel engine.

Future diesel mechanics learn to repair diesel engines. *Courtesy of Robert C. Atwood.*

PW100 turboprop jet engine. *Courtesy of Pratt and Whitney Aircraft of Canada.*

PT6 turboprop jet engine. *Courtesy of Pratt and Whitney Aircraft of Canada.*

Cross section of a jet engine.

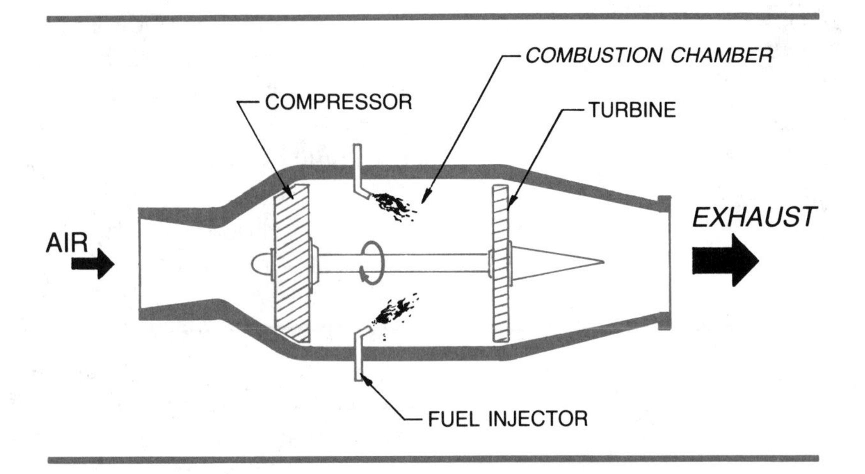

Cross sections of two types of rocket engines.

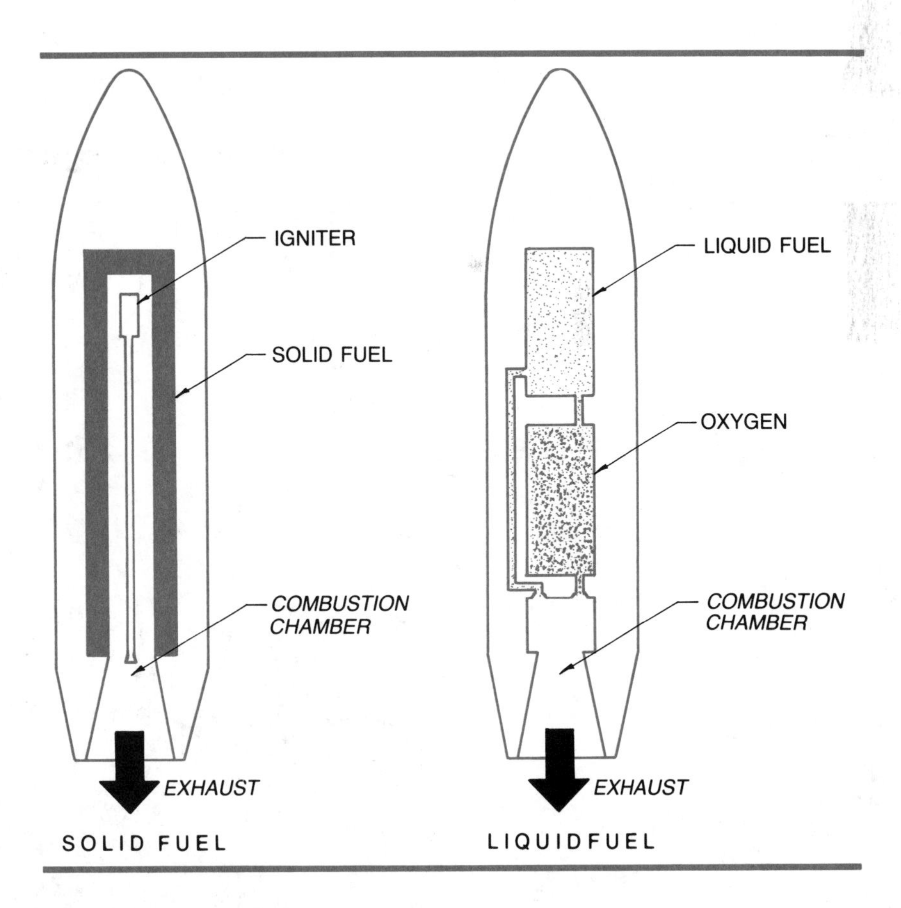

TRANSMITTING POWER

There are a number of different sources of energy and a number of different devices and processes for converting these energy sources into power. We now need to look at how this power is transmitted from where it is produced to where it is needed. But first let's look at an example of what transmitting power means and why we have to do it.

When you ride a bicycle, you use energy, power, and the transmission of power. The energy is yours. It comes from the food you eat. Your muscles convert this energy into power so that you can pedal the bicycle. But the power is not needed at the pedals. It is needed at the back wheel to turn it and to make the bicycle move. The power generated by your muscles is transmitted to the back wheel by the chain of your bicycle as it goes around the sprocket of the back wheel. Let's look at some other forms of transmitting power.

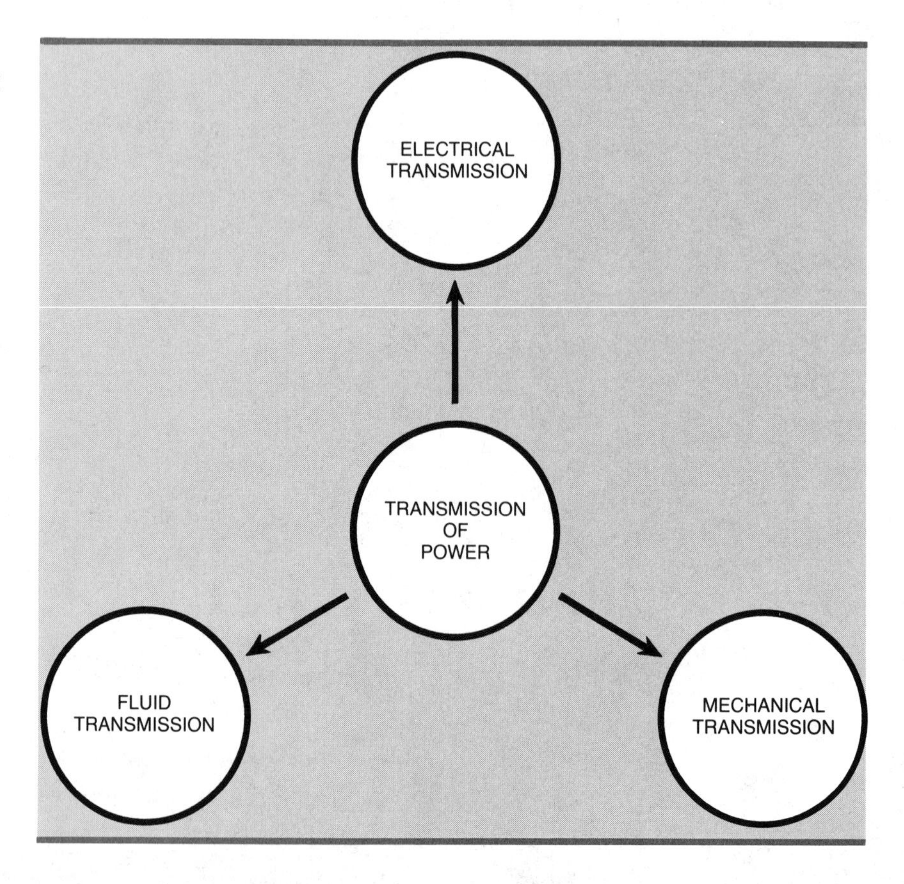

How power is transmitted from where it is generated to where it is needed.

Electrical Transmission

We use electricity in our homes, schools, offices, industrial plants, hospitals, and wherever people live and work. But before we can use it, we have to get it from where it is generated to where we need it. We do this by **electrical transmission**.

Electricity is transmitted from where it is generated to where we need it by transmission lines. You have seen them. Some people call these electrical transmission lines telephone lines. But what they are referring to are usually power lines for transmitting electricity. These are the lines on the large poles that you see along roads and highways.

Mechanical Transmission

The earlier example of pedaling a bicycle was an example of **mechanical transmission** of power. Power is transmitted mechanically by such devices as the bicycle chain, belts and pulleys, and gears.

Fluid Transmission

Fluid transmission means transmitting power by using air or some type of fluid such as water or oil. There are a number of examples of power transmitted by fluid. The hydraulic lift used in auto repair shops is an example of power transmitted by fluid. A barber chair that goes up and down is another example. Probably the best example of fluid transmission is the braking system in a car. The power applied to the brakes is transmitted to the wheels by a special brake fluid.

HOW WE USE ENERGY AND POWER

How many different types of energy do you use in a day? And how many different ways do you use energy? It is important to understand how we use energy. There are four major categories of energy use with which you should be familiar. They are electricity generation, transportation uses, industry uses, and residential and commercial uses.

Electricity Generation

Electricity generation is the largest area of energy use. We use more energy in generating electricity than in any of the other three areas. Most electricity is generated in plants that use steam turbines. In creating the steam, we use coal, oil, water, nuclear energy, and even garbage. About 30 percent of the energy used in this country is used to generate electricity.

Graph of energy consumption in this country.

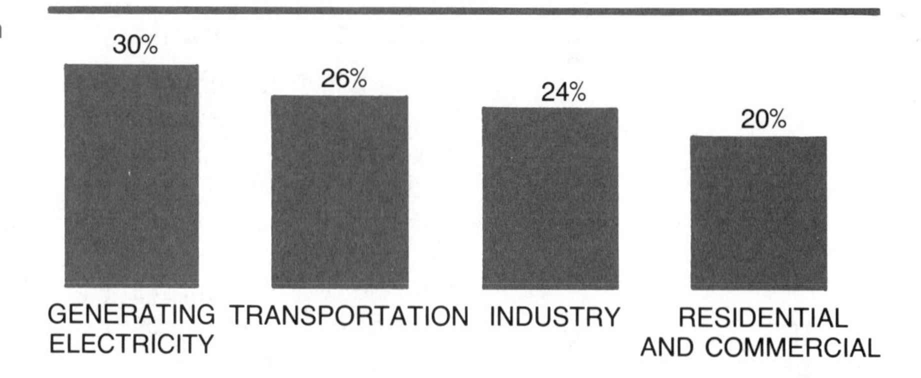

Transportation Uses

The transportation industry is second in using energy and accounts for about 26 percent of all energy used in this country. Cars, trucks, buses, motorcycles, ATVs, and various types of military vehicles use gasoline and diesel fuel. Propeller aircraft, jet aircraft, and rockets use gasoline, a special jet fuel, liquid rocket fuel, and solid rocket fuel. Various types of water vessels use wind, steam, gasoline, diesel fuel, and nuclear energy.

Industrial Uses

Industry is third in using energy and accounts for about 24 percent of all energy used in this country. It takes a lot of energy to mine raw materials, process them, and convert them into finished products. If the raw material is a tree, it takes gasoline to power the chain saw and cut it down. It takes diesel fuel to power the heavy equipment to load the felled trees on trucks for shipment to lumber mills. At the mill, it takes gasoline and electricity to power the machines and tools that cut the trees into various sizes and shapes of lumber. Think of an automobile manufacturing plant and how much energy is used there. Every type of industry uses large amounts of energy.

Residential and Commercial Uses

Last on the list in terms of using energy are the residential and commercial uses, which account for about 20 percent of all the energy used in this country. We use a lot of energy in our shelters regardless of whether they are built for residential or commercial use.

We use energy for lighting, heating, cooling, powering appliances, powering machines, powering equipment, powering tools, and for cooking. Cooking and some heating systems use natural gas. Other forms of heating and cooking use electricity, as do cooling, lighting, and appliances. Some shelters use solar energy for heating water.

CAREERS IN ENERGY AND POWER TECHNOLOGY

There are many different career fields in energy and power technology. You should be familiar with some of them. The best way to learn them is to examine them in broad terms. Some careers are scientists, engineers, managers, technicians, production workers, operators, and mechanics. The scientists, engineers, and managers are called *professional-level* workers. Technicians are *technical-level* workers. Production workers, operators, and mechanics are called *skilled-labor-level* workers.

Professional Careers

Professional careers associated with energy and power technology are the scientists, engineers, and managers. Scientists conduct research into new forms of energy, better ways of converting the existing types of energy into power, and alternative forms of energy. Scientists have discovered all of the principles that have led to the various forms of energy we have used and will use in the future. For example, the scientific principle of internal combustion led to the development of the gasoline engine.

Engineers take the principles developed by scientists and design the tools, devices, machines, and processes for applying them to everyday life. For example, engineers take the various scientific principles of solar energy and apply them to design such products as solar hot water systems.

Managers organize operations, develop operating plans and procedures, conduct evaluations, hire and supervise employees, and do many other tasks to ensure efficient business operations. Most professional-level workers have at least a college education, and many have advanced degrees.

Technical Career Fields

Technicians assist scientists and engineers in their work. There are many different types of technicians, and their jobs vary greatly. There are electrical technicians, heating and cooling technicians, nuclear technicians, and laboratory technicians.

Technicians require specialized training. Most technicians have a two-year degree from a community college or a diploma from an area vocational trade school.

Skilled-Labor Careers

The skilled workers in energy and technology are production workers, operators, and mechanics. Production workers are the people involved in manufacturing energy and power-related products. Production workers perform such tasks as welding, machining, and assembling and manufacturing automobile engines, heaters, and air conditioners.

Operators are the people who operate the many different machines, vehicles, vessels, equipment, and processes relating to energy and power. A truck driver is an operator. A forklift driver is an operator. The engineer on a locomotive is an operator. The person who runs a large electrical generating plant is an operator.

Mechanics are the people who maintain and repair the machines, tools, vehicles, vessels, and equipment relating to energy and power. Auto, diesel, and heating and cooling mechanics all fall into this category.

Production workers, operators, and mechanics all require special training that can be obtained from area vocational centers, high schools, and some community colleges. Training can also be received through the apprenticeship and on-the-job training routes.

SUMMARY

All technology areas depend on energy and power. Energy is the ability to do work. Many different sources of energy are available on earth. Power is the amount of work done in a given amount of time. Power is also the rate of energy conversion. The ability of something to convert energy is a measure of how powerful it is.

There are several different sources of energy for the power people need every day. The most important energy sources are food, sun, wind, water, fossil fuels, nuclear materials, and the earth's core.

Energy cannot be used directly. It has to be converted into some type of power. Various devices convert energy into power. The most common are human and animal muscles, turbines, steam engines, and internal combustion engines.

After energy has been converted into power, it must be transmitted from where the conversion takes place to where the power is needed. There are three broad categories of power transmission: (1) electrical transmission, (2) mechanical transmission, and (3) fluid transmission.

NEW TECHNOLOGY FOR RAILROADS

It was about 150 years ago that steam-powered locomotives were first used to carry passengers and freight. By 1960 the steam engines were gone from the rails, replaced mainly by (1) more efficient diesel powered locomotives. Other advances in railroading include the use of (2) containerized freight, sometimes called "piggyback" trailers. These containers can be transported by ship, train, and truck, making global transportation systems possible. (3) Computers are now an important part of modern railroads, too. This yardmaster is sorting cars onto different tracks. Electronic scanners in the yard track "read" coded markers on the cars, allowing them to be arranged by place of final destination. (4) The Maglev train is being planned for future use. It will ride above a single rail by means of a powerful magnetic field.

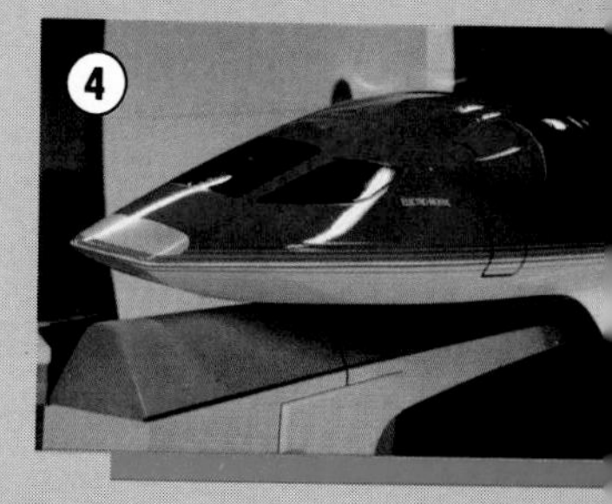

THE NAVY'S NUCLEAR FLEET

Nuclear energy is used to power many of the Navy's major warships. It allows them to travel great distances without refueling, and to attain high speeds both on the surface of the water and underneath the seas. (1) The U.S.S. Enterprise was the first nuclear aircraft carrier to join the fleet. (2) The U.S.S. Ticonderoga, a cruiser, carries Aegis guided missiles and sophisticated electronic tracking equipment. (3) The U.S.S. Michigan carries an arsenal of nuclear weapons, and can stay underwater for long periods of time.

There are four major categories of energy used: generating electricity, transportation, industry, and residential and commercial uses.

Some of the most commonly practiced career fields relating to power and energy technology are scientists, engineers, managers, electrical technicians, heating and cooling technicians, nuclear technicians, laboratory technicians, production workers, welders, machinists, assemblers, truck drivers, forklift drivers, and mechanics.

KEY WORDS

Energy
Power
Electricity
Horsepower
Fossil fuels
Uranium
Earth's core
Muscle power
Windmill
Nuclear energy
Geothermal energy
Steam engine
Internal combustion engine
Turbine
Generator
Water turbine
Gas turbine
Steam turbine
Gasoline engine
Diesel engine
Turbojet
Turbofan
Turboprop
Electrical transmission
Mechanical transmission
Fluid transmission

SOME QUESTIONS TO TRY

1. Define energy.
2. Define power.
3. Name four sources of energy.
4. Name three ways that energy is converted into power.
5. Name two ways in which power is transmitted.
6. What are the four broad categories of energy and power use in this country?
7. Name four careers commonly practiced relating to energy and power technology.

TECHNOLOGY LEARNING ACTIVITIES

1. Make a simple solar hot water system and demonstrate its use for your class.
2. Make a working model of a waterwheel that will show how the water's energy is converted into power.
3. Make a simple steam turbine that shows how the steam drives the turbine.
4. Make a working model of a windmill that will show how the wind's energy is converted into power.
5. Make a timeline of important technological developments in the area of energy and power.

CHAPTER 6

MANUFACTURING TECHNOLOGY

OBJECTIVES

After completing this chapter, you should be able to

- define manufacturing.
- list the most frequently used manufacturing materials.
- list and explain the most frequently used manufacturing processes.
- explain the design process.
- define computer-aided manufacturing and explain how it differs from traditional manufacturing.
- explain how manufactured products are distributed.
- list some of the careers associated with manufacturing technology.

Manufacturing has been around since the beginning of time. The earliest people needed tools and implements for collecting, preparing, and storing food. They needed tools and implements for building shelters, for making clothes, and for collecting and storing water. And they needed weapons for hunting and protection. People have always provided these things for themselves by manufacturing them.

WHAT IS MANUFACTURING?

Manufacturing is the process of converting **raw materials** into products. In early times, when a person converted clay into a pot or some other type of container, that was manufacturing. Now machines do most of the work of manufacturing.

As more people wanted more manufactured products, it became necessary to produce them in large amounts quickly and inexpensively. This need led to the Industrial Revolution in the latter part of the 1700s. Manufacturing moved from products made by hand to products made by machines in factories. This switch from hand tools and processes to machines is called **mechanization**.

Converting iron ore to steel in a steel manufacturing plant. *Courtesy of Bethlehem Steel Corporation.*

The Industrial Revolution also led to the important manufacturing concepts of (1) interchangeable parts and (2) mass production.

Mass Production

Before the Industrial Revolution all manufactured products were made by hand one at a time. This process was slow and inexact. It meant that of any product no two would be exactly alike. Of course, this method caused problems.

Suppose a farmer bought a handmade plow. After it was used several times, the blade wore out and broke. The farmer couldn't simply replace the blade, as we can do now. A new blade had to be specially made instead. Can you imagine having to have a part specially made every time you needed one for a bicycle, car, or television?

Such problems led to the concept of interchangeable parts and mass production. Eli Whitney is credited with the development of this concept about 1798. He thought of the idea as a way to fulfill a contract for making 10,000 rifles.

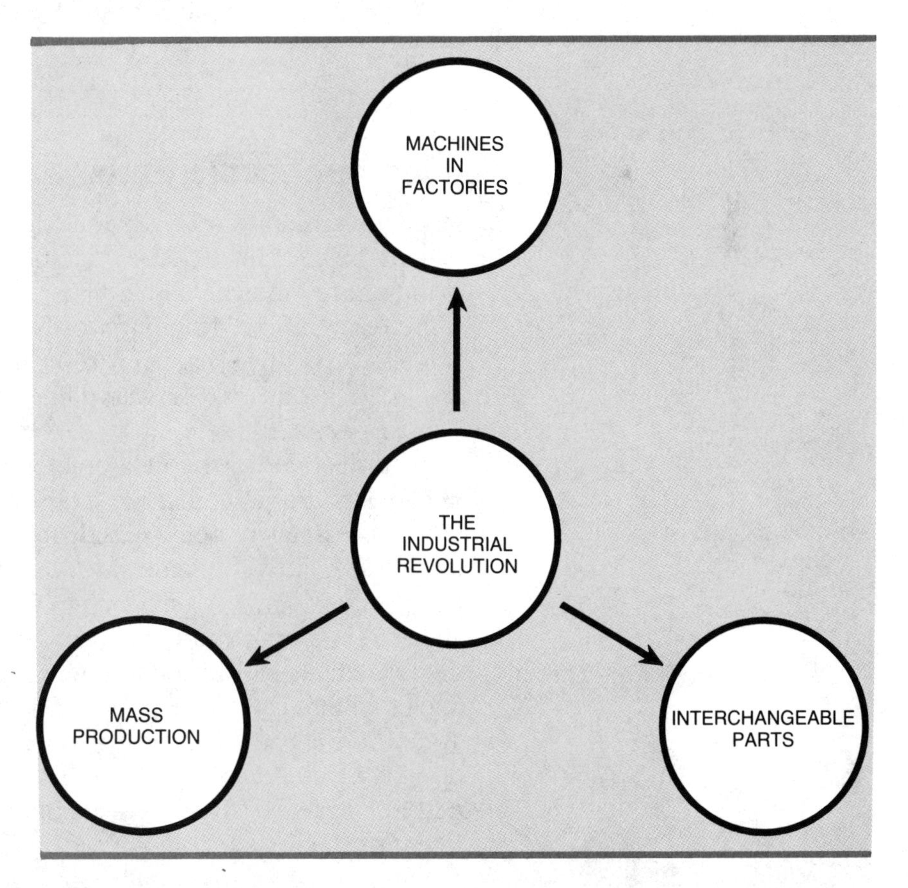

Results of the Industrial Revolution.

Rather than making 10,000 rifles one at a time, he decided to make 10,000 barrels, 10,000 stocks, 10,000 trigger mechanisms, and so on. Then he would assemble the 10,000 rifles needed. This method of manufacturing, called **mass production**, was more efficient.

A machine could be set up to make a part (the rifle barrel, for example), and it wouldn't have to be broken down and set up for a new part until every barrel had been manufactured. This method saved a lot of time and work. In the old way, the machine would have to be broken down and set up again every time a part was made.

For example, let's say Mr. Whitney's rifles each had 10 parts. With the new system of mass production, only 10 machine setups were needed—one for each part. With the old system, 10,000 × 10 or 100,000 machine setups were needed—one for each part of each rifle to be manufactured.

If it took 5 minutes to set up a machine for manufacturing a part, Mr. Whitney would spend a total of only 50 minutes setting up his machine throughout the entire job done the new way. The old way, however, would require 500,000 minutes or over 8,000 hours just setting up machines.

Interchangeable Parts

Mass production was obviously a much better approach to manufacturing, but it led to another problem. If parts were to be manufactured individually in mass quantities, each duplicate part had to be exactly alike. In other words, if Mr. Whitney reached into his box of 10,000 triggers, all 10,000 had to be exactly alike. During assembly, if one trigger was different, it wouldn't fit into the rifle being assembled.

This need led to the idea of interchangeable parts. **Interchangeable parts** are parts that are exactly the same or are at least within acceptable limits of being exactly the same. If you had to draw 10,000 circles, all exactly the same size, what would you do? You could use a template, of course. A *template* is a tool that allows you to make circle after circle exactly the same size. Whitney used such tools also, and they are still used even today in mass producing interchangeable parts. Such tools are called *jigs*. A jig is a device that holds a part and guides a tool during manufacturing processes. Tools that are specially designed to hold a part in position during manufacturing are called *fixtures*. Jigs and fixtures are important tools in mass producing interchangeable parts.

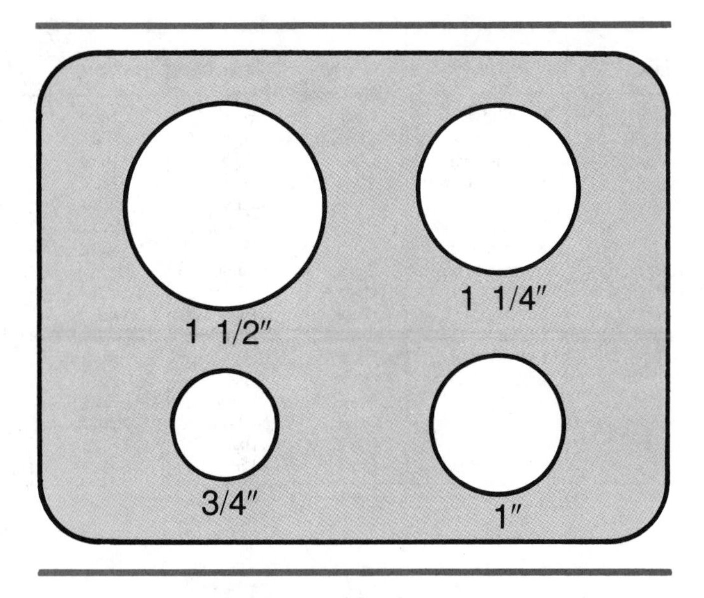

A circle and hole template used in drafting.

MANUFACTURING MATERIALS

Manufacturing is converting raw materials into products. Many materials are used in manufacturing, but they all fall into one of two groups: natural materials and synthetic materials.

Natural Materials

Natural materials are available in nature. Natural materials used in manufacturing are metal and wood. You should be familiar with the more common woods and metals used in manufacturing.

Wood Wood is a natural material because it is provided by nature in the form of trees. Wood is classified as hardwood or as softwood. The hardwoods are ash, beech, birch, cherry, elm, gum, hickory, mahogany, maple, oak, poplar, sycamore, walnut, and willow. The soft woods are cedar, fir, pine, redwood, and spruce.

Besides the lumber cut from the various types of trees in manufacturing, there are also plywood, veneer, and particle board. Veneer is thin, flat slices of wood cut from a log. Plywood is layers of veneer glued around a core. Particle board is made by mixing waste wood chips, wood flakes, and particles with plastic and pressing the mixture together under heat and pressure.

Metals Metals are natural materials because they are provided by nature. All metals are classified as ferrous or nonferrous. **Ferrous metals** contain iron. **Nonferrous metals** do not.

SPECIES	COLOR	GRAIN	STRENGTH	HARDNESS	HAND TOOL WORKING	NAILING	FINISHING	COST
Ash, White	Off White	Open	High	Hard	Hard	Medium	Medium	Medium
Beech	Light Brown	Close	High	Hard	Hard	Poor	Easy	Medium
Birch	Light Brown	Close	High	Hard	Hard	Poor	Easy	High
Cherry	Reddish Brown	Close	High	Hard	Hard	Poor	Easy	High
Elm	Tan	Close	Medium	Medium	Hard	Good	Medium	Medium
Gum	Reddish Brown	Close	Medium	Medium	Medium	Medium	Medium	Medium
Hickory	Reddish Tan	Open	High	Hard	Hard	Poor	Medium	Low
Mahogany, Honduras	Golden Brown	Open	Medium	Medium	Easy	Good	Medium	High
Mahogany, Philippine	Red	Open	Medium	Medium	Easy	Good	Medium	Medium
Maple, Hard	Cream	Close	High	Hard	Hard	Poor	Easy	Medium
Maple, Soft	Brown	Close	High	Medium	Hard	Poor	Easy	Medium
Oak, Red	Light Brown	Open	High	Hard	Hard	Medium	Medium	Medium
Oak, White	Grayish Brown	Open	High	Hard	Hard	Medium	Medium	Medium
Poplar	Yellow	Close	Low	Soft	Easy	Good	Medium	Low
Sycamore	Light Brown	Close	Medium	Medium	Hard	Good	Easy	Medium
Walnut	Dark Brown	Open	High	Hard	Medium	Good	Medium	High
Willow	Medium Brown	Close	Low	Soft	Easy	Good	Medium	Low

SPECIES	COLOR	GRAIN	STRENGTH	HARDNESS	HAND TOOL WORKING	NAILING	FINISHING	COST
Cedar	Red	Close	Medium	Medium	Medium	Poor	Easy	Medium
Fir, Douglas	Reddish Brown	Close	Medium	Soft	Medium	Poor	Poor	Medium
Fir, White	White	Close	Low	Soft	Medium	Poor	Poor	Low
Pine, Yellow	Yellow	Close	High	Medium	Hard	Poor	Medium	Medium
Pine, White	Cream	Close	Low	Soft	Easy	Good	Medium	Medium
Pine, Ponderosa	Reddish Brown	Close	Low	Soft	Easy	Good	Medium	Medium
Pine, Sugar	Light Tan	Close	Low	Soft	Easy	Good	Poor	High
Redwood	Reddish Brown	Close	Medium	Soft	Easy	Good	Poor	Medium
Spruce	White	Close	Low	Soft	Medium	Medium	Medium	Medium

Characteristics of common hard woods and soft woods. *Courtesy, from Jambro, Introduction to Manufacturing, © 1986 by Delmar Publishers Inc.*

A good way to know if a metal is ferrous or nonferrous is to see if it rusts. Ferrous metals will rust; nonferrous metals will not. Cast iron and steel are ferrous metals. Aluminum, copper, bronze, brass, lead, and zinc are nonferrous metals.

Alloys are metals made by mixing other metals together. Sometimes a product needs to be made of a metal that has certain properties, but there is no ferrous metal or nonferrous metal that has all of the desired properties. Two or more metals that each have some of the desired properties can be mixed to make an alloy that has all of the desired properties. For example, brass is an alloy of copper and zinc. The solder used in electronic circuit assembly is an alloy of lead and tin.

Wood products.

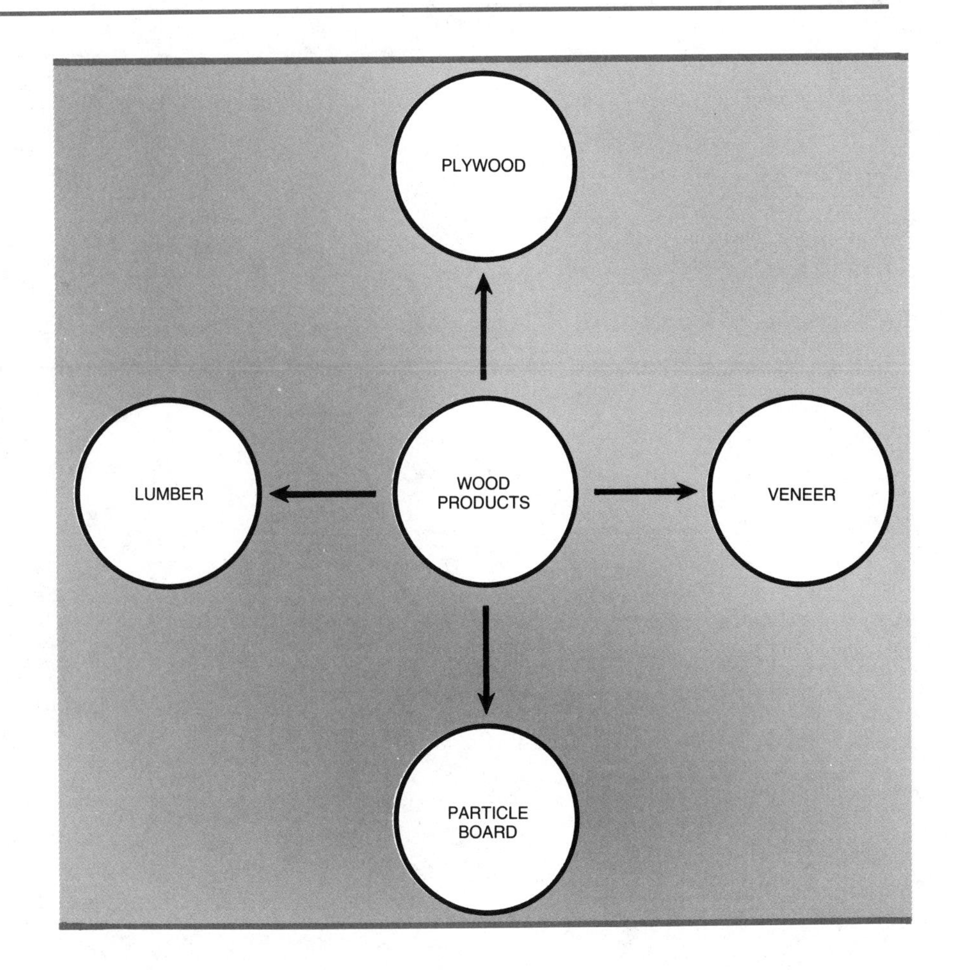

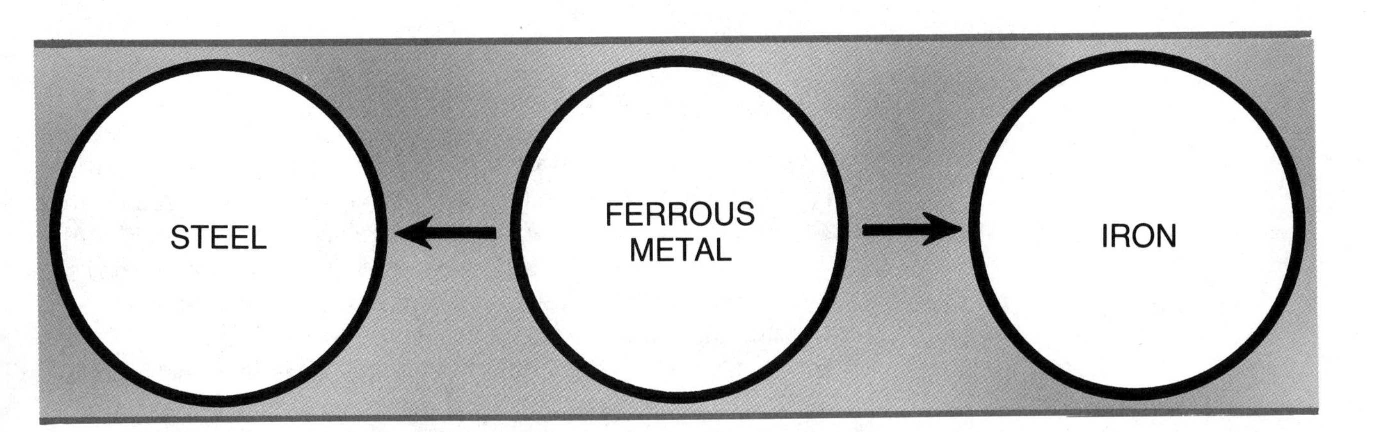

Steel and iron are ferrous metals.

Nonferrous metals.

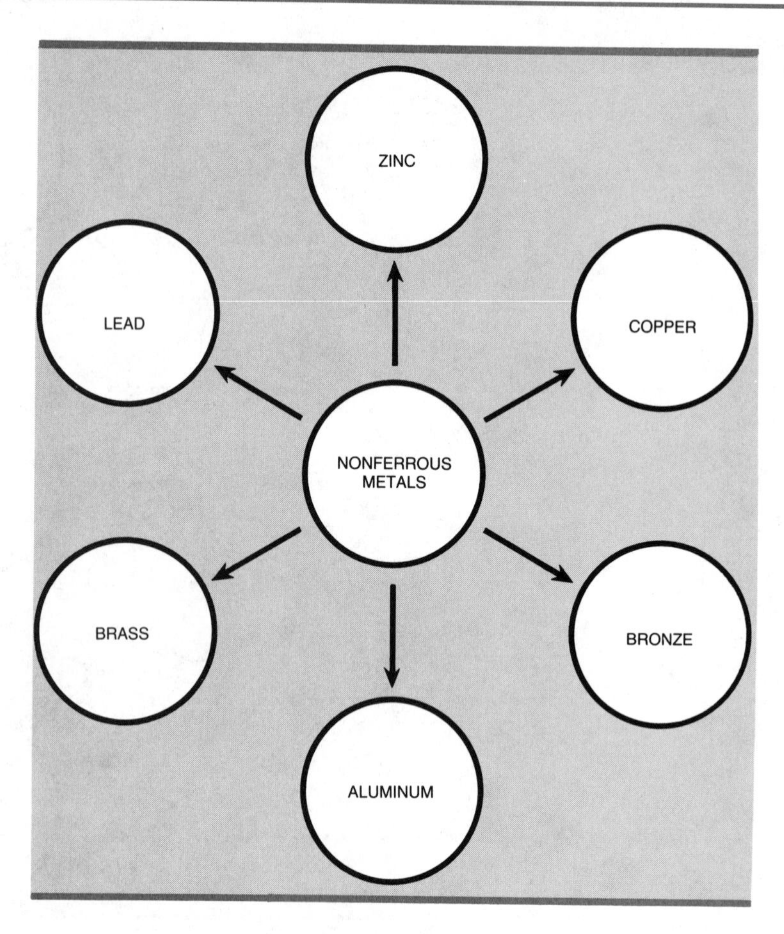

Synthetic Materials

Synthetic materials are made by people. The most commonly used synthetic material in manufacturing is *plastic*. Plastic is an important material because it can be used in so many different ways. And because it is relatively strong and lightweight, it can be used instead of wood or metal in some cases. Using plastic helps to conserve our limited natural materials, wood and metal.

Plastics are classified as *thermoplastics* or *thermosets*. Thermoplastics can be melted down and molded into other products and reused. Thermosets cannot be reused. Plastic toys are thermoplastics. Plastic dishes are usually thermosets.

THERMOPLASTICS

MATERIAL	APPLICATIONS
Acetal	Zippers Shower heads Bearings Rollers
Acrylic	Advertising signs Lenses Paints
Cellulosics	Lacquers Pens Packaging Fibers Toilet articles Eyeglass frames
Fluorocarbons	Wire insulation Pipe systems Bearings Nonstick cooking surfaces
Polyamide	Gears Bearings Fibers
Polyethylene	Packaging films Bottles Refrigeration containers
Polypropylene	Lab ware Toys Ice cube trays
Polystyrene	Novelties Latex paints Disposable products Radio and TV housings
Polyvinyl	Shower curtains Water beds Rain coats Adhesives Upholstery

THERMOSETS

MATERIALS	APPLICATIONS
Casein	Buttons Knitting needles Novelties
Epoxy	Finishes Adhesives Encapsulating
Melamine	Dishes Countertops
Phenolic	Adhesives Electrical products Appliance handles
Polyester	Auto bodies Boats Bath tubs Storage tanks
Silicone	Oils and greases Caulking Release agents Polishes
Urea	Adhesives Finishes Foams
Urethane	Adhesives Finishes Foams Chemical equipment

Common thermoplastics and thermosets and their applications. *Courtesy, from Jambro, Introduction to Manufacturing, © 1986 by Delmar Publishers Inc.*

Properties of Materials

When selecting materials for a manufacturing job, it is important for us to select materials that have the right properties for the product we want to make. Both natural and synthetic materials have seven properties that you should know: tensile strength, hardness, ductility, malleability, brittleness, elasticity, and machinability.

Tensile Strength The **tensile strength** of a material tells us how much strength is required to pull it apart. Steel has a lot of tensile strength. Plastic has much less. A substance with very little tensile strength is bubble gum. You can pull it and stretch it very easily because it has little tensile strength.

Hardness The **hardness** of a material is its resistance to denting or penetration. You may press on a steel plate as hard as you can, but you can't dent it because steel is hard. You can try to hammer a nail

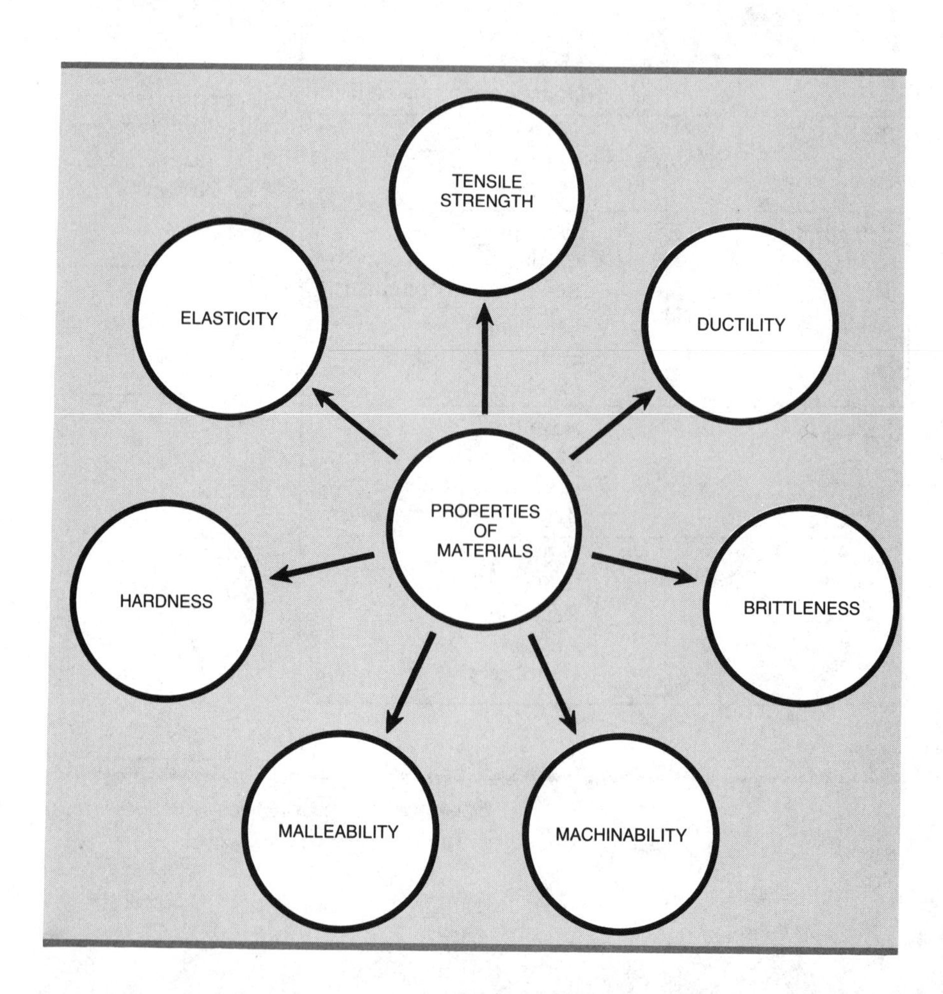

The seven properties of materials.

Steel is one of the hardest and strongest manufacturing materials. *Courtesy of Bethlehem Steel Corporation.*

through a steel plate, but the nail will bend because the steel is too hard. However, you can drive a nail into wood easily because it is not as hard as steel.

Ductility **Ductility** means that a material can be stretched without breaking. Some metals have ductility. Copper, for example, can be stretched without breaking. Wood cannot be stretched: it has no ductility.

Malleability **Malleability** means that a material can be rolled or formed without cracking. Some metals are malleable. Steel is malleable because it can be heated and hammered into shape. But wood and concrete are not malleable. You cannot heat them and pound them into shape.

Brittleness **Brittleness** means that the material breaks or cracks easily. Some forms of hard plastic are brittle.

Elasticity **Elasticity** means that the material can stretch and return to its original shape. Rubber has elasticity.

Machinability **Machinability** means that a material can be easily machined. In other words, it can be turned on a lathe, drilled, milled, planed, and so forth, without wearing out the tool.

MANUFACTURING PROCESSES

Manufacturers use many processes. The most important can be categorized as product development, product and tool design, and product and tool manufacturing.

Product Development

People in industry constantly search for new and better products to sell. This is the only way to stay ahead of the competition. For example, in the automobile industry, new products (cars, trucks, vans) and new models of existing products come out every year. The first process in the succession of processes required to make this happen is product development.

Product development has two important phases: research and development.

Research **Research** is the step-by-step discovery of knowledge. Research is carried out by scientists, who are assisted by technicians. *Pure* research means searching for knowledge without any special concern for how the knowledge will be used. *Applied* research means searching for knowledge for a specific purpose.

Most manufacturing research is applied research. For example, product development personnel are constantly trying to discover stronger and lighter metals. An ideal metal would be one that is as strong as steel but as light as plastic.

Development The **development** phase uses the knowledge discovered in research to create new products. For example, the discovery of the programmable integrated circuit led to the development of the microcomputer and the many digital products we now use every day.

Product and Tool Design

After a product has been developed, it must be designed. The ideas created in the product development phase must be converted into reality. This means that both the product itself and the jigs and fixtures (tools) needed to manufacture them must be designed. The design process is the same for both.

Design Stages In the first step of designing the product, engineers consider the product development ideas to define the problem. In the second step, they try out their ideas. This step is sometimes called *brainstorming*. Should the product be plastic? Should it be aluminum? Should it be steel? Should the product be square? Should it be round? Should it be rectangular? In the next step, engineers select the best ideas from the previous step. This collection of best ideas is usually called a *compromise solution*.

Bethlehem Steel Corporation's steel manufacturing plant of the Lehigh River. *Courtesy of Bethlehem Steel Corporation.*

The basic oxygen furnace complex of Bethlehem Steel Corporation's steel manufacturing plant. *Courtesy of Bethlehem Steel Corporation.*

An assembly line and a microcomputer manufacturing plant. *Courtesy of Commodore Computers, Inc.*

Rolling steel in Bethlehem Steel Corporation's steel manufacturing plant. *Courtesy of Bethlehem Steel Corporation.*

The design process.

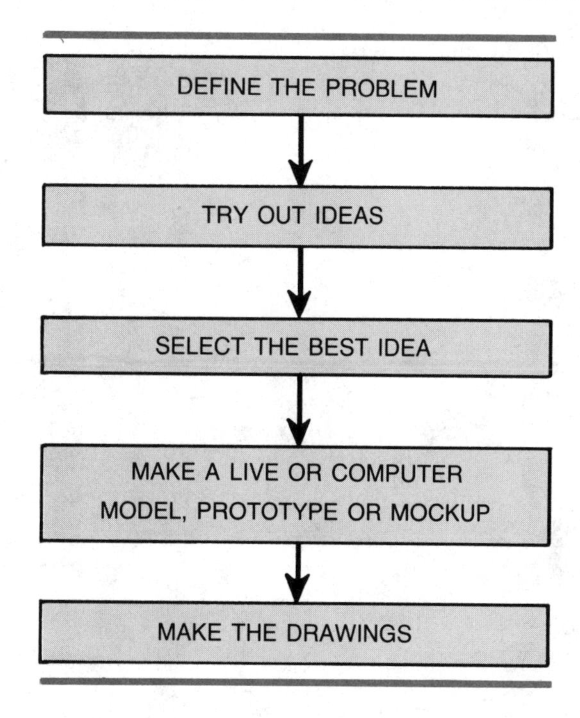

In the next step, a model, prototype, or mockup of the product is made. A *prototype* is tested to make sure it works the way it was designed to work. For example, if the product is a car, a prototype is a full-size working version of the car. It is put through numerous tests to make sure everything works the way it was designed to work.

Usually problems are found during the test and are corrected before going to the next step. When all problems have been solved, the final step in the design process—making working drawings—is undertaken. Drafting technicians make the drawings.

Modern drafting rooms are changing from traditional drafting tools and procedures (mechanical pencils, technical pens, erasers, pencil pointers, scales, and templates) to computer-aided design and drafting (CADD). CADD means the drawings are produced on a computer-based system instead of on a drafting board.

There are many CADD systems, but most contain a processor computer, disk drives, a graphics terminal, a keyboard, a digitizer, a plotter, and some type of interaction device, such as a puck or a light pen.

The same steps are used in designing the jig and fixtures or the tools needed to manufacture the product. Jig and fixture designers must know product design and manufacturing processes.

Marketing Surveys While engineers, designers, and drafting technicians are testing prototypes or mockups, marketing personnel are

A computer-aided design and drafting system. *Courtesy of Auto-trol Technology.*

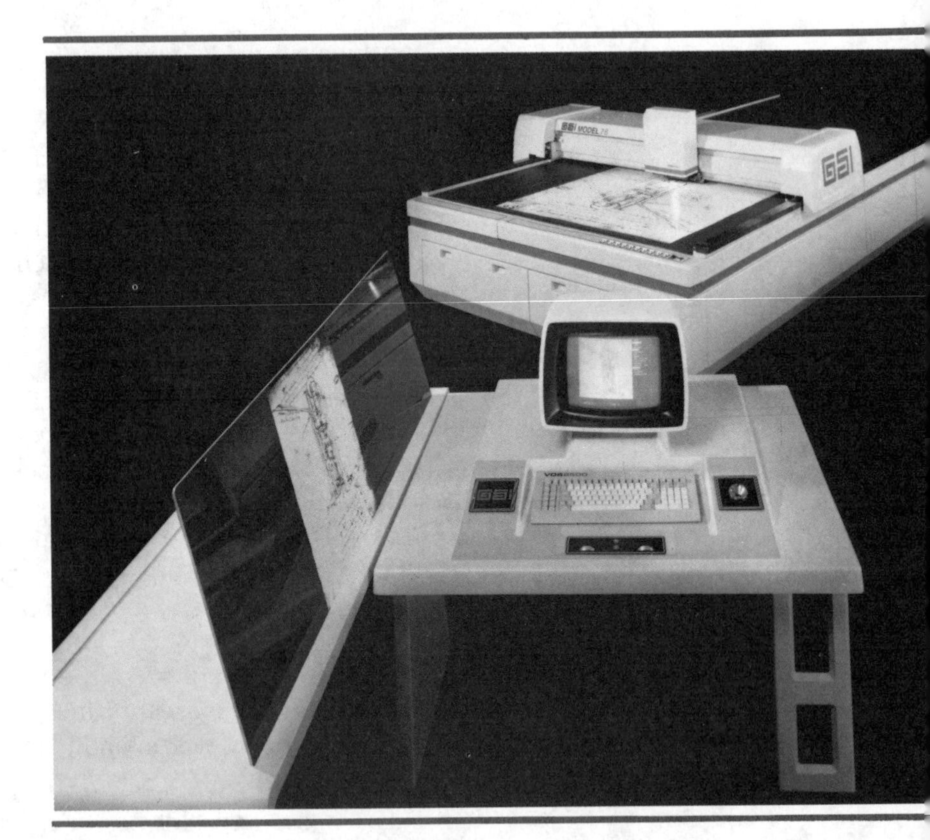

A computer-aided design and drafting system. *Courtesy of Gerber Scientific Instrument Corp.*

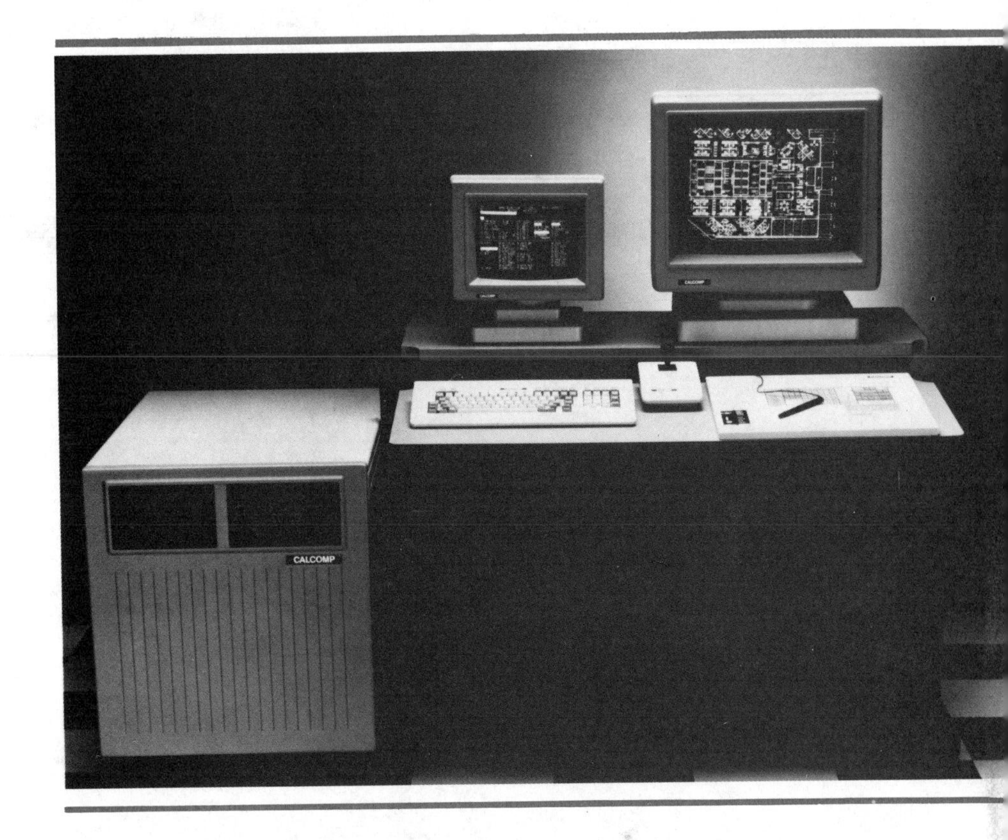

A computer-aided design and drafting system. *Courtesy of CALCOMP, Inc.*

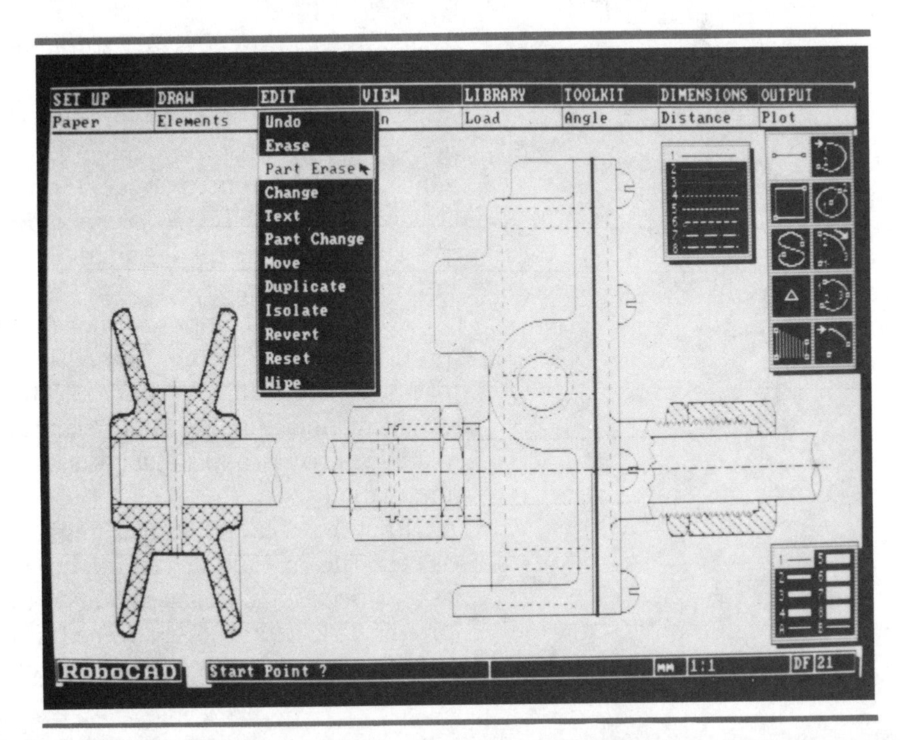

Screen on a CADD system showing how a drafter gives the system commands. *Courtesy of RoboCAD.*

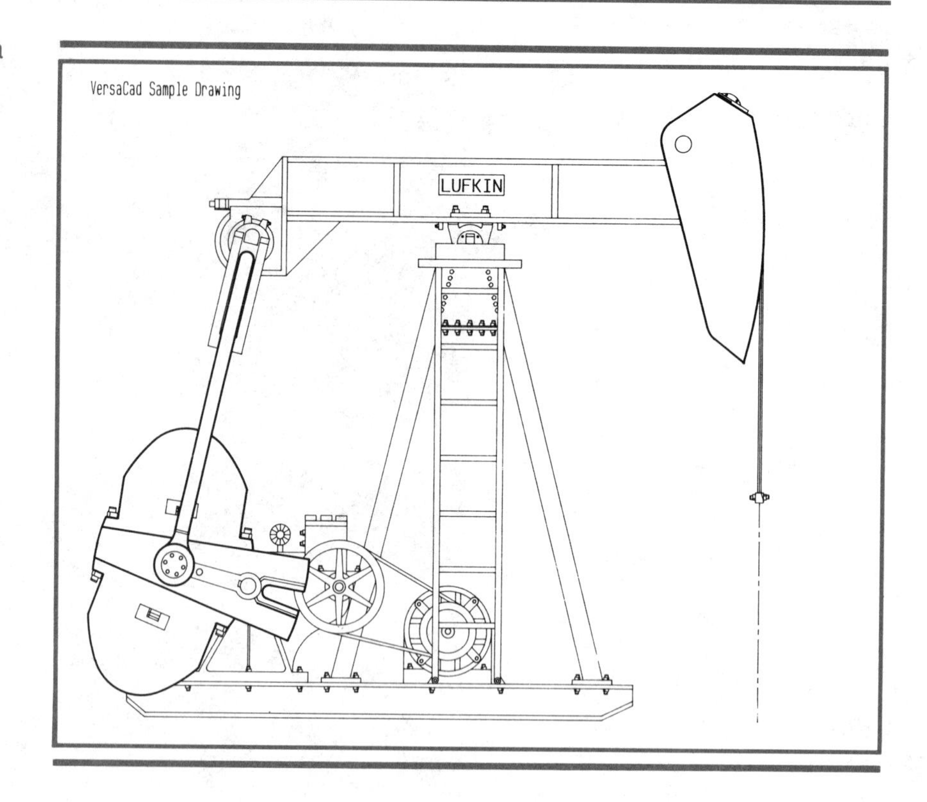

Sample drawing produced on a micro-CADD system. *Courtesy of T & W Systems, Inc.*

doing **marketing surveys**. Marketing surveys help determine how to sell the product.

A marketing survey can reveal the type of person who is most likely to buy the product, how much people will pay for the product, and consumer preferences about color, size, fragrance, shape, and flavor. This information helps salespeople do a better job of selling the product. It also alerts designers to changes that the product needs to make it more salable.

For example, suppose roller skates are going to cost $7 a pair to produce, but the market survey shows that people are only willing to pay $5 a pair. Management would halt the design process and decide not to manufacture the roller skates because it doesn't make sense to manufacture a product that costs more to produce than it will bring on the market.

In another case, the plan might have been to produce blue roller skates. The market survey might show that consumers will pay the asking price of the skates only if they are red. Management would order all skates produced in red instead of blue.

It is important to conduct market surveys before sending a design into production. Having accurate market data can mean the difference between a successful product and a failure.

Product and Tool Manufacturing

When a product has been developed, designed, and tested on the market, it is ready to undergo the manufacturing process. The manufacturing process consists of any or all of the following operations: layout and measurement, cutting, turning, planing, milling, drilling, shaping, forming, laminating, molding, casting, fastening, welding, soldering, abrading, coating, and finishing.

Layout and Measurement Before a product can undergo any other operation, it must be laid out. **Layout and measurement** involve determining the product's size and shape and preparing it for other operations. Layout involves a lot of measurement. Measurements

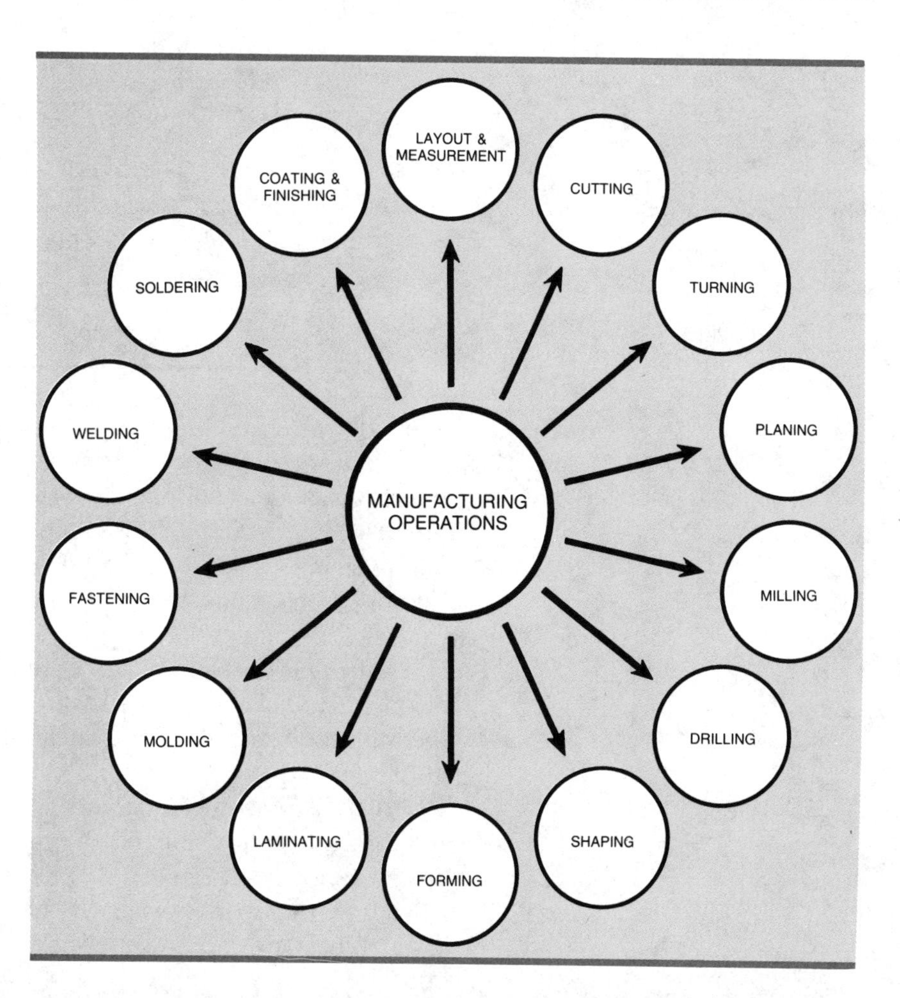

Manufacturing operations.

Making measurements with a micrometer.

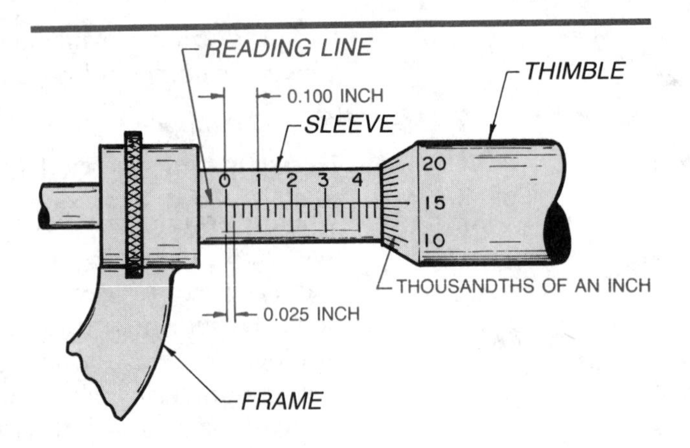

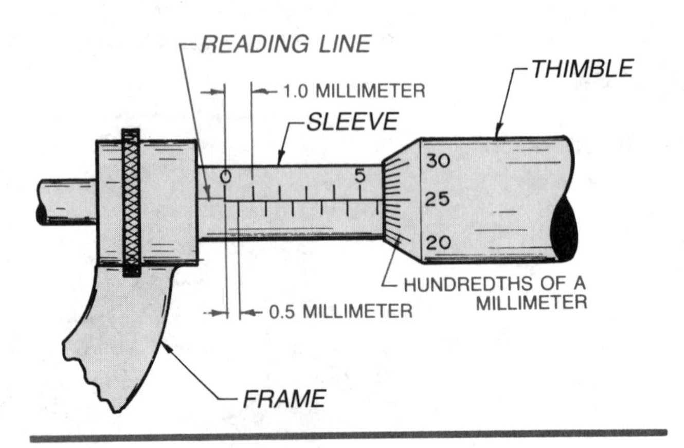

are made with rulers, tapes, scales, squares, gauges, micrometers, and numerous other devices. Layout personnel must know both the American standard and metric systems of measurement.

Cutting **Cutting** operations and manufacturing involve both sawing and shearing. *Sawing* might be accomplished manually with hand saws or mechanically with circular saws, saber saws, scroll saws, radial arm saws, and jigsaws.

Shearing is cutting a piece of paper with scissors or a paper cutter. Wood and metal can also be sheared with proper tools and machines. Shearing tools and machines include tin snips, aviation snips, bench shears, and squaring shears.

Turning **Turning** is cutting wood or metal on a lathe. The workpiece is held firmly on the machine between two spindles and turned very fast. As it turns, various types of tools are pressed against it, cutting the workpiece into the desired shape. Table legs and baseball bats are turned on a lathe.

Planing **Planing** involves leveling rough surfaces and squaring edges on wood or metal parts. It can be done manually with a hand plane, or mechanically with a jointer.

Milling **Milling** involves cutting wood or metal on a milling machine. Milling machines have different cutting attachments to perform a wide variety of tasks.

Drilling **Drilling** involves making holes in wood or metal, reaming holes, and threading. Drilling may be done manually with a hand drill or an electric drill, or mechanically with a drill press. *Reaming* is smoothing out a drilled hole to make it more accurate. *Threading* involves cutting threads into the inside of a hole to accept a screw or a bolt.

Shaping **Shaping** involves flattening or smoothing out horizontal, vertical, and curved surfaces. The work is done on a machine called a shaper.

Forming **Forming** involves converting metals into different shapes: for example, bending a flat piece of sheet metal into a length of ductwork. Forming can be done with hand tools or with machines such as the bar folder, brake, slip roll, and bender.

Laminating **Laminating** is an operation that is used to give wood extra strength. It involves bonding wood together in layers under heat and pressure so that the grain of each layer is parallel.

Molding and Casting **Molding** and **casting** are operations used to produce products in special shapes. Metals and plastics can be molded or cast. Both operations involve converting molten metal or plastic into shape using molds.

Fastening There are numerous ways to fasten materials together. Nails, screws, rivets, and nuts and bolts are the most commonly used means of **fastening**.

Welding and Soldering **Soldering** is used often in manufacturing electronic products. It involves joining metal with melted solder that quickly hardens. Soldering is done with a soldering gun.

Welding involves bonding metals or plastics together. Heat is used to melt the metals and fuse them together. There are several different types of welding: gas, spot, arc, tig, mig, and plastic welding.

Abrading, Coating, and Finishing **Abrading** means removing surface imperfections on wood and metal workpieces. Filing, grinding, and sanding are all abrading operations. **Coating** means applying protective coats to workpieces. Coats may be applied by hot dipping, electroplating, or anodizing. Electroplating is how chrome is applied to car bumpers. **Finishing** involves filling, staining, and painting.

Types of welding.

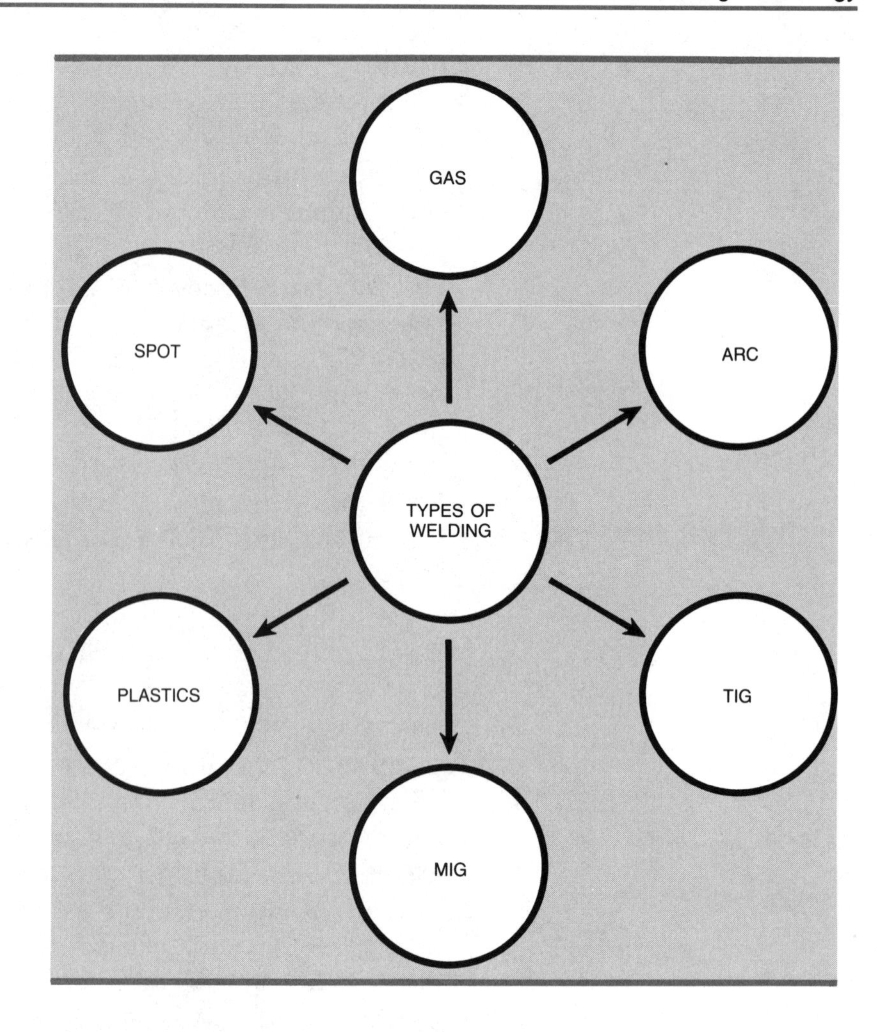

COMPUTER-AIDED MANUFACTURING

All of the manufacturing operations just explained can be done manually or mechanically. In modern manufacturing plants, many of these operations have been automated through the use of computers and robots. This automation is called computer-aided manufacturing (**CAM**).

In manual or mechanical manufacturing, people load, operate, and unload the machines that perform turning, drilling, milling, shaping, planing, and welding. In an automated manufacturing setting, robots load and unload machines. The robots get their operating instructions from computers.

Computer-aided manufacturing is a general term. It includes the many ways in which computers are used in manufacturing. Two important ways are computer-aided design and drafting (CADD) and robotics.

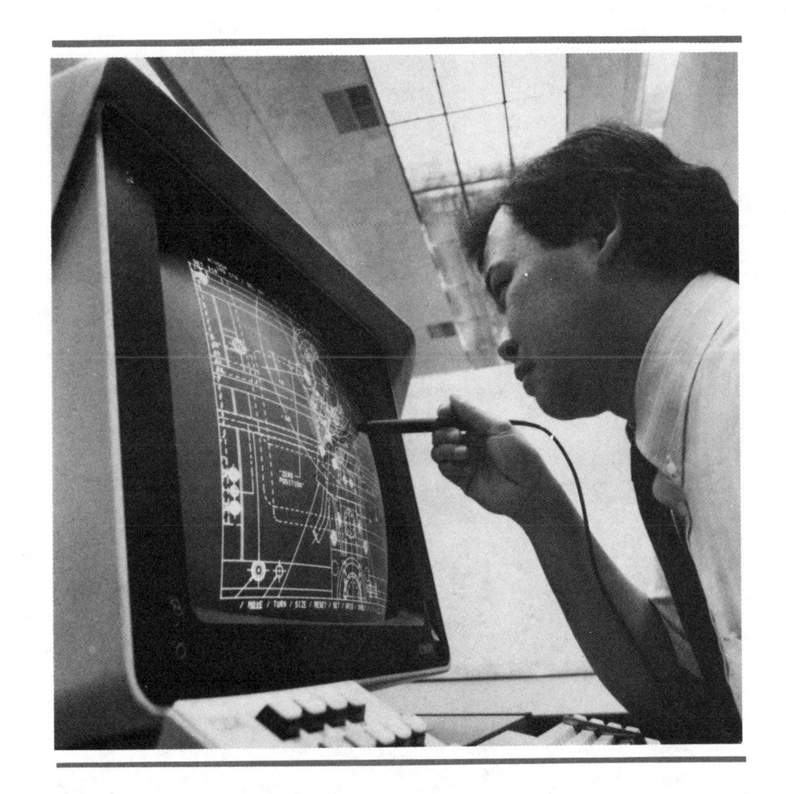

In automated manufacturing plants, the computer is used to increase productivity. *Courtesy of Cincinnati Milacron.*

In modern manufacturing plants, robots such as this G-26 series robot perform work that used to be performed by people. *Courtesy of PRAB Robots, Inc.*

Using the computer in operating machine tools is the most fundamental example of CAM. Numerical control (NC) is becoming the norm in modern manufacturing plants. The three principle forms of NC are

1. Traditional
2. Computer
3. Direct

Numerical Control (NC)

NC means controlling machining processes (or any other manufacturing process) with programs that consist of alphanumeric characters and symbols. In an NC operation, individual programs can be written for each job. This concept of programmability brings maximum flexibility to machining. NC can be applied to all of the traditional machining processes (i.e., drilling, turning, milling, etc.)

An NC system has three components: (1) the program, (2) the controller, and (3) the machine tool. The *program* is a detailed,

Three types of numerical control.

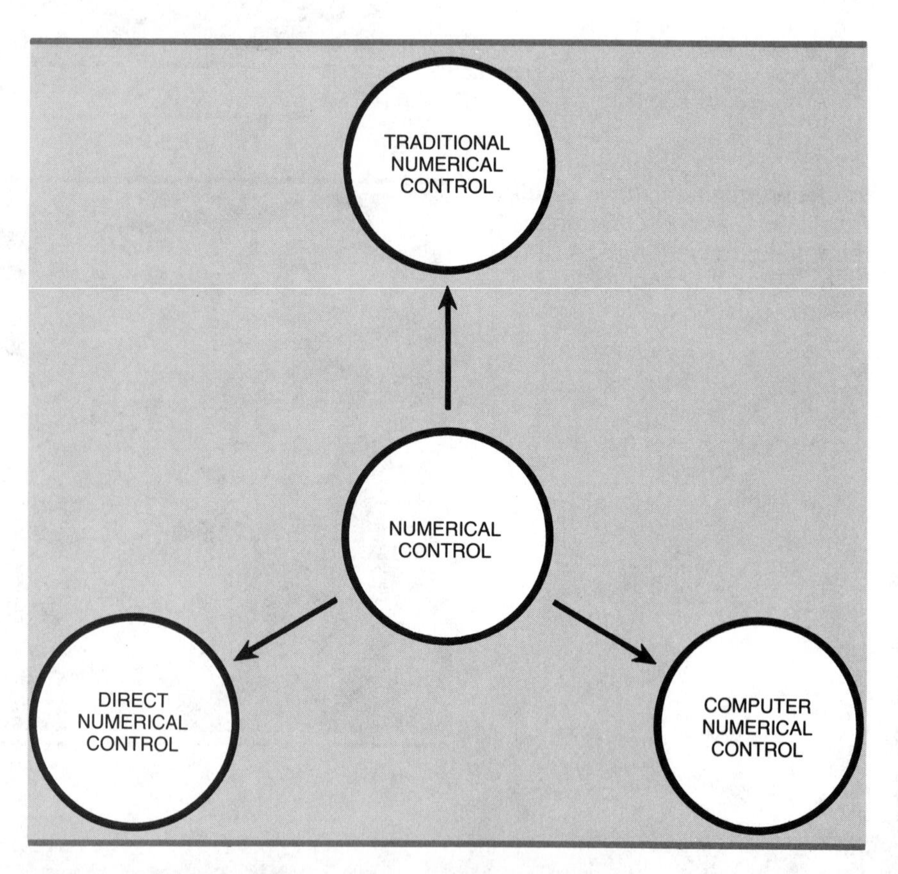

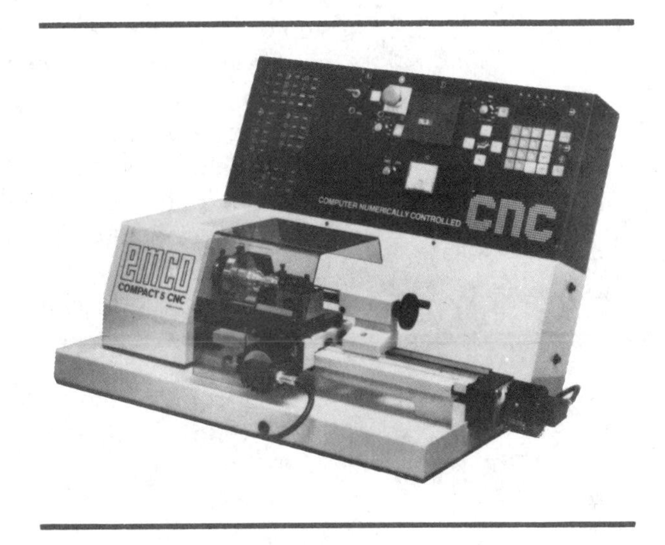

A CNC compact lathe. *Courtesy of EMCO.*

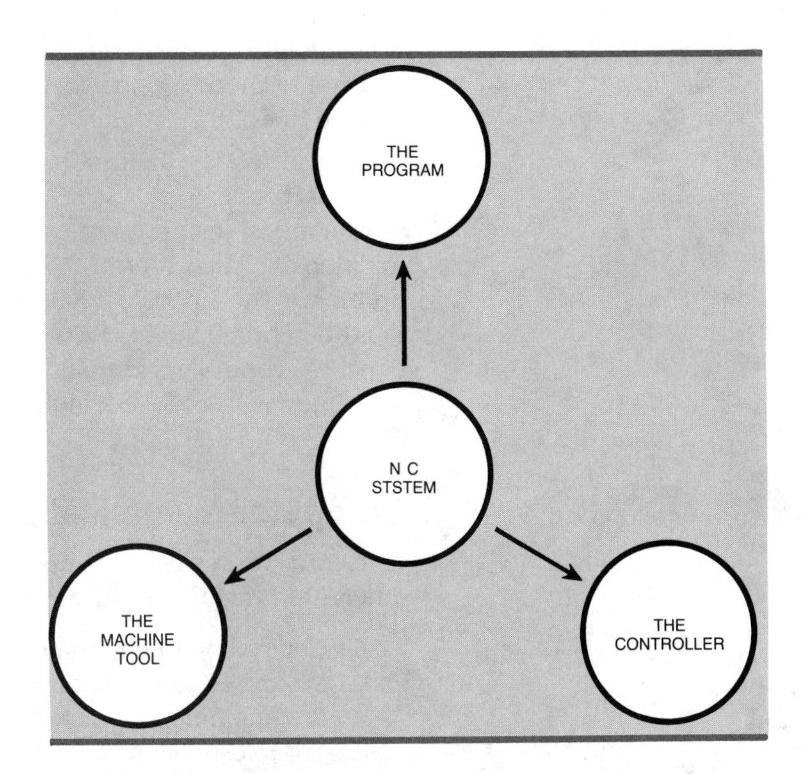

Three components of an NC system.

step-by-step set of instructions for carrying out a machine task. The *controller* is an electromechanical machine that "reads" the program, interprets the instructions, and sends the appropriate signals to the machine tools. The *machine tool* is the machine (mill, drill, lathe, etc.) that actually does the work.

The key to numerical control (and to CADD and robotics) is coordinate geometry. All positions and movements of an NC machine tool are specified by *X* and *Y* or *X*, *Y*, and *Z* coordinates contained in the program. NC machining is a four-step process:

1. Process planning
2. Part programming
3. Tape preparation/validation
4. Machining

Process Planning Process planning begins with the working drawings prepared in the CADD phase. By studying the drawings, planners develop a flowchart of the various operations that will be required in the conversion process.

Part Programming Part programming involves writing the code that tells the machine to perform the necessary operations.

Tape Preparation/Validation Tape preparation/validation involves converting the written instructions into tape form and doing a trial run to check the program for correctness. In this step "bugs" are worked out of the program before it is actually used.

Machining After the tape has been validated, the machining operations take place.

Numerical control represents a significant improvement over traditional manual control of machine tools. However, NC does have problems. The tape is a fragile medium that is prone to damage and rapid deterioration. Tape-reader technology is unreliable, and traditional NC controllers are inflexible. For these reasons, traditional NC is giving way to computer numerical control (CNC).

Computer Numerical Control (CNC)

CNC involves using a computer to solve some of the problems inherent in traditional NC. A CNC system has four basic functions:

1. Controlling the machine tool
2. Making adjustments during the machining process
3. Programming
4. Troubleshooting

Controlling the Machine Tool In CNC systems, the computer interprets the program code and sends the appropriate signals to the machine tool. The computer does not completely replace the controller. The program is still fed into the controller on tape. However, once it has been fed in, the computer stores it. Thus the program can be run many times without rerunning the tape. With traditional NC systems, the tape must be rerun each time. Therefore, the computer saves much time in NC.

Making Adjustments During Machining Another advantage of CNC over NC is that the computer allows machinists to make revisions and adjustments during the machining operation. In traditional NC, a system is locked into the speed and feed rates specified in the program. This is a problem because various machining tasks are better accomplished at varying speeds and feed rates. The only way to change them is to shut down the machine and revise the program. The computer, however, allows such changes to be made without a complete shutdown and rewrite.

Programming The computer also saves time in the writing of part programs. The differences between writing a part program in a traditional NC setting and in a CNC setting are the same differences that exist between writing in longhand and using a word processor.

Troubleshooting Finally, the computer can be used to troubleshoot maintenance problems in an NC system and decrease costly downtime.

Direct Numerical Control (DNC)

DNC is an even greater improvement. It eliminates the tape reader step. Thus, with DNC, programmed instructions go directly from the computer memory to the machine tool. DNC systems generally use one large computer or a series of smaller computers networked to one large computer to control several machine tools.

A DNC system has three major components: (1) a central computer with a large bulk memory, (2) networking lines, and (3) machine tools. The *central computer* is a large mainframe with enough bulk storage to store the programs for all of the machine tools that interface with it. The *networking lines* are telecommunication lines that allow instant two-way communication between the computer and the machine tools it controls. The *machine tools* are the same tools already discussed in the sections on NC and CNC.

The DNC system can do jobs that cannot be done by an NC or a CNC system:

Three components of a DNC system.

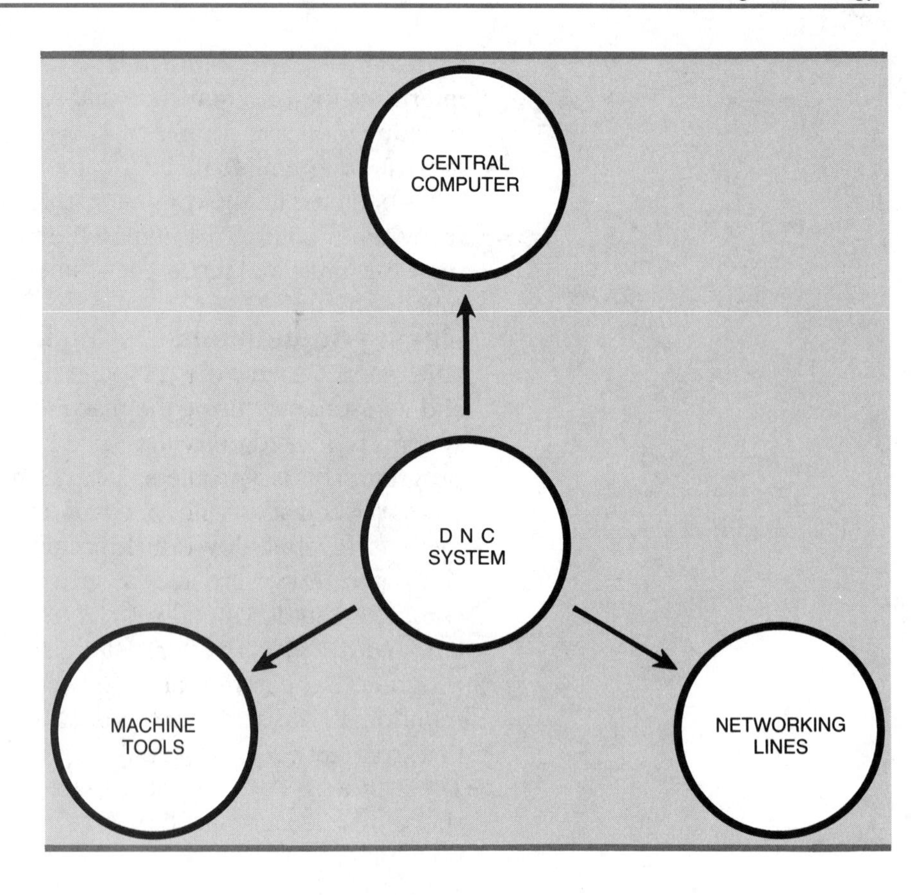

1. Numerical control without the tape
2. Storage of part programs
3. Two-way communication
4. Communications networking

Numerical Control You have already learned that the tape medium and the controller create problems in a traditional NC setting. These problems are partially reduced by CNC. Because DNC eliminates the tape and traditional controller unit altogether, the problems associated with them are partially solved.

Program Storage Mass storage of part programs is possible because of the large bulk memory of the central computer in a DNC system.

Two-way Communication DNC allows two-way communication between the computer and the machine tools. There is one-way communication with NC—from the controller to the machine tool. With CNC there is better one-way communication, but it is still one way. DNC allows continuous, immediate two-way communication. In other words, the computer can collect, store, and report data continuously throughout a machining operation. Work can therefore be closely monitored and adjustments can be quickly made.

Communications Networking Communications networking means more than just two-way communication between the computer and machine tools. With DNC, all manufacturing processes and components can be linked together for continuous, instantaneous communication. The DNC system, CADD systems, front office, shop floor, shipping and receiving, sales and marketing, and all other components can be linked together and kept up to date on the status of any project.

Even greater production gains are achieved when DNC systems are set up with one central computer and several smaller computers. The smaller computers hold active programs, so the time needed to move instructions from the central computer's bulk memory to the machine tools is reduced.

Through NC, CNC, and especially DNC, the machining of manufactured goods has been revolutionized. DNC currently represents the ultimate in the machining of manufactured goods. As technology continues to develop, better and better linkages between the various CAM components (CAD, DNC, robotics) will be achieved. This in turn, will lead to better and better productivity ratios.

Robotics

A *robot* is a programmable, multipurpose, electromechanical machine that can do jobs that have usually been done by humans. Robots are used for jobs that are boring, repetitious, tedious, or dangerous. Human interaction with **robotics** systems is usually limited to programming the robot, turning the system on and off, and maintaining and repairing the system.

A robot system has four components: (1) the manipulator, (2) the controller, (3) the power sources, and (4) the end effectors. The **manipulator** is the component that mimics the human arm and actually performs the work. A manipulator has two parts: the base and the appendages. The **controller** is a microprocessor-based device that (1) stores the operational programs for the robot, (2) issues operating instructions to the robot, (3) controls peripheral devices, (4) interfaces with the robot and provides continuous updates, and (5) interfaces with humans for receiving new programs or revisions of old ones. The three *power sources* used with robot systems are electric, pneumatic, and hydraulic. *End effectors* are special adapters that allow robots to perform difficult tasks.

THE COMPUTER REVOLUTION

At one time computers were used mainly by large businesses, the government, the military, and universities. Now computers have become a part of everyday life. (1) This mechanic is using a computer to check for engine problems in a car. (2) Computerized bank machines allow us to get cash at any hour of the day or night. (3) You can use computers to shop for products from your own living room. (4) People of all ages are learning how to use computers.

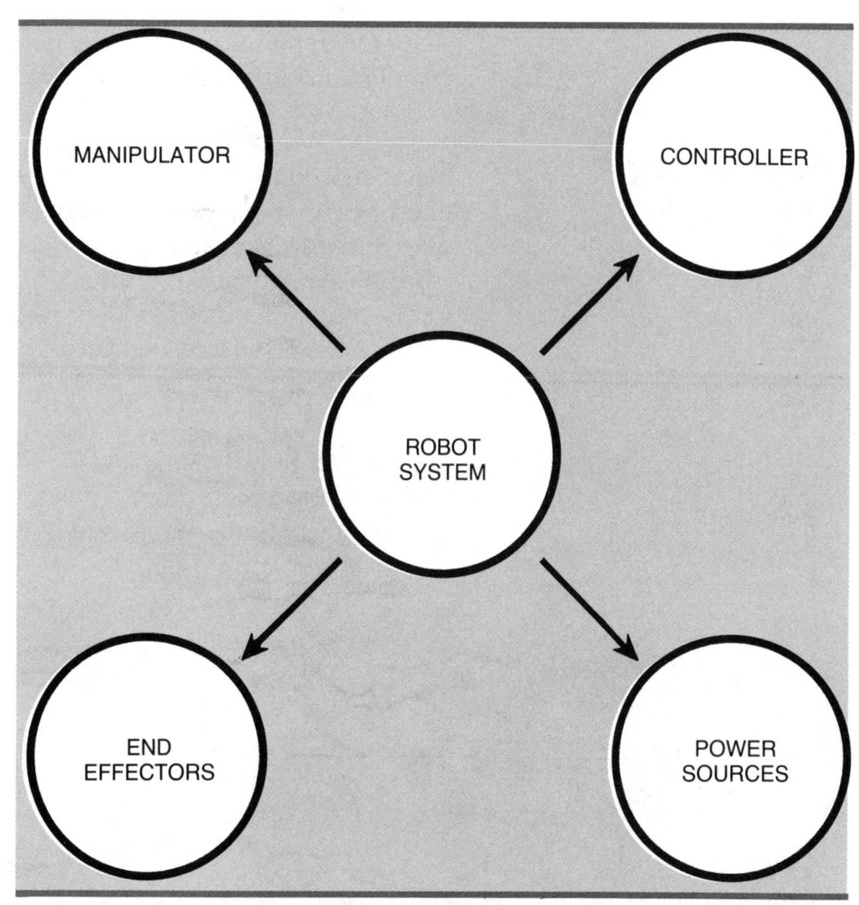

The four components of a robot system.

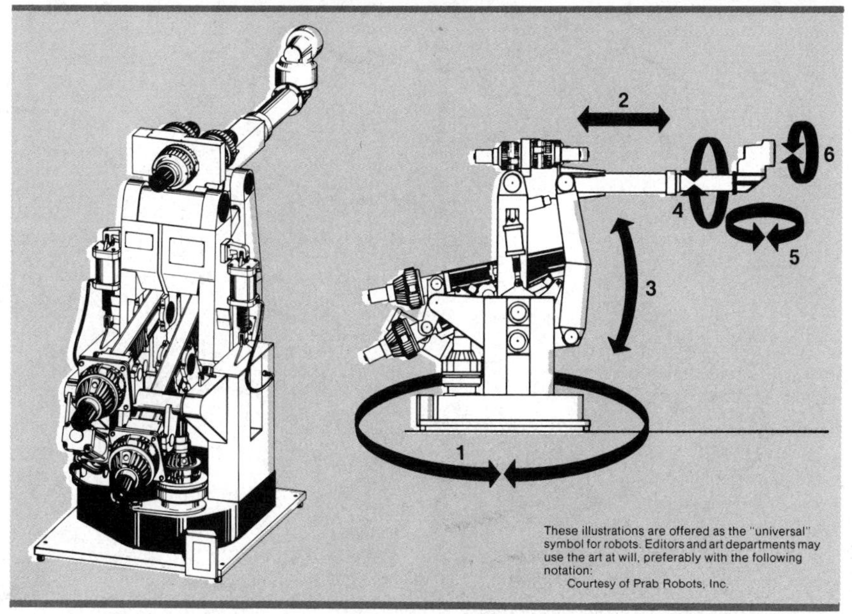

2
6
4
5
3
1
These illustrations are offered as the "universal" symbol for robots. Editors and art departments may use the art at will, preferably with the following notation:
Courtesy of Prab Robots, Inc.

The various movements of a typical industrial robot. *Courtesy of PRAB Robots, Inc.*

common applications are spray painting, welding, loading/unloading, handling materials, and assembling.

DISTRIBUTING MANUFACTURED PRODUCTS

When a product has been developed, designed, and manufactured, it must be distributed to consumers. This process is known as *marketing*. Several operations make up the marketing process. You have already learned about marketing surveys, which take place while products are tested as part of the design process. The other marketing operations are advertising, packaging, distribution, and selling.

Advertising

Advertising involves making the consumer aware of an interest in the product. Advertising comes in many forms: newspaper ads, magazine ads, radio announcements, and billboards. But the type of advertising that we see most often is the television commercial. Television

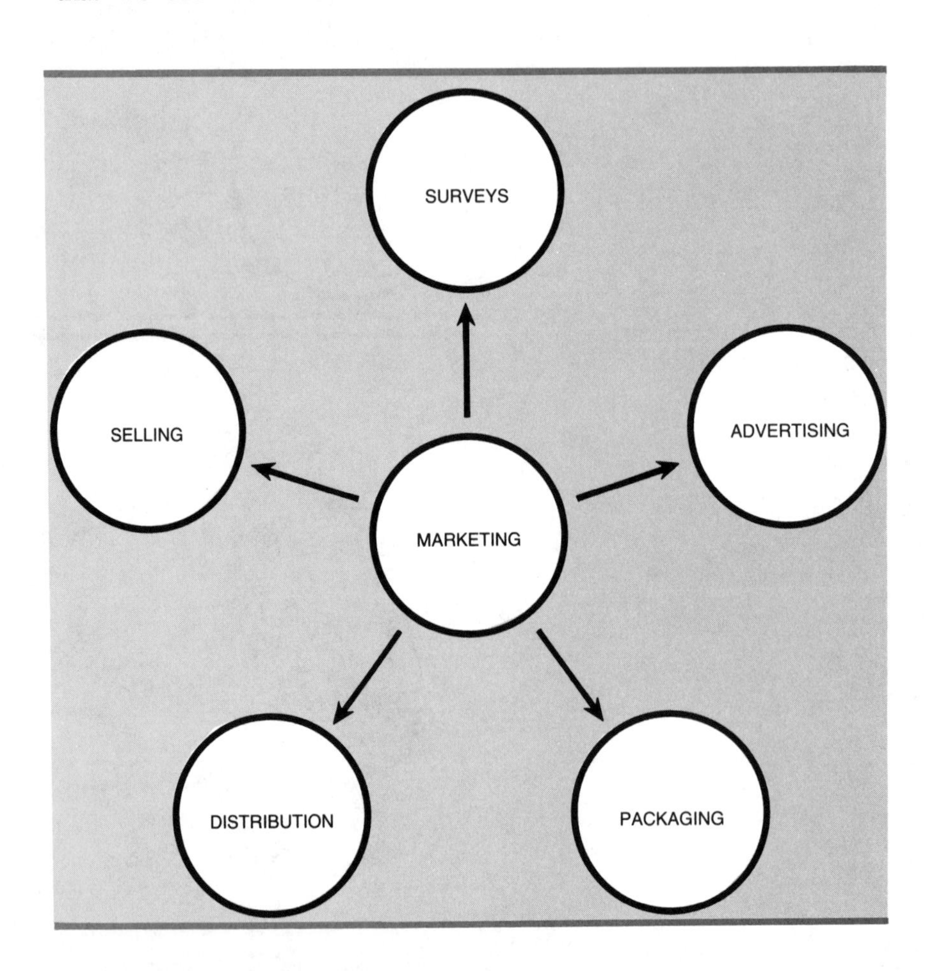

The elements of marketing.

Modeling newly produced clothing styles is a form of advertising. *Courtesy of Robert C. Atwood.*

commercials are one of the most effective forms of advertising. They are also the most expensive.

Packaging

Packaging is important in marketing a product. If the product is in an attractive, eye-catching package, it is more likely to be purchased by the consumer. A well-designed package will also protect the product and identify its manufacturer.

Distribution and Selling

Distribution means getting the packaged product from the manufacturer to the consumer and selling it. Three methods used for distributing products to the market are direct selling, indirect selling, and wholesale distribution.

Direct selling means the manufacturer sells the product directly to the consumer through sales outlets or by mail order. *Indirect selling* means the manufacturer sells the product to a retail store, which then sells it to the consumer. *Wholesale distribution* means that the manufacturer sells the product to a wholesale outlet. The wholesaler then sells the product to a retail store which sells it to the consumer. **Selling** is the final step in the marketing process. Selling means calling on the potential consumers and trying to convince them to buy the product. It is important for salespeople to know the product well, to know the needs of the consumer, and to know how to make a good impression on consumers.

SOME CAREERS IN MANUFACTURING TECHNOLOGY

Many career fields are associated with manufacturing technology. Most fall into one of the following six categories: (1) unskilled workers, (2) semiskilled workers, (3) skilled workers, (4) technical workers, (5) supervisory workers, and (6) professional workers.

Unskilled Workers

Unskilled jobs require little education and training. They include such jobs as laborer, plant and grounds worker, and janitor. These jobs include such tasks as loading and unloading, maintaining grounds, maintaining plants and the lawn, cleaning windows, cleaning offices, and running errands.

Semiskilled Workers

Semiskilled jobs require only a small amount of training, usually on-the-job training. They include such jobs as assembler, machine operator, truck driver, power equipment operator, and materials handler.

Skilled Workers

Skilled jobs require an apprenticeship or special vocational training. Skilled jobs include such things as setup worker, millwright, tool and die maker, metal fabricator, electrician, and machinist.

Categories of manufacturing careers.

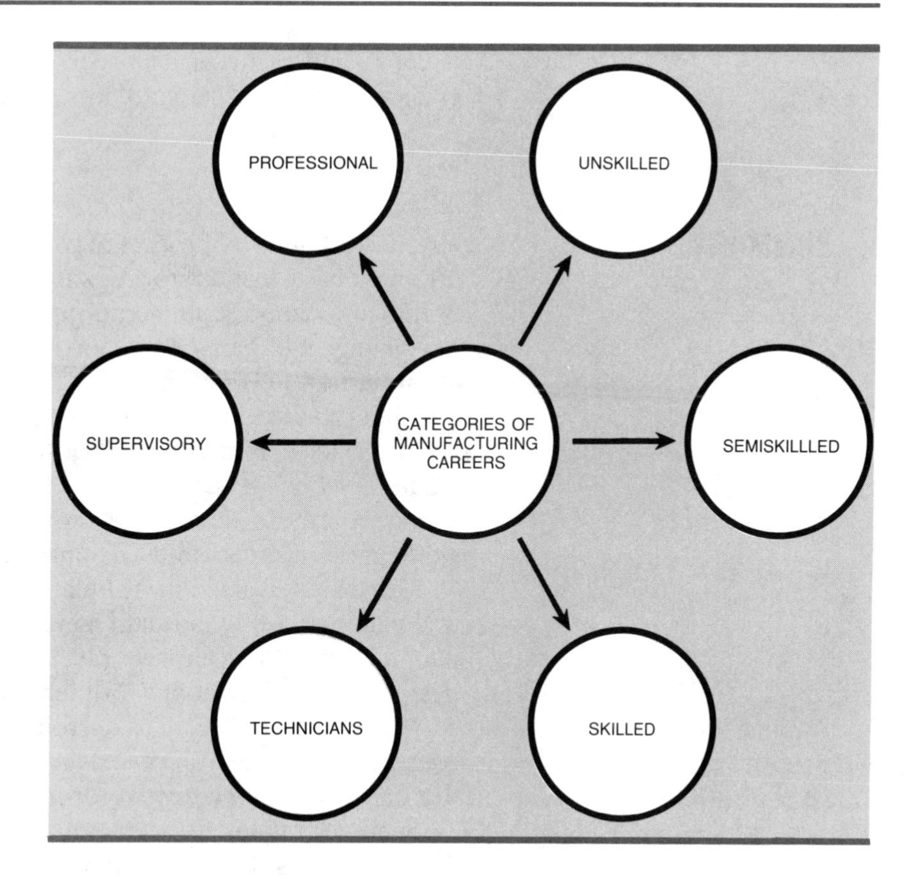

Technical Workers

Technical jobs require two-year degrees from community colleges or technical colleges. They include such jobs as designer, drafting technician, CADD technician, quality control inspector, materials tester, production controller, and lab technician.

Supervisory Workers

Supervisory jobs require at least two-year degrees from a community college or technical college. Supervisors oversee other workers to ensure that schedules are kept and deadlines are met. Supervisory jobs include production supervisor, shipping supervisor, materials inventory supervisor, and employee supervisor.

Professional Workers

Professional jobs require at least a four-year college degree. They include such jobs as scientist, engineer, accountant, computer system scientist, and manager. Managers have such titles as president, vice president, chief executive officer, administrator, and plant manager.

SUMMARY

Manufacturing is the process of converting raw materials into finished products. Early manufacturing was done by hand on a piecemeal basis. Eventually, the need for more products led to mass production and interchangeable parts. These two developments are associated with the Industrial Revolution.

Mass production involved making multiple copies of each part of a product and then assembling the finished products. Mass production led to the concept of interchangeable parts. Interchangeable parts are parts that are exactly alike or very nearly so. For mass production to work, parts manufactured on a mass basis had to be interchangeable.

The materials used in manufacturing are natural or synthetic. Natural materials are woods and metals. Synthetic materials are made by people. A material can have seven properties: tensile strength, hardness, ductility, malleability, brittleness, elasticity, and machinability.

Metals are classified as ferrous or nonferrous. Ferrous metals contain iron. Nonferrous metals do not contain iron. Alloys are metals that are made by mixing two or more metals together.

Synthetic materials used in manufacturing are mostly plastics. Plastics are classified as thermosets or thermoplastics. Thermoplastics can be melted down and reused. Thermosets cannot.

There are many different manufacturing processes. The most frequently used fall into three categories: product development, product/tool design, and product/tool manufacturing.

All manufacturing operations have traditionally been done manually by people and machines. Automation is changing that. Modern manufacturing processes are being converted to computer-aided manufacturing (CAM) and robotics. The most commonly used forms of CAM are numerical control (NC), computer numerical control (CNC), and direct numerical control (DNC).

After a product has been manufactured, it must be distributed to the market. This involves four processes: advertising, packaging, distribution, and selling.

Many career fields are associated with manufacturing technology. Some of the most widely practiced are assembler, machine operator, truck driver, power equipment operator, millwright, tool and die maker, metal fabricator, machinist, CADD technician, drafting technician, production controller, quality control inspector, scientist, engineer, and manager.

KEY WORDS

Manufacturing
Raw materials
Mass production
Interchangeable parts
Mechanization
Natural materials
Synthetic materials
Tensile strength
Hardness
Ductility
Malleability
Brittleness
Elasticity
Machinability
Ferrous metals
Nonferrous metals
Alloys
Research
Development
Marketing surveys
Layout and measurement
Cutting
Turning
Planing
Milling
Drilling
Shaping
Forming
Laminating
Molding
Casting
Fastening
Welding
Soldering
Abrading
Coating
Finishing
CAM
NC
CNC
DNC
Robotics
Manipulator
Controller
Point-to-point movement
Continuous path movement
Advertising
Packaging
Distribution
Selling

SOME QUESTIONS TO TRY

1. What is manufacturing?
2. Name two frequently used manufacturing materials.
3. Name two frequently used manufacturing processes.
4. What are the steps in the design process?
5. What is computer-aided manufacturing?
6. What are the four steps in distributing manufactured goods?

TECHNOLOGY LEARNING ACTIVITIES

1. Using a plywood backboard, make a manufacturing materials sample board. The board should contain small examples of all of the various woods, metals, and plastics used in manufacturing. Each sample should be attached to the board and labeled.
2. Make a scale model that will illustrate how a manufacturing plant works. Include all operations from research and development to distribution and selling.
3. Form a manufacturing company among your classmates. Assign different students to research and development, design, marketing, manufacturing, and distribution jobs. Each job should be clearly defined and have a written description. This is the responsibility of the students assigned as managers and supervisors. When the company has been formed and staffed, its job is to develop, design, manufacture, and market a product that will be bought by other students at your school or people in the community. Such a product could be special-event sand candles, for example. Such candles prepared for a given holiday would make a good product. Regardless of the product selected, the following steps should be completed.
   a. The research and development students develop the product.
   b. The design students design it and complete detailed sketches.
   c. While it is being designed, the marketing students conduct a market survey to determine consumer preferences and cost limitations.
   d. The manufacturing students plan the production phase, obtain all of the needed materials, and make the desired number of the product.
   e. The distribution students advertise and sell the product.
4. Make a timeline of important technological developments in manufacturing.

CHAPTER 7

TRANSPORTATION TECHNOLOGY

OBJECTIVES

After completing this chapter, you should be able to

- define transportation.
- explain significant developments in transportation technology.
- explain the components of transportation technology.
- explain the various ways that people and goods are transported.
- list some of the most widely used applications of transportation technology.
- list some careers associated with transportation technology.

For many years, transportation was one of the slowest developing technologies. It took thousands of years to develop transportation technology from animal-drawn vehicles to the horseless carriage. But in the past 100 years, transportation technology has gone from horseless carriages and sailing ships to sleek modern automobiles, jet aircraft, nuclear-powered ships, and even spacecraft.

During this century, transportation technology has developed faster than any other technology except electronics. Transportation technology has an impact on all other technology areas.

WHAT IS TRANSPORTATION?

Transportation is the technological activity that involves moving people, products, goods, and materials from one location to another. The earliest form of transportation was walking. But the need to go longer distances in less time has led to continuous developments in transportation technology.

When you peddle a bicycle, ride in a car, fly in an airplane, sail on a ship, or blast off in a spaceship, you are being transported. When water is piped from a well to your house, it is being transported. When cargo is shipped overseas, it is being transported. When cattle are shipped on a train from a ranch in Texas to a processing plant in Chicago, they are being transported. When a satellite is launched into space, it is being transported.

SOME IMPORTANT DEVELOPMENTS IN TRANSPORTATION TECHNOLOGY

Some of the most significant developments in transportation technology have been

1. The wheel
2. The domestication of animals
3. The steam engine
4. The internal combustion engine
5. The discovery of petroleum as an energy source
6. The jet engine
7. The rocket engine

The Wheel

The wheel may be the most important development in the history of technology. It came about 10,000 B.C. during the New Stone Age. The wheel allowed for the development of vehicles. Early wheeled vehicles were crude carts that could be loaded with goods and pulled by people. Using the wheel, people could transport more goods faster and with less work.

The Domestication of Animals

The **domestication of animals** was an important development in transportation technology. Wheeled vehicles made a big difference, but wheeled vehicles pulled by animals represented the real step forward in transportation technology. A team of horses, mules, or oxen could pull heavily loaded carts and other wheeled vehicles long distances and even up steep hills. Using animals in this way led to the development of the horse-drawn carriage, buckboard, and stagecoach.

Significant developments in transportation technology.

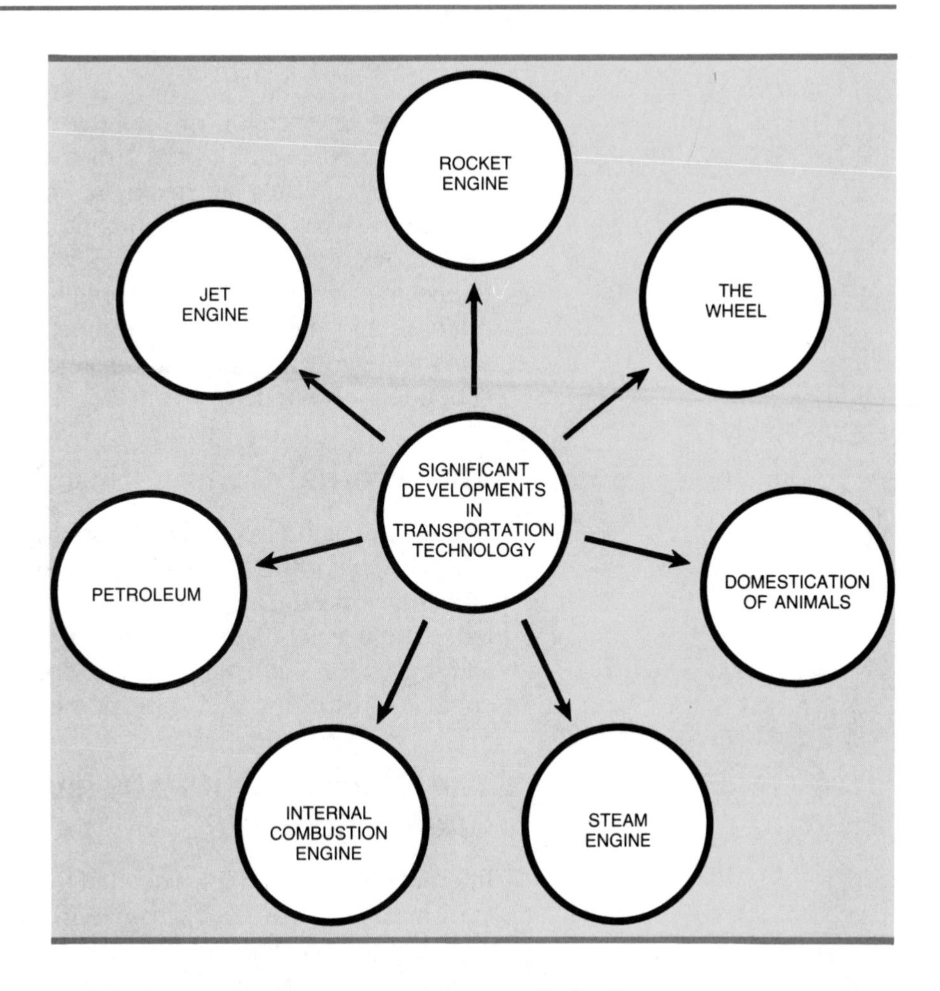

The invention of the wheel allowed for the development of wheeled vehicles.

The Steam Engine

More than one version of the **steam engine** was invented. But the one most people think of is James Watt's steam engine. Developed around 1760, Watt's steam engine was much better than the other versions. It led to the development of the steam locomotive and the steam-powered ship.

Steam-powered trains could go faster and farther than horse-drawn vehicles, and they could carry much heavier loads. Steam-powered ships were much faster and stronger than sailing ships and not as dependent on wind.

The Internal Combustion Engine

The steam engine worked well for large vehicles such as locomotives and ships, but it was too big to work in smaller vehicles. The **internal combustion engine**, invented around 1860, solved this problem and led to the development of smaller motor vehicles, such as cars, trucks, and buses. Gasoline and diesel engines are internal combustion engines. That is, the fuel is burned inside the engine.

The Discovery of Petroleum as an Energy Source

Internal combustion fuel is needed for an internal combustion engine. Fuel must be extremely sensitive to heat and must react to heat by exploding. This need led to the use of **petroleum** fuel—that is, gasoline refined from oil. The first oil well in this country was dug in Titusville, Pennsylvania, in 1859. Now, in spite of wells throughout the country, oil is imported into the United States.

The Jet Engine

The internal combustion engine was used on early airplanes, and it is still used on some aircraft. But the invention of the **jet engine** in 1936 was a major step forward in air transportation.

Jet-powered aircraft are much faster and more powerful than aircraft powered by internal combustion engines. The jet engine is now the mainstay of modern airliners and military aircraft.

The Rocket Engine

The **rocket engine** was actually invented around 1927, nine years before the jet engine. But it wasn't perfected enough for space travel

Operation of a four-cycle gasoline internal combustion engine.

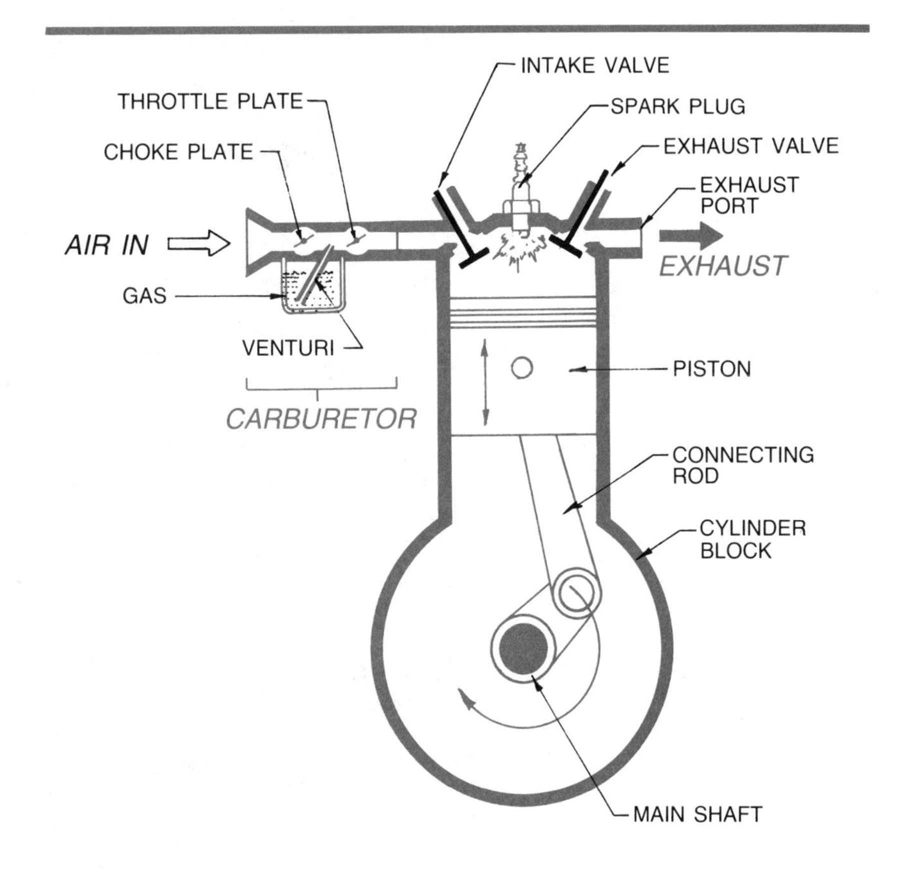

FOUR-CYCLE GAS ENGINE

until the Soviet Union and the United States began their space race in the 1950s. The rocket engine is important because it provides the thrust needed to break though the earth's gravitational pull and travel into space. The rocket engine is one of the major breakthroughs that, along with many others, eventually led to the first manned moon landing by the United States in 1969 and the space shuttle program that continues even now.

TRANSPORTATION COMPONENTS

Transportation systems have two basic components: vehicles and pathways. This is true regardless of whether the mode of transportation is ground, water, or air or whether people, materials, or goods are transported.

Vehicles do the transporting. Cars, trucks, buses, airplanes, and ships are vehicles. All these things are vehicles because they transport people, goods, or materials.

Some examples of vehicles.

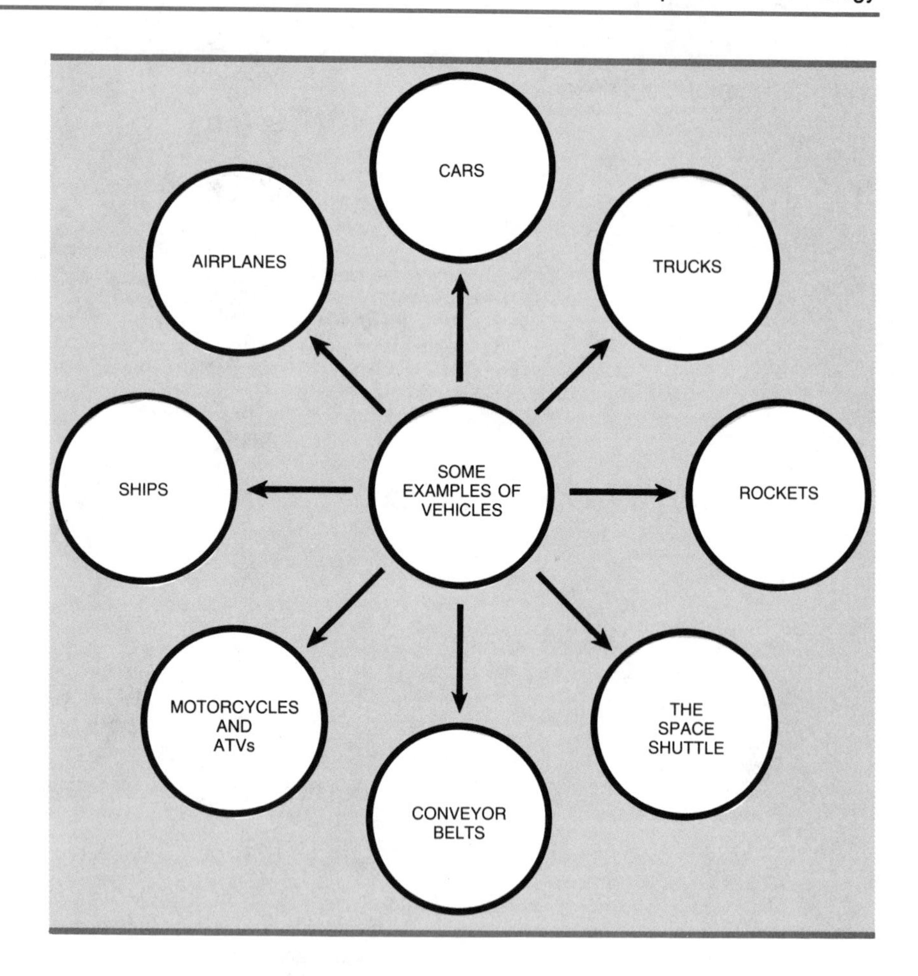

Vehicles use **pathways** to get from one place to another. Pathways are roads, streets, highways, seaways, airways, tracks and canals. Land transportation depends on roads, streets, and **highways**. We follow such pathways when traveling in a land vehicle. Maps showing the roads to take are important when people go on a vacation or travel to an unfamiliar place.

The oceans and seas are marked into **seaways** or **shipping lanes**. A ship must stay in the right shipping lane. Shipping lanes help avoid accidents and make a ship easier to find in case of an accident or late arrival.

The sky is marked off into **airways** that all aircraft must follow. Airways keep thousands of planes organized and help ensure that planes don't crash into each other. Without airways, midair collisions would happen much more frequently.

Trains and subways move on pathways called *tracks*. Train and

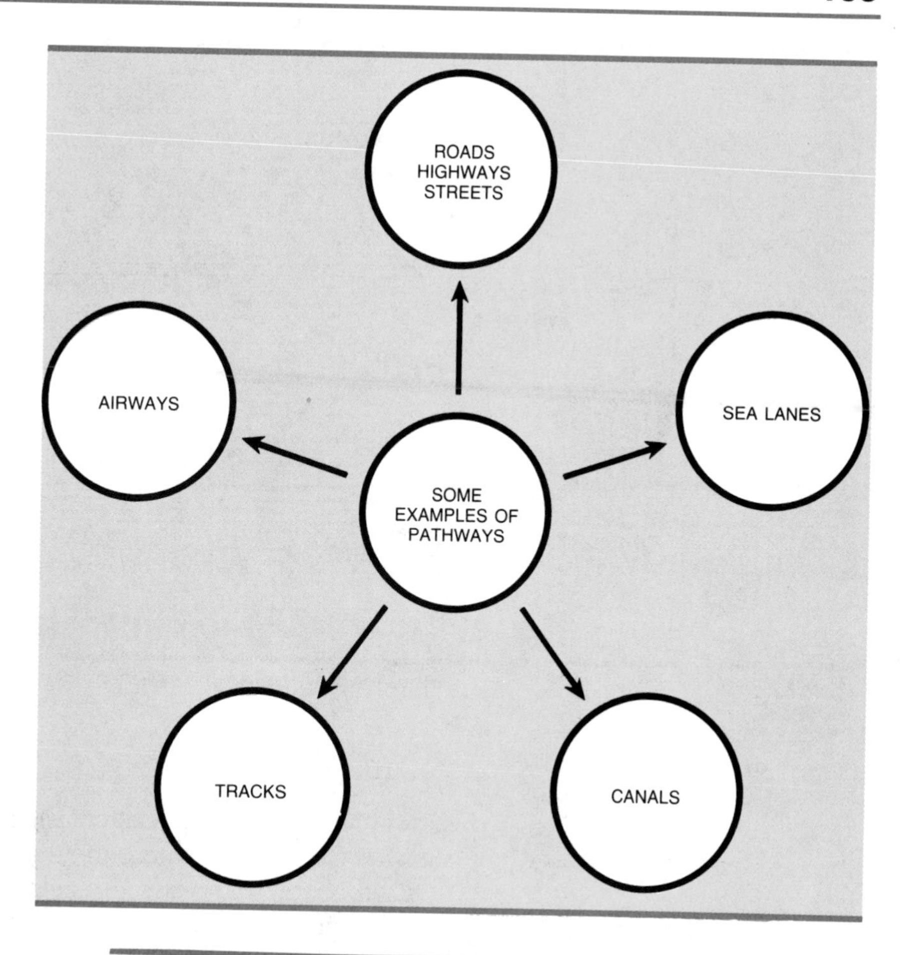

Some examples of pathways.

This intercoastal canal is an example of a pathway. *Photo by Deborah M. Goetsch.*

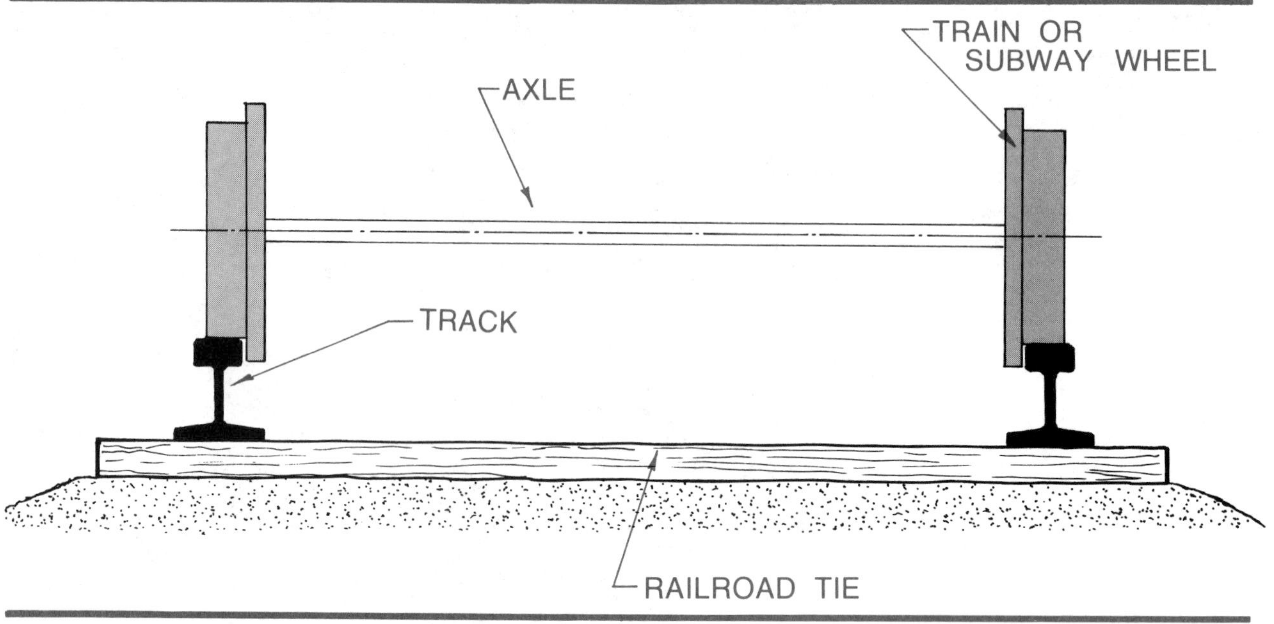

Subways and trains move on pathways called tracks.

subway wheels are especially designed to fit over tracks. Trains and subways can only go where the tracks lead.

Canals are waterways that allow smaller vessels and barges to transport goods inland. Often canals are waterways built to link natural bodies of water such as lakes, rivers, oceans, seas, or bays.

TRANSPORTING PEOPLE AND GOODS

Our society is very mobile. In other words, we are always on the move. The national and international economies and defense forces depend on mobility. That is, transportation must be available to move people on land, in the air, on the water, and even through space.

Land Transportation

Land transportation has come a long way since the earliest people had to walk from place to place. Now we can walk, peddle a bicycle, ride a motorcycle, drive a car, ride in a subway, or ride on a train. Goods can also be shipped in these ways.

Because of the many ways we can travel on land, our horizons are much broader today than in the old days when people had to walk or ride animals. We can more often visit friends and relatives in other cities or states. We can go to a school that is miles away from our

Commercial airliners such as this Boeing 747 are one of the most common types of air transportation. *Courtesy of Boeing Aircraft Corporation.*

Automobiles such as this midengine futuristic sports car being studied by Chrysler Pacifica's design staff are the most common form of ground transportation. *Courtesy of Chrysler Corporation.*

house. We can shop at stores outside our neighborhood. We can take vacations to distant places. We can live in Maine and eat fruit grown in California. We can live in Kansas and eat fish caught off the Florida coast. We can travel by air any city in this country in just a few hours if we fly, or in just a few days if we drive.

ATVs such as this Polaris Trailboss 250 are becoming a common and popular form of transportation. *Courtesy of Polaris.*

Motorcycles such as this Sport Glide FXRT and Sport Glide FXRD Grand Touring Edition have long been popular forms of ground transportation. *Courtesy of Harley Davidson.*

Sport Glide™ FXRT

Sport Glide™ FXRD Grand Touring Edition

Streamlined diesel trains are a popular and common form of land transportation, especially in European countries.

One of the oldest but still popular forms of water transportation is the sailboat. *Photo by Deborah M. Goetsch.*

The Chrysler Laser Atlas and Satellite system will provide drivers of the future with space-age vehicle navigation capabilities. *Courtesy of Chrysler Corporation.*

The first Opel Kadett was introduced in 1936. Compare the earlier photograph with the more recent Kadett GTE. *Courtesy of General Motors.*

The first Chevrolet was built in 1911. Compare it with the modern Chevrolet Corvette. *Courtesy of General Motors.*

Compare the 1908 Buick (top) with this 1983 Buick Skyhawk Limited Edition. *Courtesy of General Motors.*

Compare the 1908 Cadillac (top) with the 1983 Cadillac Sedan DeVille. *Courtesy of General Motors.*

Modern motorcycles such as Wide Glide FXWG and Electra Glide FLHT are very popular forms of ground transportation.
Courtesy of Harley Davidson.

Wide Glide® FXWG

Electra Glide® FLHT

One of the most popular and most recent forms of ground transportation is the ATV (all terrain vehicle) such as this Polaris Trailboss 250. *Courtesy of Polaris.*

Many students still ride to school on the school bus. *Photo by Deborah M. Goetsch.*

Water Transportation

Water transportation includes canoes, rowboats, sailboats, tugboats, motorboats, cabin cruisers, paddle wheelers, ferries, fishing trawlers, tankers, freighters, submarines, ocean liners, troop transports, aircraft carriers, and many other types of watercraft.

Many people that live near rivers and lakes still depend on small boats for transporting themselves and their goods. Sailing is a popular sport among people who live near lakes, bays, or the ocean. Many of the goods that people need are transported by freighters and tankers. Much of our oil and natural gas is shipped to this country in tankers. Fishing trawlers transport their harvest from the sea.

Troop carriers, submarines, aircraft carriers, and many other oceangoing vessels carry the people, material, and equipment needed by the military for national defense. Ocean liners are a popular form of transportation from continent to continent or country to country. Every day, thousands of people and thousands of tons of goods are transported over the water.

The Vikings traveled around the world as long ago as A.D. 800 in ships like this. *Courtesy of Spizzirri Publishing Company, Inc.*

Pirate ships like this used to sail the seas searching for other ships to rob.

Commercial fishing and shrimping boats such as these are common in coastal areas.
Photo by Deborah M. Goetsch.

Barges like this, being completed at the Ambridge, Pennsylvania, plant of the United States Steel Corp., can carry as much as 650 tons of various types of materials. *Courtesy of United States Steel Corporation.*

Air Transportation

People and goods are transported through the air by propeller-driven aircraft, from small one- or two-seat models to large military cargo carriers. Jet aircraft also come in the same sizes. Propeller and jet aircraft are called *fixed-wing* aircraft.

Another form of air transportation is *rotary-wing* aircraft or helicopters. Helicopters range in sizes from small one-person models to large models capable of lifting a load as heavy as a large truck.

Air transportation is critically important to everyday operations in business, industry, government, and national defense. Every day, men and women in business, industry, and government depend on air transportation to get them to important meetings in other cities, states, and countries. The military depends on air transportation for transporting troops and supplies to trouble spots quickly and efficiently. Without modern methods of air transportation, the economy would grind to a halt.

The Wright Brothers' *Flyer* is the world's first pilot-controlled, engine-powered aircraft to have achieved flight. On December 17, 1903 at Kitty Hawk, North Carolina, the plane, piloted by Orville Wright, flew for a total of 12 seconds. *Courtesy of Spizzirri Publishing Company, Inc.*

Tiny ultralight aircraft such as this one are becoming popular with amateur flyers. *Photo by Deborah M. Goetsch.*

Helicopters range in size from small one-person models to large models capable of lifting heavy loads. *Courtesy of Ken Elliott.*

Modern jet aircraft are one of the most important forms of transportation to the military. *Courtesy of Ken Elliott.*

NASA's space shuttle is perhaps the most famous and well-known form of space transportation.
Courtesy of Spizzirri Publishing Company, Inc.

Space Transportation

For the past 20 years, people have been coming closer and closer to the day when traveling though space will be as common as normal air travel. One big difference, of course, is that to get into space, you have to break through the earth's gravitational pull. A powerful rocket is needed to blast space vehicles into space.

Right now, NASA's space shuttle is the most advanced form of space transportation. It enters space piggybacked on two large booster rockets. It can transport people and goods, such as satellites that need to be launched and left in space. It can reenter the earth's airspace and land like a normal aircraft.

APPLICATIONS OF TRANSPORTATION

Transportation can be grouped according to how people use it. There are three major groups: commercial transportation, military transportation, and general-purpose transportation.

Commercial Transportation

Air

Commercial transportation is provided by companies whose business is transporting people and goods for a price. Airline companies provide this transportation. A person can buy a ticket and fly from an airport in one city to an airport in another city. There are over 500 commercial airports in this country. Over 200 million commercial airline tickets are sold every year.

Although much of a commercial airliner's business involves transporting people, they do carry a large amount of goods or freight. For example, much of the mail (letters and packages) is shipped by air. Of course, airmail delivery is faster, but it is also more expensive. Is there an airport in your town?

Ground

Commercial ground transportation includes buses, trains, shuttles, and taxicabs. Buses and trains serve the same purpose as airliners. Their routes connect cities so that people can buy tickets and be transported from one place to the other. Two well-known commercial bus companies are Greyhound and Trailways. Is there a bus station in your town? Is there a train station in your town?

Commercial airliners such as this Boeing 757 can transport hundreds of people and hundreds of tons of freight each day. *Courtesy of Boeing Aircraft Corporation.*

Stagecoaches were one of the first forms of commercial ground transportation. *Courtesy of Spizzirri Publishing Company, Inc.*

The cable car originated as a form of commercial transportation in San Francisco in 1873. Cable cars like this were powered by an endless moving wire cable with a stationary engine. *Courtesy of Spizzirri Publishing Company, Inc.*

The New York Central Railroad's Engine 999 was one of the most famous of all commercial railroad engines. In the late 1800s, it held the international ground speed record of 112½ miles per hour.
Courtesy of Spizzirri Publishing Company, Inc.

Most airports have shuttle transportation services available. Shuttles transport people from one part of the airport to another, from the airport terminal to the parking lot, or from the airport terminal to a hotel or motel. Shuttle vehicles might be subway trains, elevated trains, vans, or buses.

Taxicabs are cars or vans especially adapted for commercial transportation. A taxicab driver will take you almost anywhere you want to go for a set fee. They usually charge by the mile. A special meter keeps track of how far the taxicab has traveled.

Water Commercial water transportation comes in three main forms. Ocean liners transport people overseas and on shorter vacation trips. Freighters transport cargo from various ports. For example, automobiles manufactured in Japan for sale in the United States are transported on freighters. Tankers transport oil and natural gas across the oceans and seas. Oil drilled in the Middle East is shipped to the United States in tankers.

COMPUTERS COME IN ALL SIZES

In 1945 ENIAC, the first electronic computer, was completed. It was a huge machine, measuring over 100 feet in length. Today's computers are compact in comparison, and can calculate far more information in shorter amounts of time. (1) This Cray super computer is among the largest and most powerful now manufactured. (2) Minicomputers are used by many businesses. (3) The microcomputer is now found in many homes and schools. (4) Lap-sized computers are small enough to be carried just about anywhere.

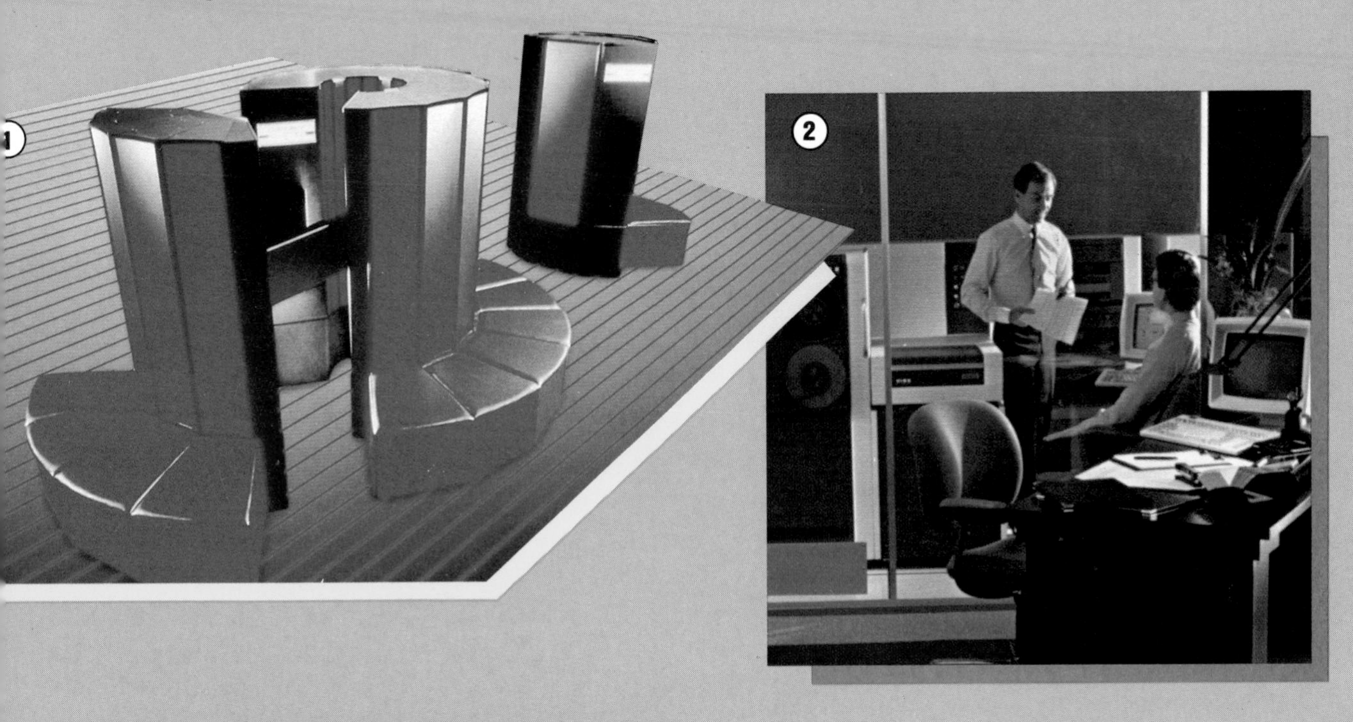

COMMUNICATION BREAKTHROUGHS

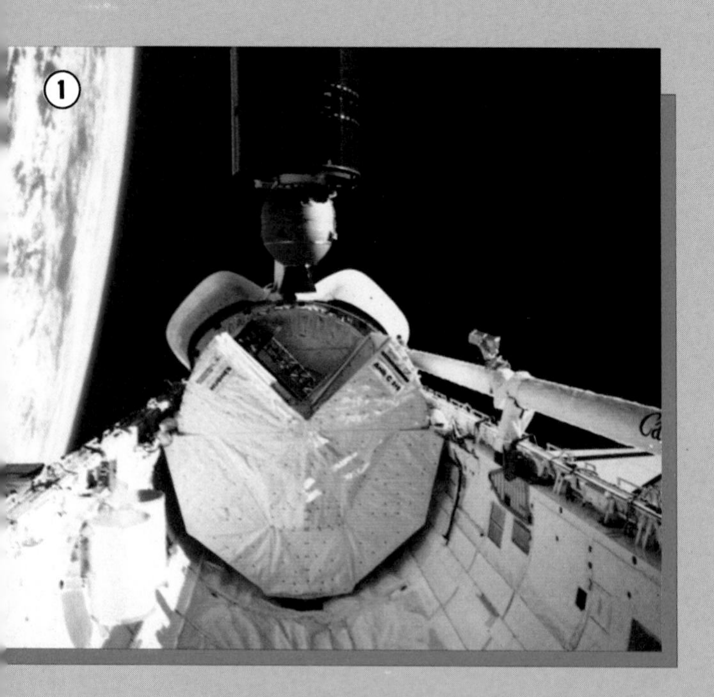

Many different technologies have been brought together to change the ways we communicate with other people and with machines, too. (1) Advances in space transportation have made it possible to put communications satellites into orbit around the earth. This one, the Telesat-I, is being launched from the space shuttle Discovery. (2) On earth, special "satellite dishes" send out and receive signals from space. (3) People can now speak to computers and be understood. This man is inputting data to the computer by means of a newly developed speech command system.

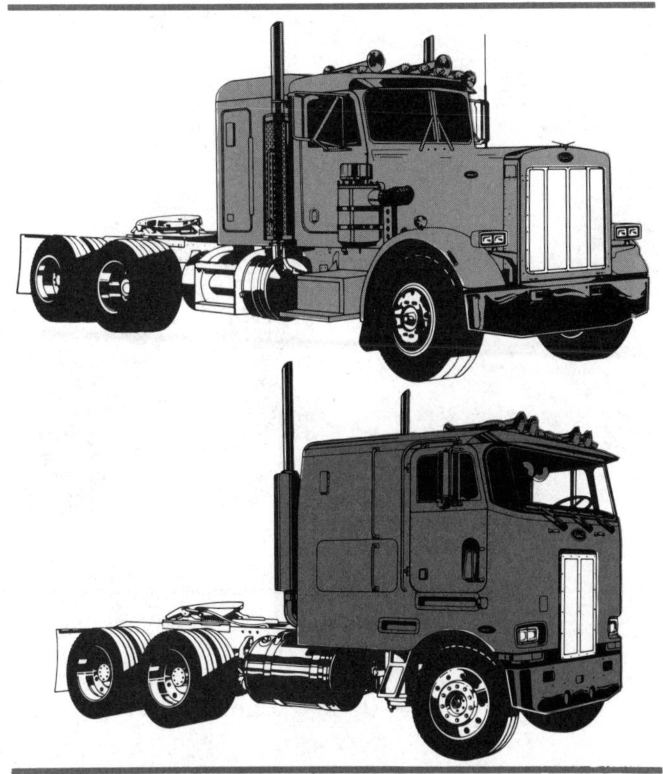

Commercial trucking is an important form of ground transportation. *Courtesy of Peterbuilt.*

Large oil tankers are an important form of commercial water transportation for transporting oil and petroleum products.

Freighters like this are an important form of commercial water transportation for transporting nonpetroleum goods.

Military Transportation

Air **Military transportation** is transporting military goods and personnel. Military air transportation has two major forms: fixed-wing and rotary-wing aircraft. **Fixed-wing aircraft** are fighters, bombers, reconnaissance planes, and transport planes. Only transport aircraft are designed specifically for people and goods. Some military fixed-wing transport aircraft resemble commercial airliners. Others are especially designed to transport military necessities such as trucks, jeeps, and tanks.

Military **rotary-wing aircraft**, or helicopters, are like fixed-wing aircraft in that some are used to transport people and goods and some have special purposes. Helicopters are particularly useful for getting troops out of places where airplanes cannot land and trucks cannot go.

Modern jet aircraft like this F16 fighter bomber are important forms of military transportation. *Photo by Deborah M. Goetsch.*

Modern military depends on both fixed-wing and rotary-wing aircraft. *Courtesy of Ken Elliott.*

The Boeing B17 *Flying Fortress* was the most versatile American bomber used in World War II. It saw extensive service in both the European and Pacific theaters. *Photo by Deborah M. Goetsch.*

The B25 bomber was an important aircraft in World War II. It was able to take off in very short distances, such as from the deck of an aircraft carrier. *Photo by Deborah M. Goetsch.*

Small but fast fighter aircraft such as this P47 fighter bomber and torpedo plane were very important in the Pacific theater of operations during World War II. *Photo by Deborah M. Goetsch.*

Jet aircraft like this F100 fighter bomber replaced slower propeller aircraft. *Photo by Deborah M. Goetsch.*

The F104D Lockheed star fighter-bomber. *Photo by Deborah M. Goetsch.*

The U. S. Air Force F-84 jet fighter. *Photo by Deborah M. Goetsch.*

Jet fighters like this F-85 fighter bomber saw extensive action during the Korean War. *Photo by Deborah M. Goetsch.*

The F-16 fighter bomber is one of the most modern forms of military aircraft in the world. *Photo by Deborah M. Goetsch.*

Ground Military ground transportation comes in many forms. The most frequently used are jeeps and trucks. Other vehicles such as tanks are also used for transporting people on occasion.

Water Military water transportation comes in many forms. Some of the most common forms are aircraft carriers, submarines, battleships, cruisers, destroyers, and transport ships. These, and a number of other types of vessels transport the people and goods of the military.

During World War II and the Korean War, one of the most important forms of ground transportation was the tank. *Photo by Paul T. Laushey.*

Armored personnel carriers like this are an important form of ground transportation to the military. *Photo by Paul T. Laushey.*

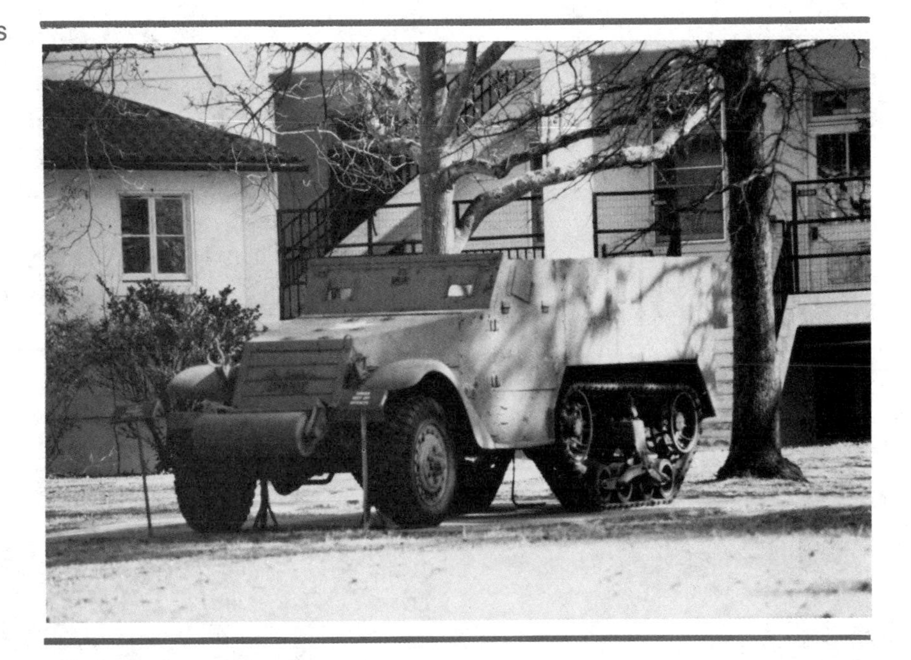

Armored half-track trucks like this saw a lot of action in the European theater of operations during World War II. *Photo by Paul T. Laushey.*

Armored personnel carriers like this are still an important form of military ground transportation. *Photo by Paul T. Laushey.*

General-Purpose Transportation

General-purpose transportation is that used for personal transportation and recreation. The family car, your bicycle, or motorcycle, are types of general-purpose transportation. Some of the transportation for business, industry, and government is general purpose. For example, private airplanes used to transport company officials, company cars, and pleasure craft, such as sailboats and cabin cruisers, are forms of general-purpose transportation.

Motorcycles like the Sportster 883XLH and Sportster 1100XLH are examples of general-purpose transportation vehicles. *Courtesy of Harley Davidson.*

Sportster® 883 XLH

Sportster® 1100 XLH

All-terrain vehicles like this Scrambler 250 are a new and popular form of general-purpose ground transportation. *Courtesy of Polaris.*

The most common and popular form of general-purpose transportation is the automobile. Americans drive more automobiles and small trucks and vans than any other people in the world. *Photo by Deborah M. Goetsch.*

Small private aircraft like this are an example of general-purpose air transportation. *Photo by Deborah M. Goetsch.*

SOME CAREERS IN TRANSPORTATION TECHNOLOGY

Many career fields are associated with transportation technology. As in all technology areas, there are different levels of jobs. Jobs available in transportation technology are grouped according to ground, air, water, and space transportation. In each area, there are professionals, technicians, and skilled workers.

Ground Transportion Careers

Most careers in ground transportation fall into four major categories: professionals, operative, sales, and maintenance and repair.

Professional Professionals are the people in charge of commercial transportation companies, such as train, bus, and taxicab companies. They have titles such as president, vice president, manager, and mid-manager. Professionals usually have college degrees.

Operative Operatives are the people who operate the vehicles—for example, **locomotive engineers**, bus drivers, and taxicab drivers. The amount of training required varies. Taxicab drivers require the least, and train engineers the most.

Sales Sales personnel are responsible for helping people buy tickets and plan their routes. Sales personnel ususally have a high school or vocational school diploma.

Maintenance and Repair Maintenance and repair workers keep the cars, trucks, buses, and trains in top condition, and repair them when they are not. **Auto mechanics**, auto body repair workers, and diesel mechanics are in this category. Maintenance and repair workers need specialized training. They can get it through on-the-job training, through an apprenticeship program, or from a vocational school, trade school, comprehensive high school, or community college.

Air Transportation Careers

Most careers in air transportation fall into five major categories: professionals, technicians, sales, service, and maintenance and repair.

Professionals Professionals are the managers in charge of the airline companies and airports. They are also the **pilots** and navigators who fly the planes.

Technicians Technicians are the air traffic controllers who monitor all of the airplanes that are taking off, flying, and landing, to make sure they stay in their airways. Air traffic controllers require a college degree or special training from a technical school. Many air traffic controllers receive their training in the military.

Sales Sales workers are the ticketing agents who help people plan their flights, make reservations, and buy their tickets. Ticketing agents usually have a high school or vocational school diploma and special on-the-job training.

Service Service workers are flight attendants and cargo handlers. **Flight attendants** on commercial airliners help passengers to find their seats, serve refreshments, and see to the comfort and safety of passengers during flight. Flight attendants complete special training, usually provided by the airline they work for.

Cargo handlers load airplanes ready to take off and unload those that have landed. Cargo handlers usually have a high school diploma.

Maintenance and Repair Maintenance and repair workers are the aircraft power and frame mechanics. They keep airliners in top condition, and repair them when they break down. Aircraft power mechanics work on the engines of airplanes. Aircraft frame mechanics work on the bodies of airplanes. **Aircraft power and frame mechanics** must have extensive specialized training that can be received through apprenticeship, on-the-job training or from a vocational school or community college. They must also be licensed by the Federal Aviation Administration.

Water Transportation Careers

Most water transportation careers fall into four categories: professional, service, sales, and maintenance and repair.

Professionals Professionals are the managers who run the freight, tanker, and ocean liner companies. They are also the officers, navigators, and engineers who pilot the ships. Most professionals in water transportation have a college degree and a high level of specialized training.

Service Service workers make up the crews of ships and docks. Their jobs involve loading and unloading cargo. They require little formal education and training, but they must be in good physical condition and be able to do hard physical work.

Sales Sales workers are the ticketing agents who help people plan trips on an ocean liner, make reservations, and buy tickets. Most sales workers have a high school or trade school diploma. Some have community college degrees.

Maintenance and repair Maintenance and repair workers are primarily **marine diesel and equipment mechanics**. They maintain and repair the diesel engines in ships and the equipment used for loading and unloading cargo.

Space Transportation Careers

Although space transportation is still a relatively new field, many people work in it. Most people work for NASA and are classified as professionals, technicians, and administrative workers. Professionals are the research scientists, design engineers, and top-level managers. Technicians are computer programmers and operators who work on the supercomputers used to monitor space flights. Administrative workers are the secretaries, clerks, and office workers who support the technical and professional workers.

SUMMARY

Transportation is the technological activity that involves moving people, products, and materials from one place to another. The earliest form of transportation was walking. The most recent form of transportation is NASA's space shuttle. The most significant developments in transportation technology have been the invention of the wheel, the domestication of animals, the steam engine, the internal combustion engine, the discovery of petroleum as an energy source, the jet engine, and the rocket engine.

Any transportation system has two basic components: vehicles and pathways. Cars, trucks, buses, airplanes, and ships are all vehicles because they transport people, goods, or materials.

Vehicles use pathways to get from one place to another. Roads, streets, highways, seaways, airways, tracks, and canals are pathways.

Transportation can be grouped according to how people use it. The three major groups are commercial transportation, military transportation, and general-purpose transportation.

Some of the most commonly practiced career fields relating to transportation technology are locomotive engineers, bus drivers, taxicab drivers, pilots, navigators, air traffic controllers, flight attendants, dock workers, ticketing agents, and maintenance and repair workers, including such fields as marine diesel repairers, auto mechanics, bus mechanics, truck mechanics, air power and frame mechanics, and marine equipment mechanics.

KEY WORDS

Transportation
Domestication of animals
Steam engine
Internal combustion engine
Petroleum
Jet engine
Rocket engine
Vehicle
Pathway
Shipping lane
Seaway
Airway
Highway
Commercial transportation
General-purpose transportation
Military transportation
Rotary-wing aircraft
Fixed-wing aircraft
Pilot
Locomotive engineer
Flight attendant
Marine diesel and equipment mechanics
Marine equipment mechanic
Aircraft power and frame mechanics
Auto mechanic

SOME QUESTIONS TO TRY

1. Define transportation.
2. Name three important developments in transportation technology.
3. Name the components of a transportation system.
4. Name three applications of transportation technology, and give examples of each.
5. Name four commonly practiced career fields in transportation technology.

TECHNOLOGY LEARNING ACTIVITIES

1. Make a large poster showing and illustrating the most important developments in the history of land transportation.
2. Make a large poster showing and illustrating the most significant developments in the history of water transportation.
3. Make a large poster showing and illustrating the most important developments in air transportation.
4. Make a large poster showing and illustrating the most important developments in space transportation.
5. Make either detailed sketches or live models of the following:
   a. An automobile in the year 2025
   b. A jet airplane in the year 2025
   c. A train in the year 2025
   d. A space orbiter in the year 2025
6. Make a scale model of a small community with a complete transportation system as you think it will be in the year 2025. Transportation system should include ground, water, air, and space transportation, vehicles, pathways, and facilities.

89.99
99.99

PART THREE

THE INFORMATION COMMUNICATION TECHNOLOGIES

CHAPTER 8

INFORMATION PROCESSING TECHNOLOGY

OBJECTIVES

After completing this chapter, you should be able to

- define information processing.
- explain how the computer is used in processing information.
- list some of the more common computer applications in information processing.
- list several other computer applications.
- list some of the careers associated with information processing.

When people talk about processing, they usually think of processing materials. Cooking raw vegetables or meat so that they can be eaten is processing. Refining crude oil into gasoline is processing. Cutting a tree into lumber is processing. But there is another type of processing that is becoming more and more important—information processing.

We often hear that we are living in the information age. If this is true, then the processing of information is important technology.

WHAT IS INFORMATION PROCESSING?

When we process anything, we collect raw materials, perform some kind of conversion on them, store them, and finally use them. For example, we process crude oil. It is collected by drilling, converted into gasoline by refining, stored in large tanks, and finally used in automobiles. This same type of process applies to information. **Information processing** is to collect it in raw form, record it, sort it, classify it, perform calculations relating to it, store it, and retrieve it. "Data" and "information" are sometimes used in the same way, but they are not the same thing. **Data** are raw facts. **Information** is data that have been processed in some way to make them useful to humans.

Through research, scientists collect information about medical problems. The information is then organized, categorized, and written up in a report or a book. The report or book stores the information.

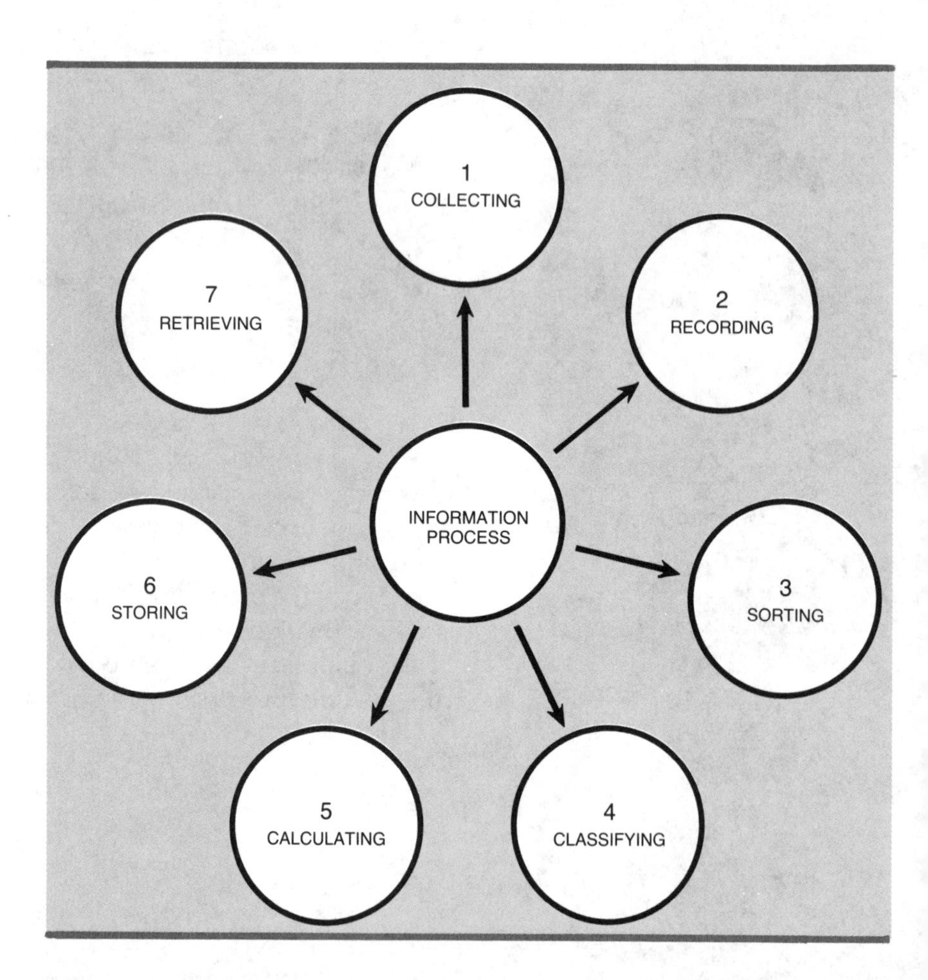

Elements of information processing.

When a medical doctor must treat someone who is ill, he or she uses the information stored in the book or report.

For thousands of years, people have relied on this type of approach to information processing. Many developments in information processing have occurred. The various mediums we write information on have developed from stone tablets to clay tablets to papyrus to high-quality paper. Tools we write information with have developed from the wood stylus to quill pens to lead pencils to fountain pens to ballpoint pens to typewriters to word processing systems.

All of these things help make it easier to collect, record, store, and retrieve information. But none of these developments allowed us to process information as fast as we needed to.

NEED FOR BETTER INFORMATION PROCESSING

Every day that people have been on earth, they have collected more and more information. Now we have so much information that it is hard to keep up with it all. Think of all the information that is stored in libraries, in books in our homes, and in office filing cabinets.

By the mid-1900s, people had collected so much information that processing—especially the storage and retrieval steps of processing—had become a problem. Let's use the example of a medical doctor again to illustrate this point.

Pretend you are a doctor in the year 1930. One of your patients complains of a new kind of back problem you have never seen. You don't know what to do, so you go to the nearest medical library to do some reading and research on back problems. Maybe some other doctor has treated the problem and written about it. If so, the information will be stored in the medical library. Your problem is retrieving it.

First, you walk or drive to the library. Then you go to the card catalog and search through the cards on medical books until you find books covering back problems. When you find the right card, you search through the shelves to locate the right book. Of course, you cross your fingers and hope the book is not checked out to someone else already.

Let's say you were able to find the book. Now you must sit down and read until you find a back problem that has symptoms like those of your patient. Assume it takes two hours to find the information. All of this time your patient has been sitting in your office with severe back pain. Clearly, you need a better way to process information, especially the retrieval step. Problems like this led to the development of the computer.

THE COMPUTER

In the previous example, if you had had a computer the information could have been retrieved in a matter of seconds. Today's medical doctors can tie an office computer into the American Medical Association's computer network. When they need information about a problem they simply type in the problem and in seconds the information they need appears on the computer screen.

Computer Defined

A **computer** is an electronic machine that has storage, logic, and mathematical capabilities and has been designed to perform tasks at extremely high speed. A computer has four distinguishing characteristics.

1. Computers perform all operations electronically.
2. Computers have an internal storage capability.
3. Computers receive operational instructions from stored programs.
4. Computers can modify programs and executions by making logical decisions

A modern computer system.
Courtesy of Sperry Corporation.

A computer is an electronic machine. *Courtesy of Commodore Computers.*

Development of Computers

Historians trace the first computer all the way back to the Chinese abacus in 3000 B.C. However, the first computer that really fits our

Important dates in the development of computers.

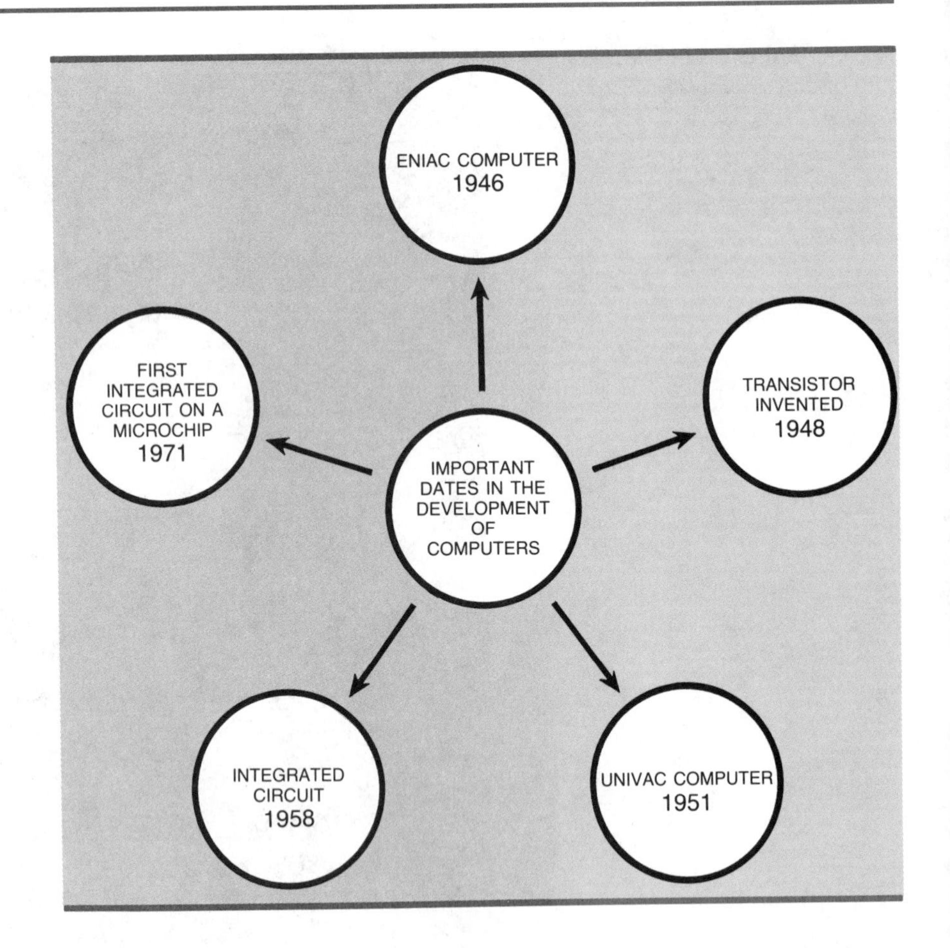

definition was developed in 1946. It was called **ENIAC**, which stands for Electronic Numerical Integrator and Computer.

ENIAC was a slow, huge monster compared to modern computers. It ran on vacuum tubes. It is now displayed as a historical artifact in the Smithsonian Institution in Washington, D. C.

In 1948, the **transistor** was invented. Transistors would do the same things as vaccum tubes but they were much smaller. They allowed a computer to be developed that was as powerful as earlier models but much smaller.

The next important step in the development of computers was **UNIVAC**. UNIVAC was a large computer used by the United States Census Bureau. UNIVAC was faster and more powerful than ENIAC, but it was still big and slow by today's standards.

In 1958, the invention that paved the way for today's small powerful computers came about. That invention was the **integrated circuit** (IC). The IC could do the same things as transistors, but it was even smaller. It allowed computers to be even smaller but still be powerful.

Finally, in 1971, scientists learned how to put ICs on tiny microchips. Then the computer revolution really took off. An IC on a microchip no bigger than your fingernail can now contain as many electronic components as the huge ENIAC computer.

With the development of the microchip came the **microcomputer**. Now we have several different sizes of computers used in information processing.

Sizes of Computers

You have to be careful about categorizing computers by size. Today we have computers that will fit on a desk, but they are more powerful than older models that will fill a whole room.

However, you should know that in general terms the sizes of computers are microcomputers, minicomputers, **mainframe computers**, and **supercomputers**. The smallest is the microcomputer. The largest is the supercomputer.

Microcomputers are those we use as personal and desktop computers. Some microcomputers are so small they can fit in a shirt pocket. Many computers are medium-sized computers. They are used like small mainframes. Several terminals can be tied into one minicomputer.

Mainframe computers account for most of the information processing that takes place every day. The reason is that most companies, especially large companies, use mainframe computers. Numerous terminals can be tied into one mainframe computer.

Supercomputers are the largest and most-used computers. There aren't very many of them, although the number is growing. The NASA space program uses a supercomputer to help launch, track, and monitor spacecraft.

Because they make it so easy to store and retrieve information that has been collected and recorded, computers are one of the most important developments in the history of information processing. Generally speaking, the more powerful the computer, the more information it can store and the faster the information can be retrieved.

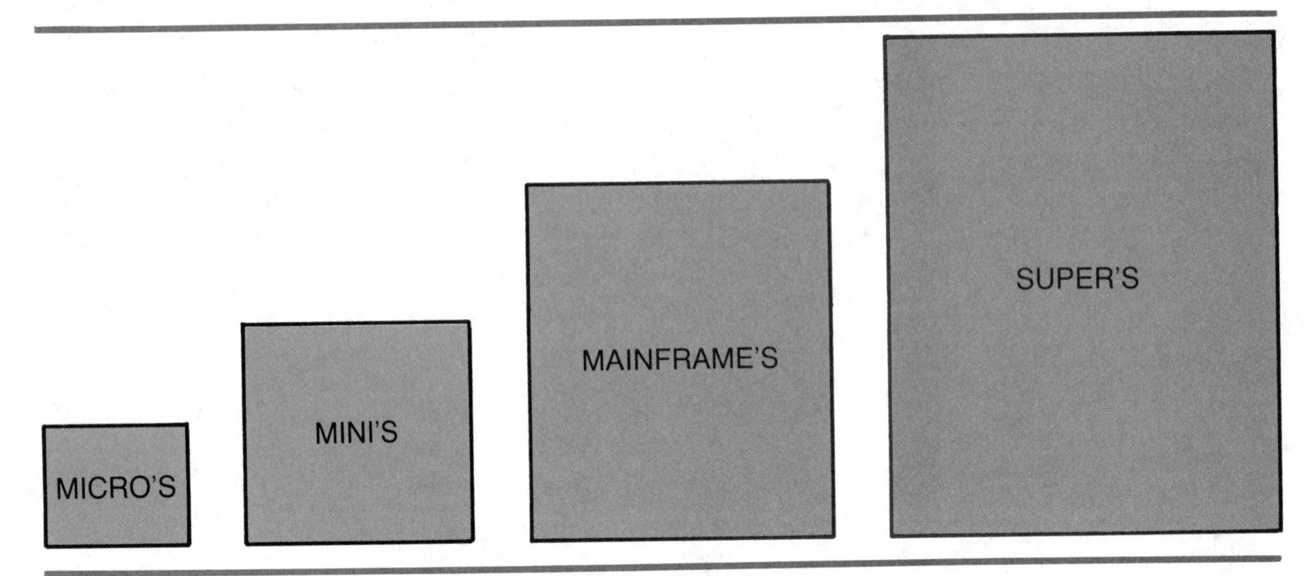

The sizes of computers.

Micro-, mini-, mainframe, and supercomputers. *Courtesy of Sperry Corporation.*

The Macintosh microcomputer.
Courtesy of Apple Computers, Inc.

The Apple IIc microcomputer.
Courtesy of Apple Computers, Inc.

The Tandy 600 portable computer. *Courtesy of Tandy Corporation.*

The Tandy 1000 personal computer. *Courtesy of Tandy Corporation.*

The Tandy Model 200 microcomputer. *Courtesy of Tandy Corporation.*

The Tandy 100 portable computer. *Courtesy of Tandy Corporation.*

The SMC-70 microcomputer.
Courtesy of Sony Corporation.

The Burroughs XE-550 minicomputer. *Courtesy of Burroughs Corporation.*

The V Series of mainframe computers. *Courtesy of Burroughs Corporation.*

The A-15 mainframe computer. *Courtesy of Burroughs Corporation.*

The A-3 mainframe computer.
Courtesy of Burroughs Corporation.

The integrated scientific processor supercomputer.
Courtesy of Sperry Corporation.

Computer Systems

People don't just use computers for processing information; they use computer systems. Computer systems are made up of hardware and software. **Hardware** consists of devices and equipment. **Software** consists of programs, documentation, and manuals. Together, hard-

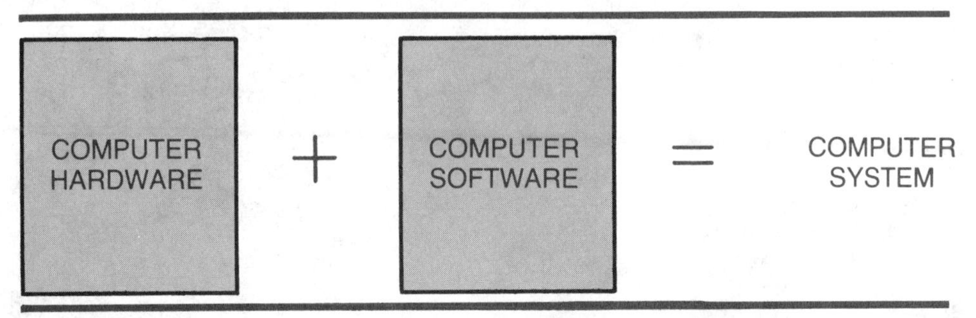

The components of a computer system.

A complete computer system. *Courtesy of Commodore Computers, Inc.*

A complete computer system. *Courtesy of Burroughs Corporation.*

ware and software allow a computer system to perform its three basic functions: **input**, **processing**, and **output**.

In the input stage, people put data information into the computer. In the processing stage, data is sorted, categorized, and stored in the memory or on a secondary storage device, such as a floppy disk or a magnetic tape. In the output stage, the information is taken out of the computer in a form usable to humans. Information displayed on a terminal screen or on a paper printout is output. Information displayed on a screen is called soft copy. Information displayed on paper is called hard copy.

Computer Hardware The various items of equipment and devices collectively known as hardware include the central processing unit (**CPU**) and memory, secondary storage devices, and input/output devices.

The CPU with memory is the computer in a computer system. It has three parts:

1. The *control unit* controls the computer system according to the instructions of the programs stored in memory.

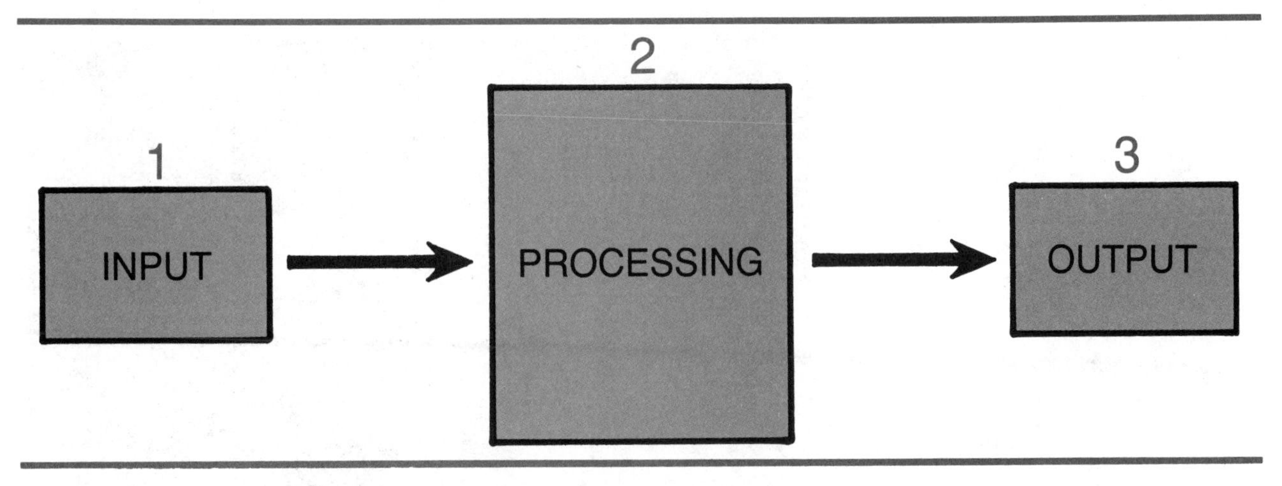

Basic computer operations.

Computer hardware.

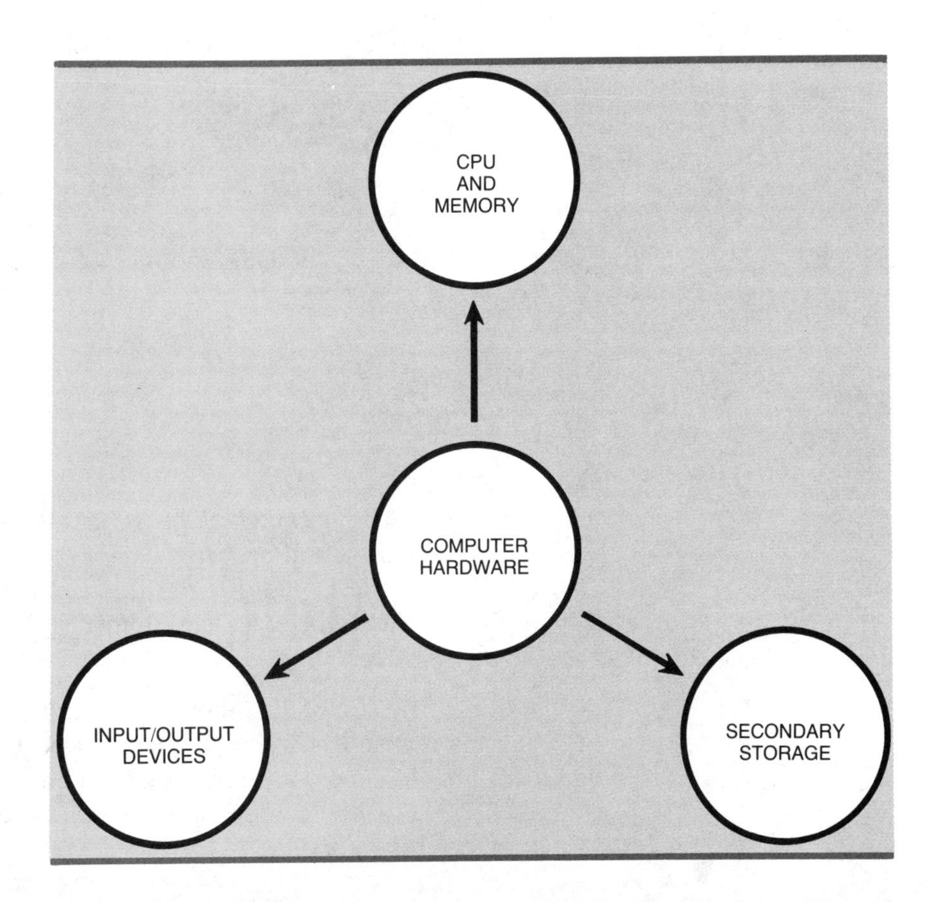

The central processing unit of a computer system.

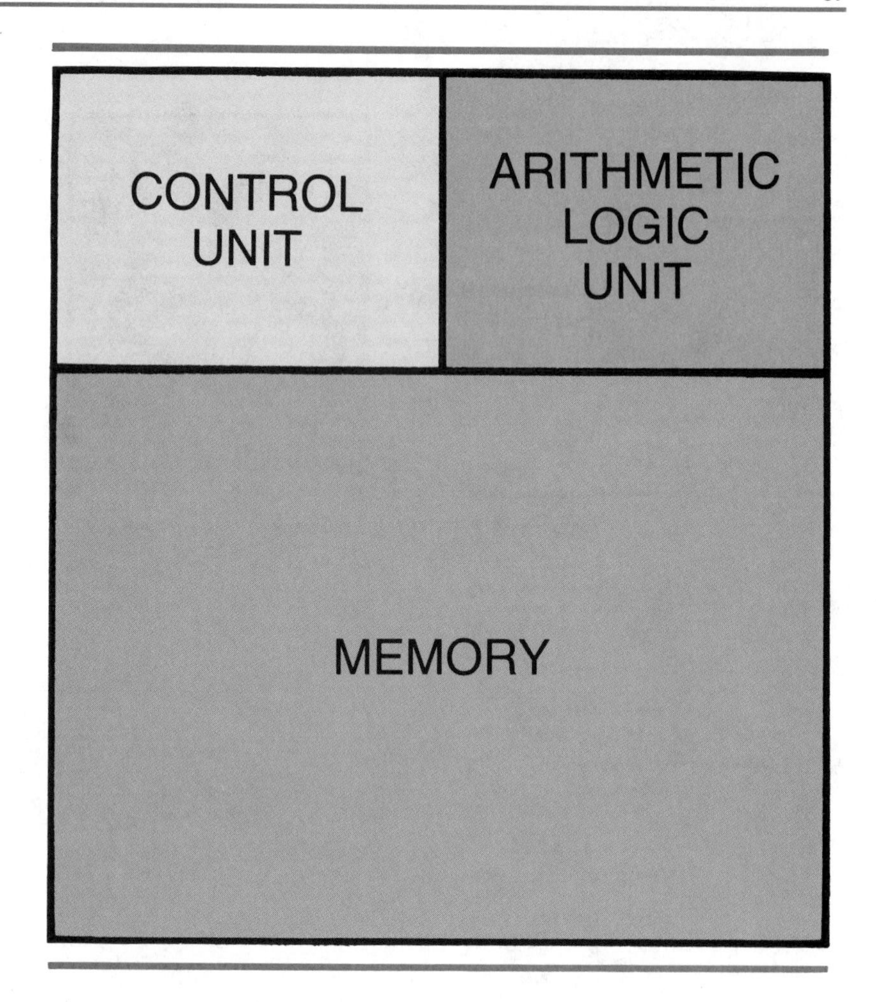

2. The *arithmetic/logic unit* (ALU) does all mathematical calculations and prepares data.
3. The *memory* is the heart of the CPU. **Computer programs** and the data and information being processed are stored there.

The size of the memory determines how much information can be stored and how fast it can be processed. Microcomputers have the smallest memories, supercomputers the largest. The memory in the CPU is called random-access-memory (RAM). RAM is sometimes called *on-line* memory.

Secondary Storage No matter how large the CPU is, it is never big enough. **Secondary storage**, called *off-line* storage, is usually necessary. It is called off-line storage because it is not instantly available in memory. It can be stored off-line, away from the computer, on a shelf.

Secondary storage devices are disks, diskettes, and magnetic tape reels. To keep the RAM from getting full, users can store information on a disk, diskette, or magnetic tape. When the information is needed, the disk or tape drive is used to read the information into the RAM a little at a time. When the information is no longer needed, the disk, diskette, or magnetic tape reel can be taken out of the drive and stored off-line.

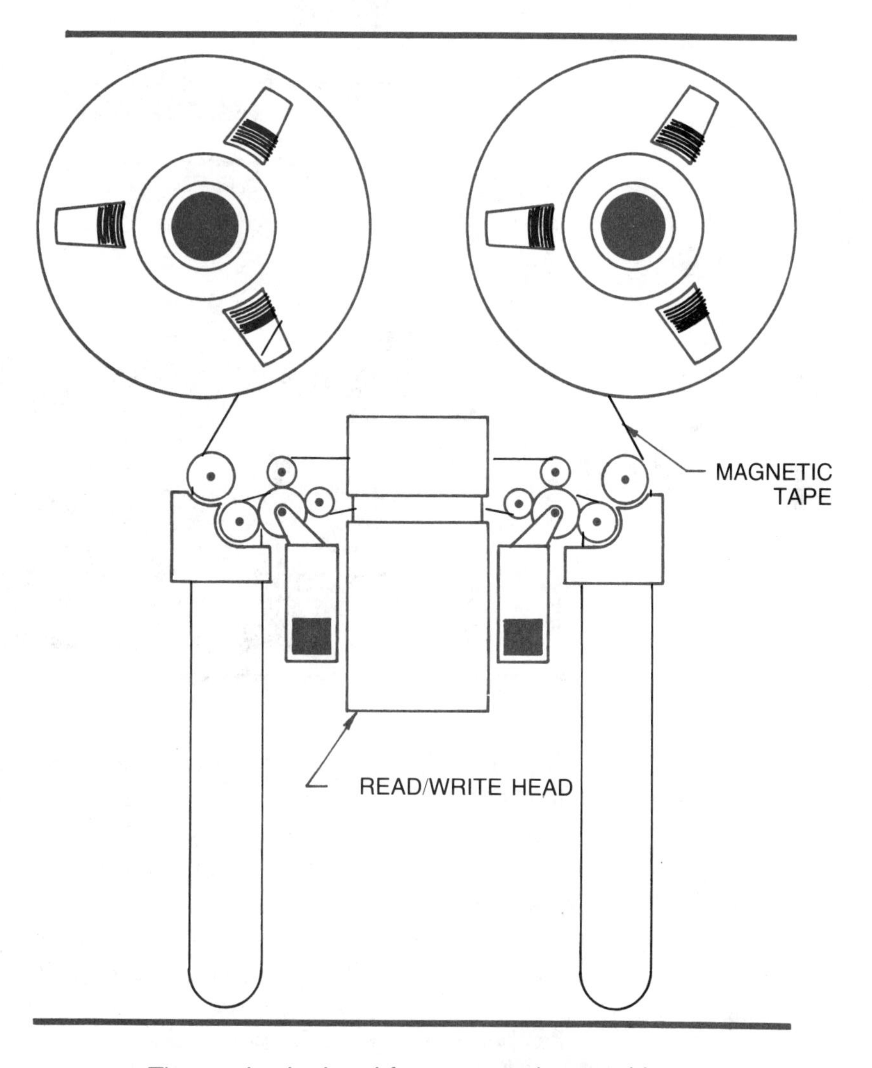

The read write head for a magnetic tape drive.

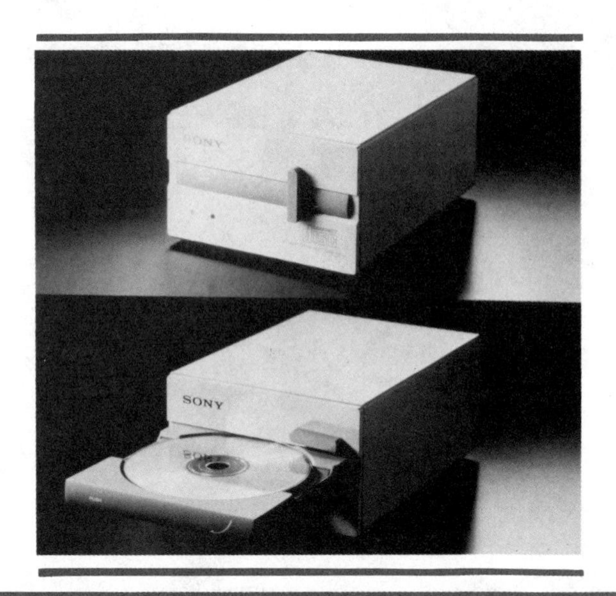

CD ROM drive unit CDU-1.
Courtesy of Sony Corporation.

Writable disk drive unit. *Courtesy of Sony Corporation.*

Input/Output Devices **Input/output devices** let people put data into the computer and take information out of it. Some devices are input devices only. That is, they can only be used to put data into the computer. Some devices are output devices only, and some are input and output devices.

The most frequently used input device is the *keyboard*. The most frequently used output device is the *display terminal*. When you can read information on the screen of a display terminal, that information has been outputted. Output on a display screen is called *soft* output.

The next most frequently used type of output device is the *printer*.

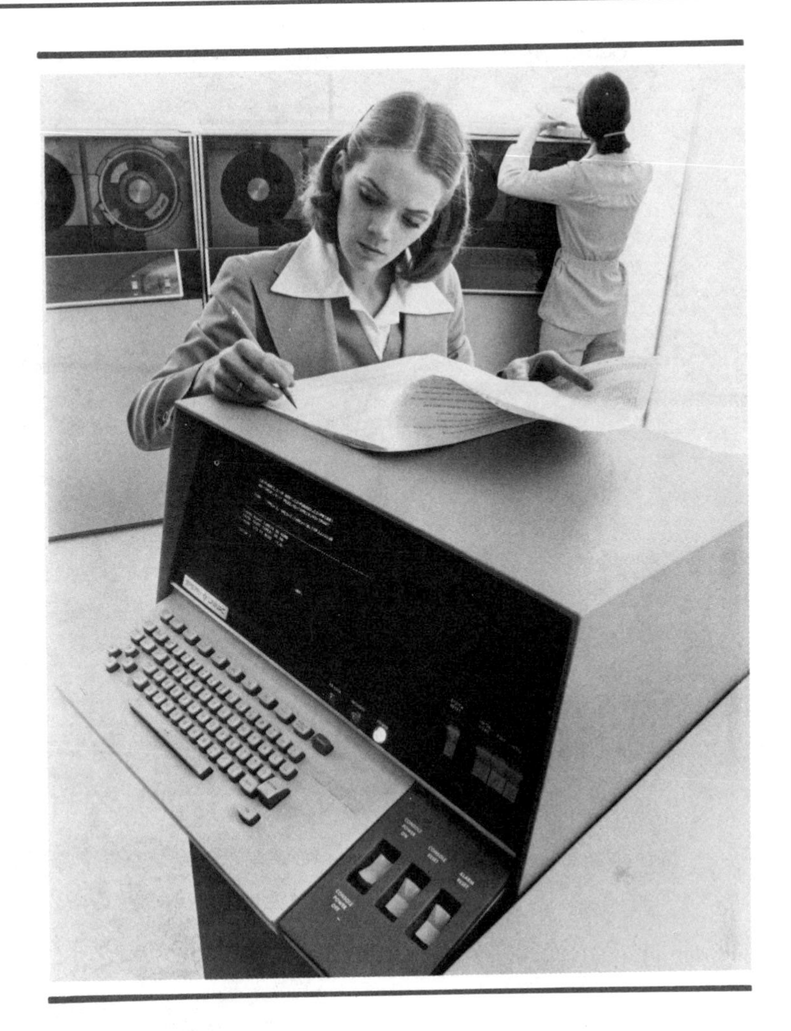

A terminal with a keyboard and display screen is an input/output device. *Courtesy of Sperry Corporation.*

The printer outputs information on paper, which is called a computer printout. A printer that has a keyboard is considered an input/output device because you can input information using the keyboard and get a computer printout as output.

Computer Software Computers can expand our capabilities and process information more than any previous development, if they are programmed to do so. *Programs* are part of the application software packages that sometimes come with computers. Applications packages contain the programs that allow computers to do different kinds of work. Frequently used applications packages include programs such as accounting, simulation, graphics, CADD, CAM, statistics, education, and word processing. People instruct the computer through programs. Computer programs are entered into the main memory of a

A printer is an output device.
Courtesy of Sperry Corporation.

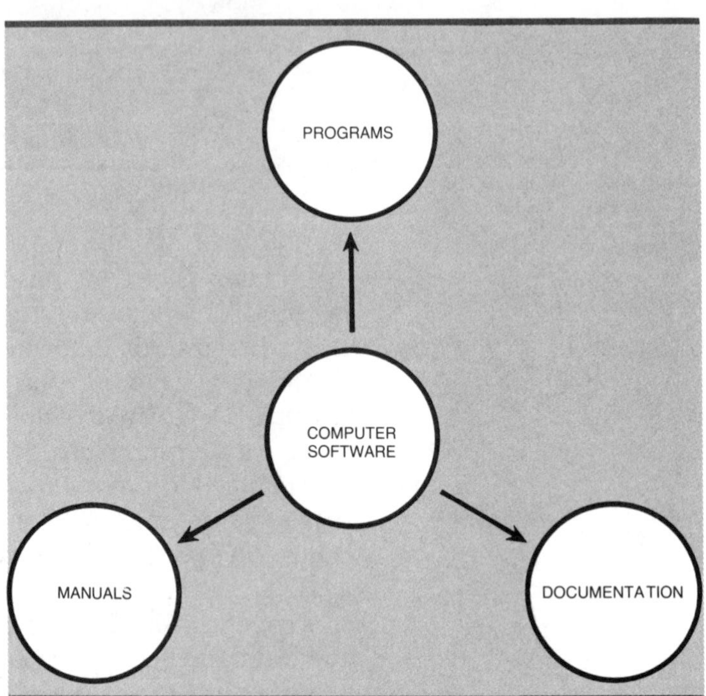

Computer software.

Students learn to use computers in grade school. *Courtesy of Robert C. Atwood.*

computer system and stored. Once stored, they tell the computer what to do.

Computer programs are the most important element in a software package. The other elements include manuals, which explain how to use the programs and actual listings of the programs.

It takes special training to be able to write computer programs. They must be written in languages that computers understand. Computer programmers are people who know one or more of these languages and are able to write programs using them.

A number of different computer programming languages are widely used. Some are for general use, and some are for very specific uses. Let's examine some of the most widely used computer programming languages.

Progamming Languages The programming languages used most are

- machine
- assembly
- FORTRAN
- COBOL
- BASIC

Widely used computer programming languages.

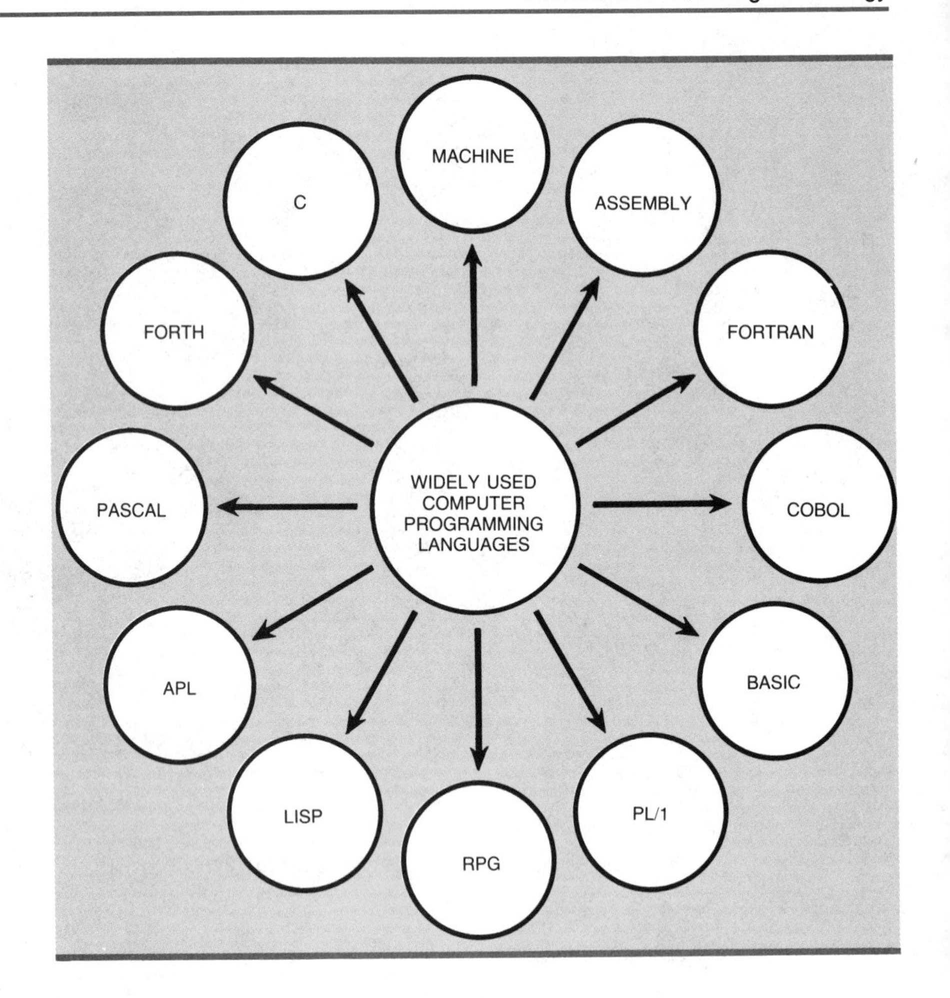

- PL/1
- RPG
- LISP
- APL
- PASCAL
- FORTH
- C

Machine language was the first computer programming language used, but it isn't used much for programming any more. A machine language program consists of 1's and 0's. It is a difficult language to learn and one in which errors are easily made. **Assembly language** was developed in the early 1950s to solve some of the problems with machine language. Instead of just 1's and 0's, assembly language used a combination of letters and numbers. Hence, programming in assembly language is much easier than programming in machine language.

FORTRAN is short for FORmula TRANslator. It is a popular programming language for scientists and engineers. FORTRAN is an arithmetic-oriented language. Consequently, FORTRAN programmers must be good in math.

COBOL is short for COmmon Business Oriented Language. It is a programming language especially designed for business applications. Since its development in 1959 and 1960, COBOL has become one of the most popular programming languages.

BASIC is short for Beginners All-purpose Symbolic Instruction Code. Because it is simple and easy to use, it has become the most popular of all the programming languages. Many of the programs written for microcomputer applications are written in BASIC.

PL/1 is short for Programming Language 1. PL/1 is a general-purpose language that combines the characteristics of FORTRAN and COBOL. However, it is too sophisticated to work on microcomputers. Consequently, its use has been limited.

RPG is short for Report Program Generator. It is a special programming language for producing business reports. An RPG program reads like a description of the report that is to be generated.

LISP is short for LIST Processor. It was developed in the early 1960s for processing nonnumeric data such as sentences and formulas. As "artificial intelligence" becomes more developed, LISP will be more widely used.

APL is short for A Programming Language. It was orignially developed as an aid in teaching mathematics. APL has only limited use because it requires a special keyboard for writing programs.

PASCAL was developed in the early 1970s as a teaching language. PASCAL is a popular language for use on micrcomputers, but it requires much more memory than BASIC, which has limited its use somewhat.

FORTH is the name given to programming designed as a fourth-generation computer language. It is low-level language designed for use on microcomputers. Its most common application is industrial process control.

C is a low-level programming language designed to be used on any computer. This is making it a popular language. It is still a relatively new language, but its popularity is growing rapidly. C has been used in making the special effects for movies such as *Star Trek*. As computer graphics applications grow, so will the use of C.

COMPUTER APPLICATIONS IN INFORMATION PROCESSING

The computer has revolutionized the world of information processing. Using computers, people are able to collect, record, store, and retrieve information in a matter of seconds. The computer is used in many areas of information processing such as business reports and accounting. But one area that really illustrates the advantages of using the computer in information processing is in word processing.

Word Processing

Word processing is the manipulation of words, sentences, paragraphs, and documents for the purpose of communication. When people write a letter in longhand or type a report, they are doing word processing. For many years, the principle tools of word processing were pencils, erasers, and the typewriter. However, in recent years, the computer has changed all of that.

Word processing is now the most common application of microcomputers. The word processing system is rapidly overtaking pencils, erasers, and the typewriter, and it is easy to see why. Let's say Mr. Jones is an old-fashioned politician who is getting ready to run for reelection. As part of his campaign, he wants to send a letter to the 500 most important voters in his district. He needs to do so in a hurry because his opponent, Ms. Smith, plans to do the same thing.

Computers are used for personal budget and finance matters in the home. *Courtesy of Commodore Computers.*

Computers are used for producing business reports and accounting. *Courtesy of Sperry Corporation.*

Mr. Jones sits down, writes a nice long letter, and gives it to his secretary, Ms. Barnes. He says, "Ms. Barnes, type up 500 originals of this letter for my signature." Ms. Barnes is shocked! It will take her days to type 500 original letters. But Mr. Jones insists on originals. He is too experienced a politician to send copies to important voters, so Ms. Barnes starts typing.

In the meantime, his opponent, Ms. Smith, a modern politician, has written her letter, too. She gives it to her secretary with the same instructions. But being a modern person, her secretary, Mr. Andrews, has a word processor. He doesn't have to type 500 letters. He just types one, stores it, enters the 500 different names and addresses, and tells his word processor to print out 500 originals.

Before poor Ms. Barnes can finish 20 letters, Mr. Andrews has finished all 500 and mailed them. Because her letters got to the important voters first, Ms. Smith wins the election! Mr. Jones belatedly decides he needs to learn more about word processing.

Computers are used on large farms to keep records of dairy cattle. *Courtesy of Sperry Corporation.*

Computers are used for teaching and learning in the classroom. *Courtesy of Commodore Computers.*

Computers are used for learning in the home. *Courtesy of Commodore Computers.*

Development of Word Processing

The first major development in word processing came in 1437 with the invention of movable type printing. The Gutenberg press marked the first time that the letters on a printing press could be changed. It worked like this. Blocks representing letters and numbers were locked into a frame. The blocks were inked, and the frame was pressed against the paper. In this way, one page of print at a time could be produced.

The next major development was the invention of the *typewriter* in the early 1700s. This development was important because the typewriter was small enough and inexpensive enough to be widely used. Early versions of the typewriter were manual. Then around 1940 the electric typewriters were developed. They have since undergone continual improvements, which continue even today. But by the 1970s microcomputer-based word processing systems were rapidly replacing typewriters in office settings.

Development of word processing.

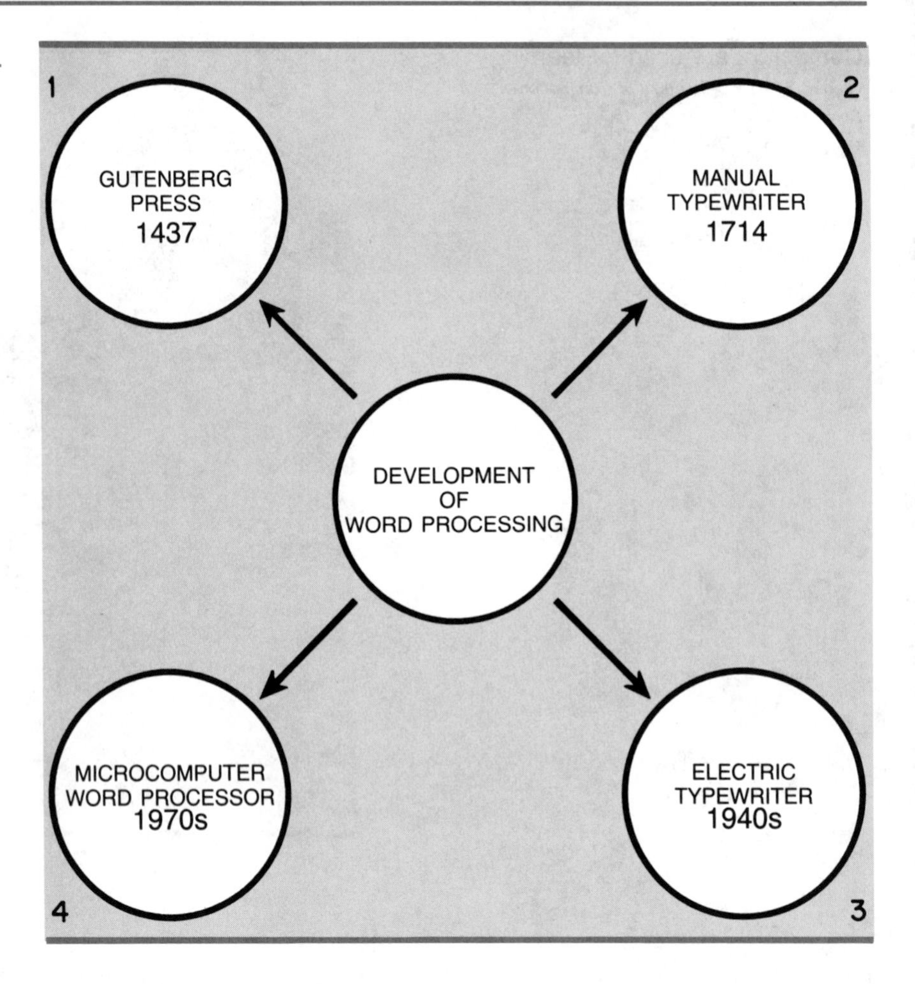

A Word Processing System

A word processing system consists of both hardware and software. Although there is a computer in all word processing systems, there are different hardware configurations for different systems. Some word processing systems are dedicated configurations. In other words, they do nothing but word processing. Then there are micro-, mini-, and mainframe-based configurations. Regardless of how they are configured, word processing hardware configurations include a computer, keyboard, secondary storage, display terminal, and printer. The computer is used for storing the word processing program and the information and data being processed.

The keyboard is used for entering data and responding to prompts built into the program. The secondary storage unit holds the disk that

Word processing hardware.

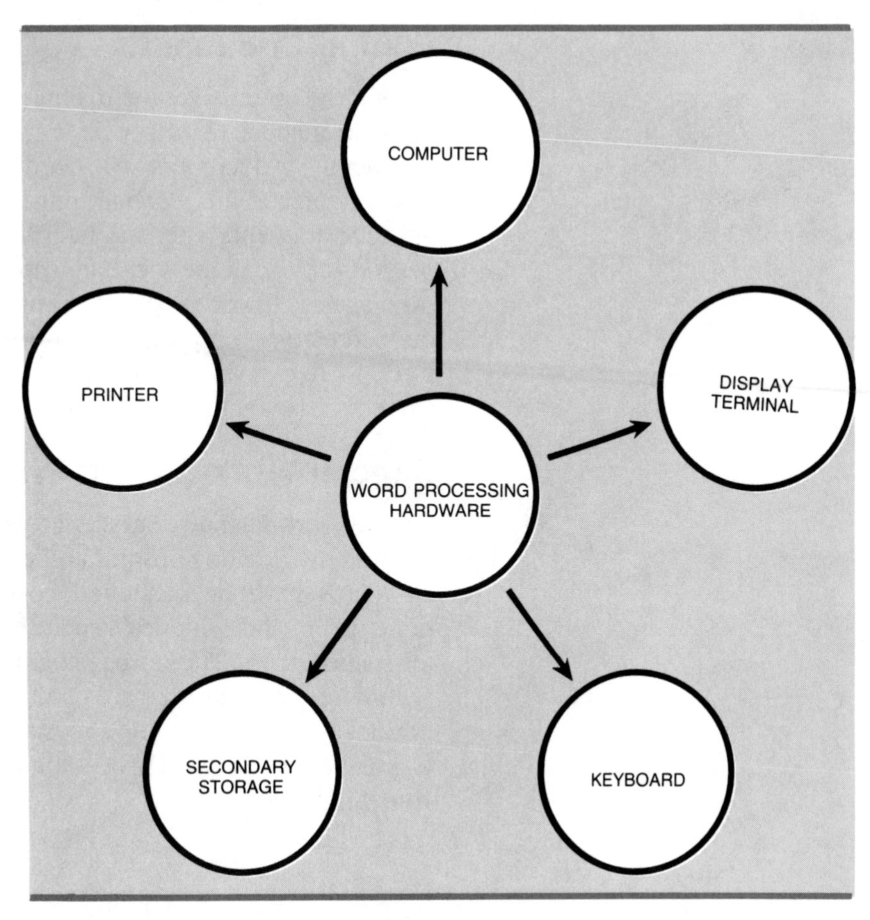

contains the word processing program and another disk for storing the letters, memorandums, and reports being worked on.

The display terminal lets the user see what is being worked on. If working on a letter, all or part of the letter will be displayed on the terminal screen. The printer prints the finished product.

Advantage of Computerized Word Processing

The earlier example of the politicians, Mr. Jones and Ms. Smith, illustrated the advantage of computerized word processing. It is much faster and more efficient than manual word processing because

1. Documents can be easily edited.
2. Special programs cut down on time-consuming work.
3. Documents can be stored for reuse.

Editing Documents

Editing documents means making changes, revisions, and/or corrections. In manual word processing, this is a time-consuming process. Frequently, a document will need so many changes that it has to be thrown out, and the secretary must start all over again. For example, if a three-paragraph letter has been typed and paragraphs 1 and 2 have been reversed, all the secretary can do is tear up the letter and start all over again. However, with a computerized word processing system, the secretary simply displays the letter and gives a few commands through the keyboard. In seconds, the paragraphs are put in the correct order by the computer.

Special Work-Saving Programs

Special work-saving programs are available with most word processing systems. Spelling programs check the documents to make sure that all words have been spelled correctly. Grammar programs check punctuation and sentence structure. Style programs check documents for syntax errors. Thesaurus programs help writers find antonyms and synonyms for words that are used repeatedly. These special programs cut out most of the editing work required on documents. Wouldn't you like to have a word processor and some work-saving programs for your English class?

INFORMATION PROCESSING CAREERS

Information processing careers can be classified as professional, technical, operative, and maintenance/repair.

Professional Careers

Professional careers in information processing include managers and system scientists. Managers oversee the operation of large computer centers that serve organizations such as schools, banks, businesses, and industrial firms.

System scientists design computer systems for schools, banks, businesses, and industrial firms. This work involves selecting hardware, selecting software, and getting the system up and running. Management and systems scientists usually require a college degree.

Technical Careers

The best-known technical career is computer programmer. Programmers develop software packages and special programs in one or more programming languages. Programmers require special training from a community college or a vocational school. Some programmers have a college degree.

Operative Careers

Operative workers operate computer systems. Such jobs include computer operator and word processor. Operatives are not computer programmers. Computer operators take in information processing requests, load programs, retrieve the desired information, and run printouts. Word processing operatives produce letters, memorandums, reports, and other printed documents on word processing systems. Operatives have special training from a community college, vocational school, or on the job.

Maintenance/Repair Careers

Computer maintenance and repair is a growing field. Like all machines, computers must be maintained and occasionally repaired. Computer repair workers must have special electronic training, which can be obtained from a community college, vocational school, or special schools provided by companies that manufacture computers.

SUMMARY

Information processing involves collecting, recording, sorting, classifying, storing, and retrieving information. Data and information are not the same thing. Information is data that have been processed. The need for better ways of processing information led to the development and use of the computer.

A computer is an electronic machine that has storage, logic, and mathematical capabilities and can perform certain tasks at extremely high speeds. A computer has four distinguishing characteristics: (1) performs all operational instructions electronically; (2) has an internal storage capability; (3) receives operational instructions from stored programs; and (4) can modify program executions by making logical decisions.

The first real computer was ENIAC. It was developed in 1945 and currently is displayed in the Smithsonian Institution. Major developments leading to the small, powerful computers were the transistor in 1948 and the integrated circuit in 1958. When scientists, in 1971, learned how to make integrated circuits very small and place them on tiny silicon wafers, the computer revolution began in full force.

From the smallest to largest, computers are classified as microcomputers, minicomputers, mainframe computers, and supercomputers. A complete computer system consists of hardware and software. Hardware is equipment. Software is programs and manuals. The three steps in computer systems are input, processing, and output.

Applications of the computer in processing information include accounting, business reporting, personnel records, registration and admission records, and word processing. Word processing is one of the most widely practiced applications.

Some of the most widely practiced are career fields relating to information processing: system scientists, managers, computer programmers, computer operators, word processors, and computer repairers.

KEY WORDS

Data
Information
Information processing
Computer
Hardware
Software
Computer program
ENIAC
UNIVAC
Transistor
Integrated circuit
Microcomputer
Minicomputer
Mainframe computer
Supercomputer
Processing
Output
CPU
Secondary storage
Input/output devices
BASIC
Assembly language
Machine Language
FORTRAN
COBOL
PL/1
LISP
APL
PASCAL
Input
FORTH
C
RPG

SOME QUESTIONS TO TRY

1. Describe information processing.
2. What is the difference between *data* and *information*?

3. Describe a computer.
4. Name the four characteristics of a computer.
5. What is hardware?
6. What is software?
7. What are programs?
8. Name three programming languages.
9. What is word processing?.

TECHNOLOGY LEARNING ACTIVITIES

1. Contact your school secretaries. Do they use typewriters or word processing systems for typing and other office duties? If they use processors, find out the following:
   a. What type of word processing system or systems do they use?
   b. What software package or packages do they use?
2. Conduct some on-site research and find out all the different ways that computers are used at your school. Make a chart showing all of the different applications.
3. Each student in class should select one occupation and determine all of the different ways computers are used in that occupation. Make a list of all the different ways for each occupation, and see who has the longest list.
4. Visit a computer store and look at
   a. The inside of a microcomputer
   b. All input/output devices that go with a computer
   c. The type of secondary storage devices used with the computer
   Make a report of your visit and give a presentation to the class. Use a lot of illustrations.
5. Make a timeline of important technological developments in information processing.

CHAPTER 9

GRAPHIC COMMUNICATION TECHNOLOGY

OBJECTIVES

After completing this chapter, you should be able to

- define graphic communications.
- list some of the significant developments in graphic communications.
- explain several ways that graphic communication is produced.
- list some of the careers associated with graphic communication.

Communication is a broad term used to describe the many different ways people send, transmit, and receive messages. There are several different ways to do this: verbally, electronically, and graphically. *Verbal* communication is people talking with each other. *Electronic* communication involves using electronic devices and equipment and sending, transmitting, and receiving messages. Television and radio are examples of electronic communication. This chapter will deal with graphic communication.

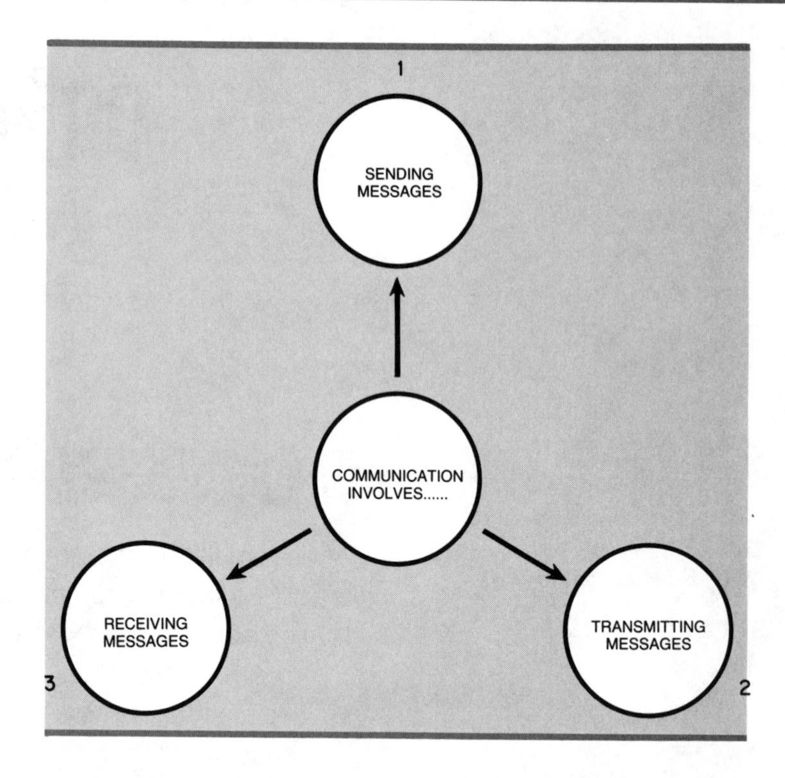

Communication involves sending messages, transmitting messages, and receiving messages.

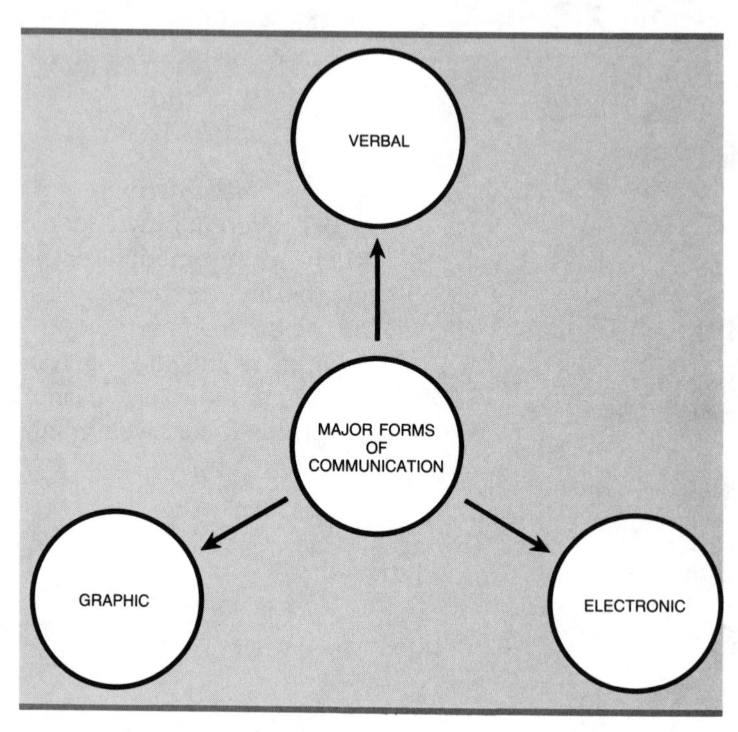

Major forms of communication.

WHAT IS GRAPHIC COMMUNICATION?

Graphic communication involves communicating visually through letters, numbers, symbols, characters, pictures, and other visual images. The images on television are visual, but they are not normally considered graphic communication because they are *dynamic*—they move. Graphic images are *static*—they don't move.

This book is a form of graphic communication; so is a newspaper or a magazine. Billboards, road signs, and bumper stickers are also examples of graphic communication. T-shirts with pictures and words are also examples of graphic communication. Do you have any T-shirts with messages on them?

All of these examples are forms of graphic communication because they use static numbers, letters, symbols, characters, and images to communicate. Graphic communication is a very old technology.

DEVELOPMENT OF GRAPHIC COMMUNICATION

There was graphic communication even before people had alphabets and written languages. The pictures drawn on the walls of caves in prehistoric times were examples of graphic communication. The

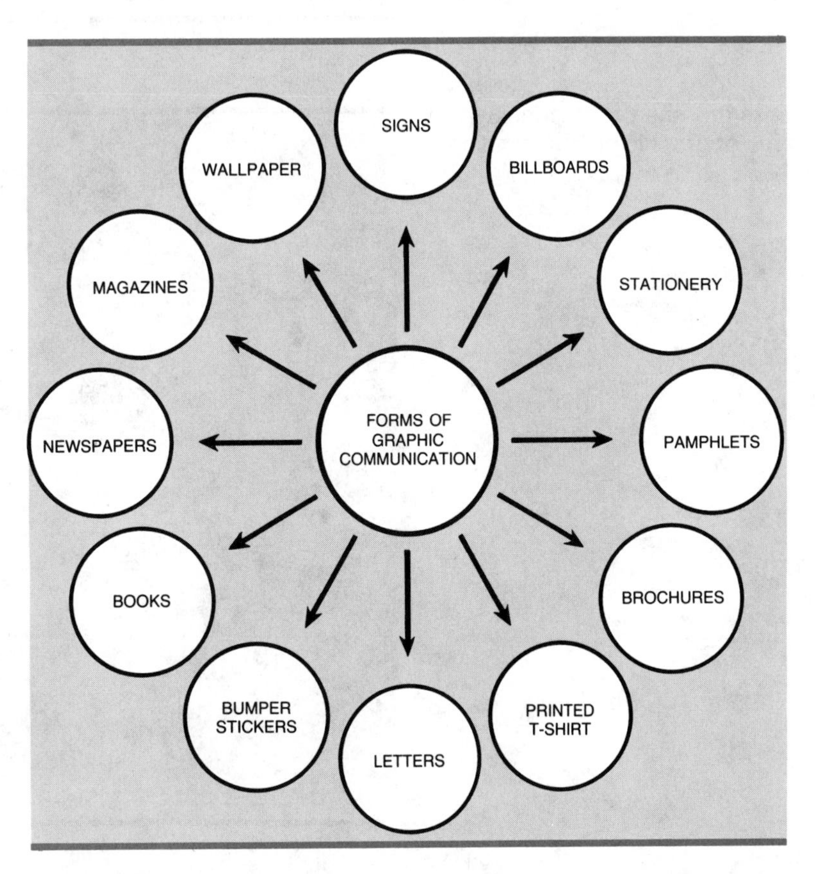

Forms of graphic communication.

Billboards are a form of graphic communication. *Photo by Deborah M. Goetsch.*

Books are one of the principal forms of graphic communication. *Photo by Deborah M. Goetsch.*

This bulletin board display is an example of graphic communication. *Courtesy of Robert C. Atwood.*

This window display is an example of graphic communication. *Courtesy of Robert C. Atwood.*

Egyptians were the first to develop an alphabet. The symbols in the early Egyptian alphabet were known as **hieroglyphics**.

Our alphabet and numbering system is based on those developed by the Greeks and Romans. Have you ever seen Roman numerals? The Romans wrote the numbers one through ten as follows:

I	II	III	IV	V	VI	VII	VIII	IX	X
1	2	3	4	5	6	7	8	9	10

Some of the most important developments in graphic communication have been stamping, movable type printing, offset printing, photography, and phototypesetting.

Stamping

Alphabets led to written communication in the form of books. Before 1400, every copy of every book was made by hand. How would you feel if you had to turn in 10 copies of a five-page report and you couldn't use a copy machine, a word processor, or even carbon paper? It would take hours to make 10 copies by hand. This is exactly what people had to do in the early days of graphic communication.

The first breakthrough in solving this problem was the process of **stamping**. Stamping was a simple but time-consuming process. Each page of a book was duplicated by carving the words and pictures into the face of a flat piece of wood. This piece of wood became a stamp. The stamp was then coated with ink and pressed against a page of paper.

It took a long time to make a stamp for each page in a book, but once the stamps were made, any number of copies of books could be turned out rapidly. Although it is one of the oldest of graphic communication processes, stamping is still used. Signature stamps, date stamps, address stamps, and many other types of stamps are still used every day. Have you ever used a stamp? What kind of stamp have you used?

Movable Type Printing

Stamping was much better than hand copying, but it still was slow and difficult. It took a long time to carve a stamp for each page of a book. Problems with stamping led to the development of **movable type printing**. Johannes Gutenberg became famous for printing a Bible using movable type in the mid 1400s.

Various types of stamps are still commonly used.

An early hand-operated printing press.

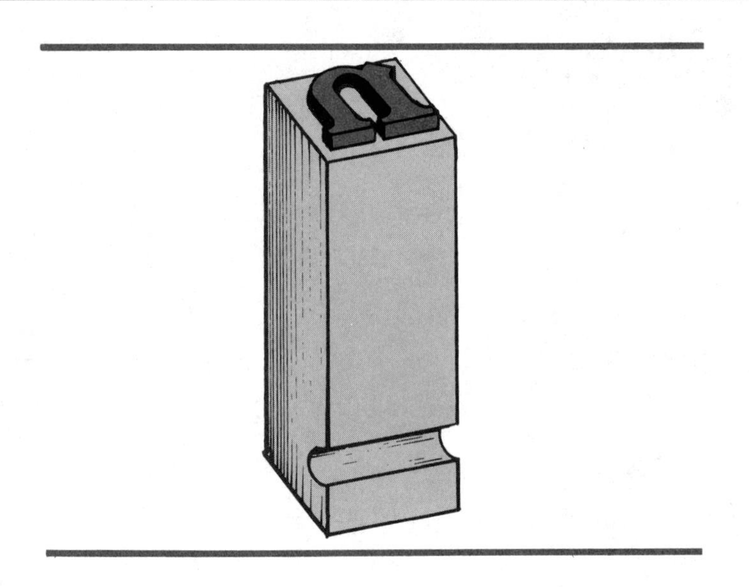

Cast-metal relief character used in movable-type printing.

With movable type printing, individual letters are cast or carved. Then they are attached to a frame as needed. Each letter can be used over and over again. If 100 copies of a book are needed, the letters are set for page 1. Then 100 copies of page 1 are made. Then the letters are reset for page 2 and the copies of page 2 are made. Then the letters are reset for page 3, and so on, until the entire book has been copied.

Movable type was a major advance over stamping. It was continually improved over the years and was a major form of printing for several hundred years. In some print shops, you can still see old movable type presses.

Offset Printing

Offset **lithographic printing** was the next major step in graphic communication. This type of printing involves making copies of an inked master. The master is called a *plate*. The printing machine has three cylinders: the plate, blanket, and impression cylinders. The plate is rolled around the plate cylinder.

As the plate cylinder revolves, it transfers the image to a rubber blanket on the blanket cylinder. As paper passes between the blanket cylinder and the impression cylinder, it receives the image. **Offset**

printing is still a popular way for producing graphic communication products such as books, magazines, and newspapers.

Photography

A very effective and efficient method for capturing static images came about with the invention of **photography**. Today people use photography frequently. There is black-and-white film, color film, large cameras, small cameras, and even instant cameras. Almost anybody can take pictures, and most people do. However, in the early days of its development, photography was cumbersome and expensive. Only experts could understand the complicated process.

Today, people can buy film in rolls, but in early days of photography, they didn't have film. Instead, heavy metal and glass plates coated with a special light-sensitive material were used. The first uncomplicated, portable camera was developed by George Eastman. It was called a Kodak camera, and it used flexible film, similar to that used today, instead of the old metal and glass plates. The Kodak camera led the way in the development of the wide-scale use of cameras by professionals and amateurs.

Photography quickly became an important means of graphic communication. In addition to stories, newspapers and magazines could now carry photographs. They formerly had to rely on artists to make drawings. Books also began to include photographs along with words. Photography is now an important part of graphic communication. Large process cameras are used to photograph materials that are to be used in books, newspapers, and magazines. The photograph becomes part of the master used for making offset copies.

If you ever have an opportunity to visit a large printing company or a newspaper publisher, ask to see the process camera and how it works. Also ask to see an offset press in operation. Photography and offset printing are very important in the mass production of several forms of graphic communication.

Phototypesetting

Phototypesetting combines the processes of photography and typing with the advantages of the computer in setting type. In phototypesetting, special type fonts are loaded into a typesetter. The typesetter console contains a keyboard, a computer, disk drives, and a display screen. A *font* is a special style of type that can be set in a variety of point sizes or heights. The font is contained on a length of photographic film. When the film or font is placed into the typesetter, everything typed will appear in that style.

Q-5000 typographic composition system. *Courtesy of Compugraphic Corporation.*

6 POINT
ABCDEFGHIJKLMNOPQRSTUVWXYZABCDEFGHIJKLMNOPQRSTUVWXYZABC
abcdefghijklmnopqrstuvwxyzabcdefghijklmnopqrstuvwxyz $1234567890

8 POINT
ADEFGHIJKLMNOPQRSTUVWXYZABCDEFGHIJKLMNOPQRS
abcdefghijklmnopqrstuvwxyzabcdefghijklmnopqrs $1234567890

10 POINT
ATUVWXYZABCDEFGHIJKLMNOPQRSTUVWXYZ
atuvwxyzabcdefghijklmnopqrstuvwxyz $1234567890

12 POINT
ABCDEFGHIJKLMNOPQRSTUVWXYZABC
abcdefghijklmnopqrstuvwxyzabcdef $1234567

14 POINT
ADEFGHIJKL aghijklmno $890123

18 POINT
AMNOPQR apqrstuvw $45678

24 POINT
ASTUVW axyzab $9012

30 POINT
AXYZA acdefg $345

36 POINT
ABCD ahijk $678

42 POINT
AEF almn 901

48 POINT
AGH aop 234

60 POINT
AIJ aqr 5

72 POINT
AK ast 6

Various font sizes and styles used in typesetting. *Courtesy, from Karsnitz, Graphics Arts Technology, © 1984 by Delmar Publishers Inc.*

As a typesetter types, the characters are stored on a disk just as in word processing. Striking a key is equivalent to taking a photograph of that character. When the product is completed, a special paper is taken from the typesetter and developed. The set type is then arranged as needed.

Phototypesetting represents the current state-of-the-art in typesetting and is a major advantage over other methods of typesetting.

PRODUCING GRAPHIC COMMUNICATION

Signs, billboards, stationery, pamphlets, brochures, letters, printed T-shirts, bumper stickers, and wallpaper are all forms of graphic communication. But the major forms of graphic communication are books, magazines, and newspapers. Producing these things requires four stages:

1. Design
2. Image creation
3. Production
4. Finishing

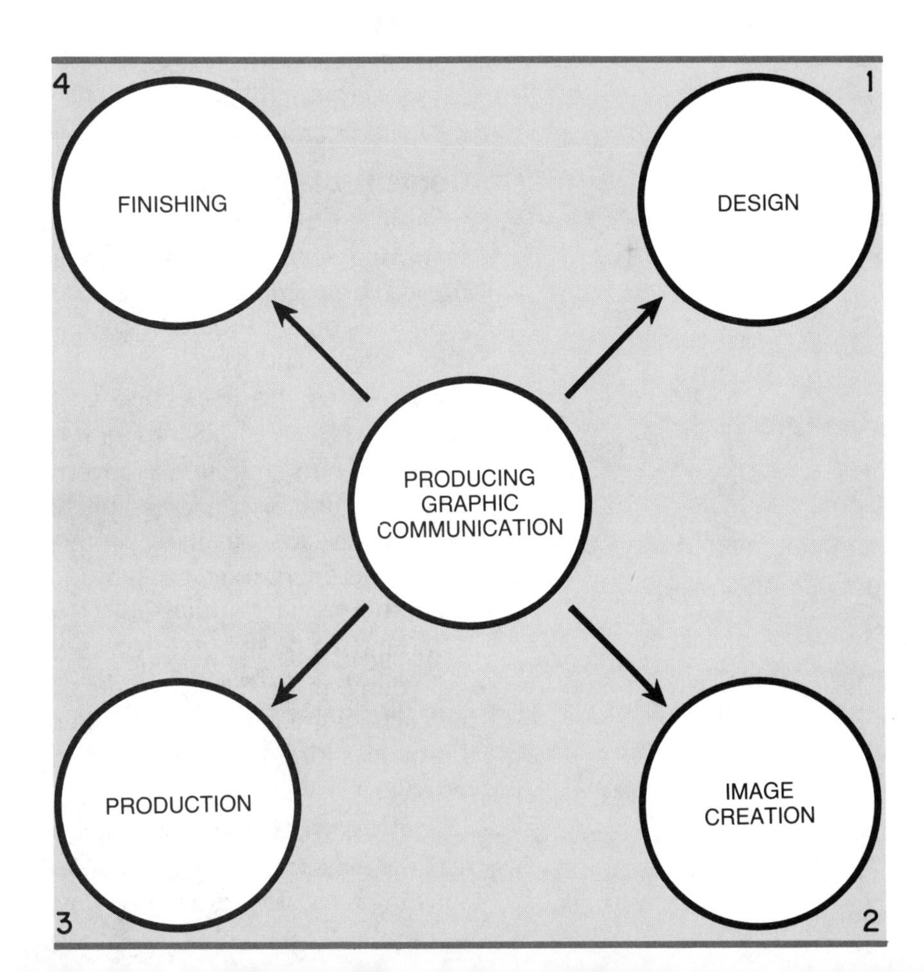

Producing graphic communication.

Design

Design involves planning the communication form so that it will actually communicate effectively. Design involves arranging words, photographs, drawings, graphs, charts, and so forth, in a pleasing, effective way.

If the communication form is a book or a magazine, the pages have to be planned carefully. The words and artwork (photographs, charts, drawings) must be arranged for the best effect in communicating the message. If the words on a page refer to a photograph, the photograph should be placed on that page or on the adjacent page.

If the communication form is a newspaper, headlines have to be carefully written and positioned. Photographs and drawings must be placed in such a way as to attract the reader's attention to the accompanying story.

Actually the first function of design is to determine how to communicate the message. Four considerations are involved:

1. Content of the message
2. Purpose of the message
3. Nature of the audience
4. Location of the reader

Content of the Message What is to be communicated? Is the message in advertising, a newspaper story, or just information? Answering such questions helps plan the content for the greatest impact. For example, if the message is a story about some newsworthy event, the content should be planned to cover the five *w*'s: who, what, when, where, and why.

Purpose of the Message Why is the message being communicated? What is its purpose? Is it to grasp the reader's interest? Is it to inform? Is it to educate? An advertisement meant simply to generate interest will be brief and eye-catching. A news story meant to inform will contain facts but little depth. A book designed to educate will contain pertinent facts, background, and supportive material, and an in-depth treatment of the subject.

The Nature of the Audience Who is the intended audience? Is it old, young, male, female, student, professional, college graduate, or high school graduate? Is the audience large or small? The answers to these questions help designers decide how to communicate the message.

Location of the Reader Where will the readers be when they read the message? Will they be in a library, in a move theater, at home watching television, driving down the highway, in a classroom, in an

Considerations in deciding how to communicate a message.

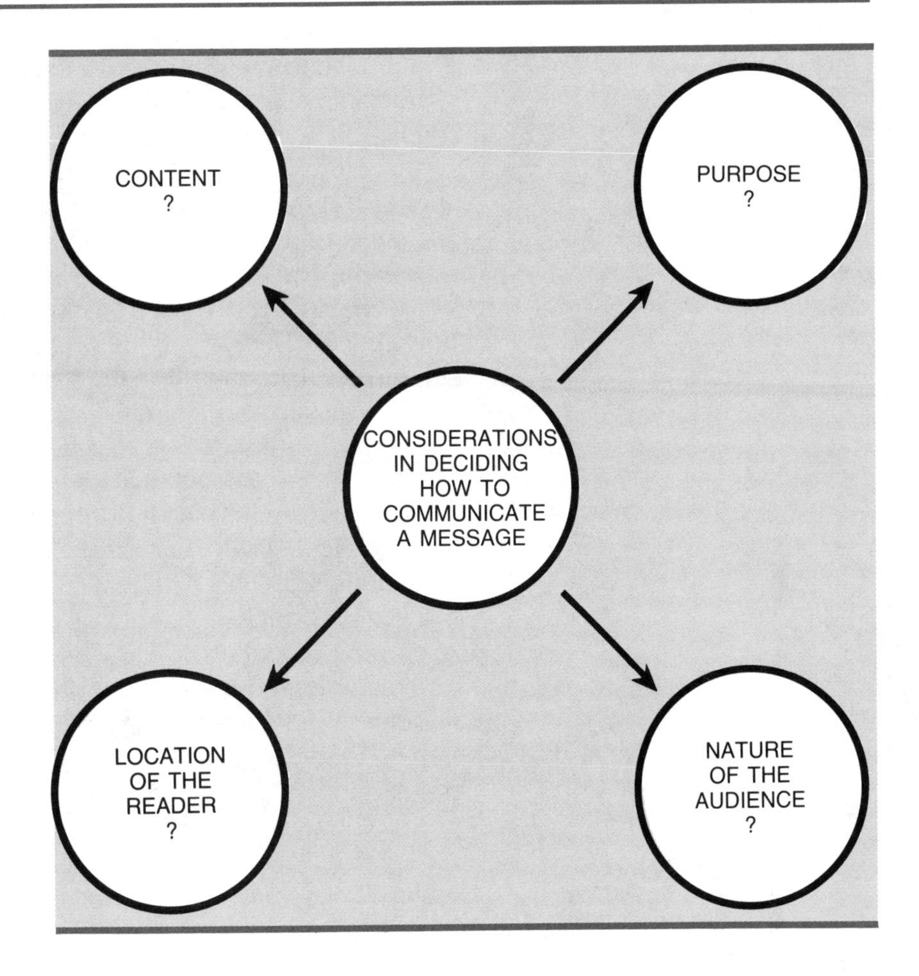

airplane, or in an office? The answers to these questions help designers decide how to communicate the message.

After the content, purpose, audience, and location of the reader have been determined, designers can select the communication form to be used. The form chosen might be a billboard, T-shirt, bumper sticker, brochure, or book. The final step in the design stage involves answering three questions:

1. What should be the quality of the finished product?
2. What quantity will be needed?
3. What will be the cost of the communication product chosen?

Quality of the Product Does the product need to be high quality with glossy colored pictures, or can it be a less-impressive but also less-expensive black-and-white approach? For example, if the product is an announcement for a store that will be closed on a given day, a simple, inexpensive, black-and-white format will probably suffice. But if the story is going to be just one of several stories being run by several stores to attract shoppers to a big sale, a more eye-catching approach might be taken. This approach might use multiple colors and fonts especially arranged to catch the reader's eye to beat the competing ads.

Quantity of the Product How many of the finished products will be needed? This is an important question because the quantity produced will be a factor in the total cost of the product. If the product is a newspaper with a circulation of 100,000, at least 100,000 will be needed. If it is a billboard, maybe only one or two will be needed.

Cost of the Product This question is the last one that designers must answer, but it is the most important. It is a function of everything in the design stage that has preceded it. How will the message be communicated? What will be the quality of the finished product? How many copies will be made? All of these things will help determine the cost of the product.

What the product costs the producer determines what it will cost the buyer. If a book costs $10 to produce, it will probably cost the buyer $14 to $16. Since it can be assumed that the buyer will want to pay as little as possible for the finished product, it is important to produce it as inexpensively as possible while maintaining the desired quality in effectively communicating the message. This is the purpose of the design stage.

Image Creation

Graphic images are made up of letters, symbols, numbers, and pictures. What do you see when you look at a page in a book, magazine, or newspaper? You see letters that make words that make sentences that make paragraphs, and so on. You see numbers, you see symbols, and so forth. And you see pictures. The pictures might be photographs, drawings, graphs, or charts. All of these things together form graphic images that communicate a message. Producing a graphic image is called *composing*. Composing graphic images can be done in many ways, but they all fall into one of two categories: (1) hot composition and (2) cold composition.

Hot composition does not mean that the materials used in composing the image are hot. Hot composition is any form of composition that uses cast-metal relief characters to produce the images. It is called hot composition because the metal used in casting characters had to be melted before it could be cast.

Cold composition is all other forms of image creation, including stamping, drawing, handwriting, typing, ruboff characters, clip art, and phototypesetting. Cold composition accounts for most of the image creation done in graphic communication technology.

Stamping Stamps and stamp pads are still used for creating certain images that are used over and over again. Date stamps, signature stamps, and address stamps are commonly used examples. But stamps can be made for any image that is to be created again and again. Some teachers have stamps made that say such things as "GOOD JOB," "TRY HARDER," or "YOU WILL DO BETTER NEXT TIME." Some people have stamps made in the likeness of an animal, a company logo, or a school motto. Stamping is one of the oldest forms of image generation.

Drawing Probably the oldest method of image generation is *image drawing*. Of course, drawing has come a long way since cave dwellers created images on cave walls. Drawing for the purpose of graphic communication is usually called *illustration* or *technical illustration*.

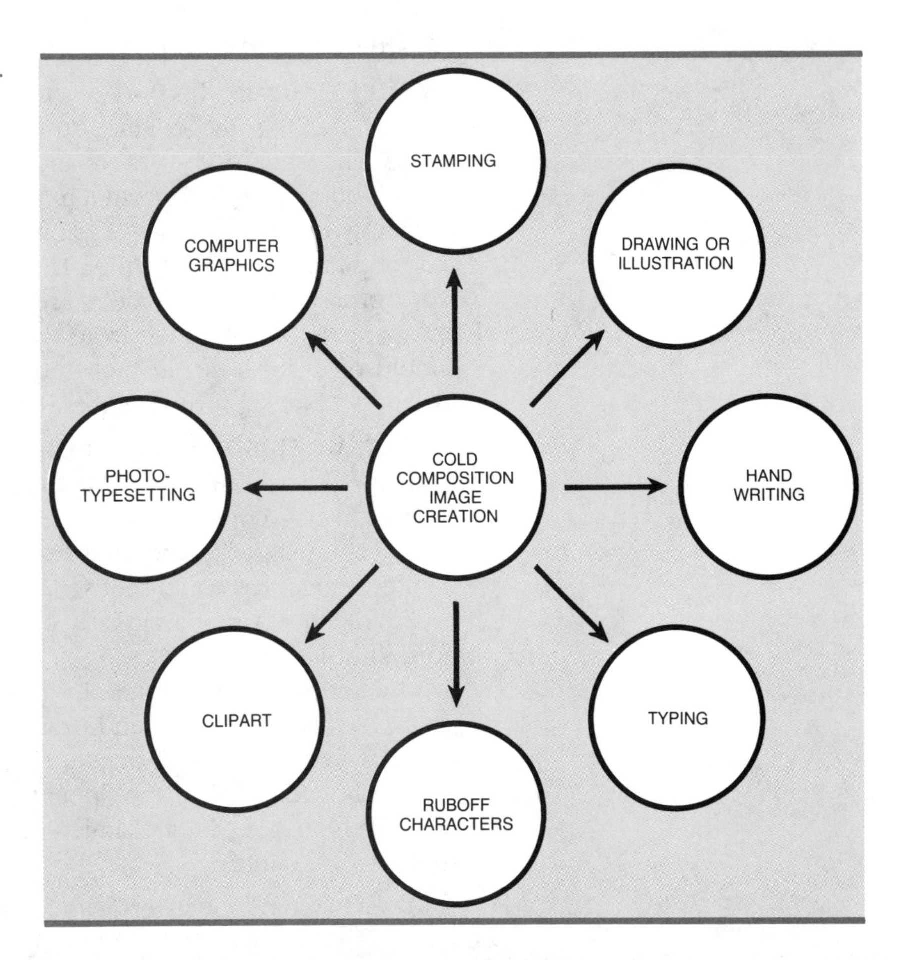

Cold composition image creation.

Sometimes a photograph cannot adequately convey the message. For example, if a picture is needed to show a cross section of an automobile engine, a photograph won't work. But a clear, detailed drawing, complete with labels can be made that will serve the purpose. In other cases, photographs may not be available of the subject. An example of this would be when a picture is needed of something that existed before the invention of photography, but now no longer exists. A dinosaur or a viking ship are examples.

Handwriting Handwriting is still a frequently used means of creating images. Forms of handwriting such as calligraphy, lettering, and cursive writing are also still important ways of creating graphic images.

Calligraphy is a general name that includes several different styles of especially "fancy" writing. You know what printing is: you learn to print in the first and second grades. *Lettering* is a free form of printing that allows for individual style. Architects are known for free-form hand lettering. *Cursive writing*, sometimes called longhand, is what you learned after you knew how to print. Different forms of handwriting are still popular on posters and greeting cards.

Typing *Typing* has become one of the most efficient and productive ways of creating images that are made up of numbers, letters, and special characters. Typewriters and the printers in **word processor** systems allow people to form letters, numbers, and special characters by simply pressing buttons. Typing is fast and efficient.

Another advantage of typing is consistency. With a typewriter or a word processor, the characters are exactly the same every time they are made. Do you know how to type? Typing is something everyone should be able to do. In high school and college, students will be assigned papers and reports that have to be typed.

Ruboff Characters There are special ruboff or dry transfer sheets made for almost every occasion. The purposes of ruboff characters are speed and convenience. There are ruboff sheets for letters, symbols, trees, cars, people, plants, bushes, and almost every other type of graphic image that could be used.

Ruboff sheets are particularly handy for people who have little or no skill in lettering or drawing. If you are making a poster and your hand lettering is not very good, ruboff letters are the answer.

Architects, illustrators, and drafters who must occasionally create pictorial drawings of buildings use ruboff trees, plants, cars, people, and bushes to enhance the appearance of the drawing and to save time. Even people who are skilled in lettering and drawing use ruboff sheets to save time.

Clip Art Like ruboff sheets, clip art serves two purposes: speed and

Vertical Gothic freehand lettering.

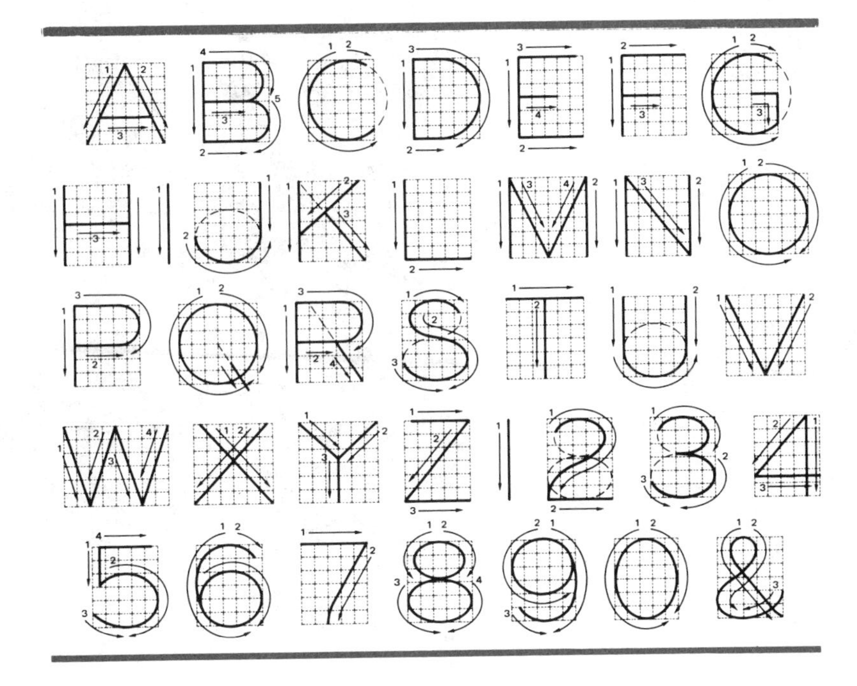

Slanted Gothic freehand lettering.

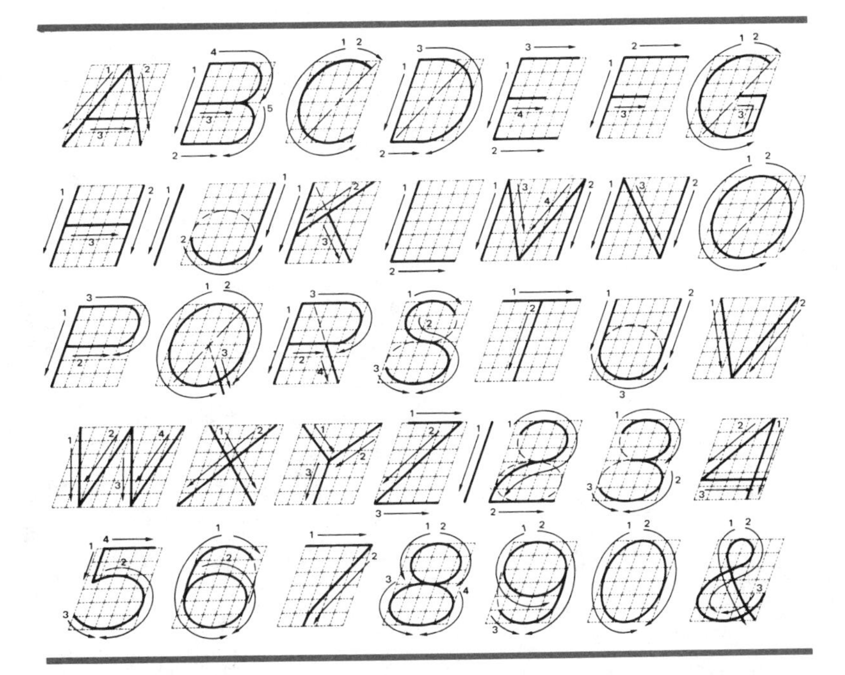

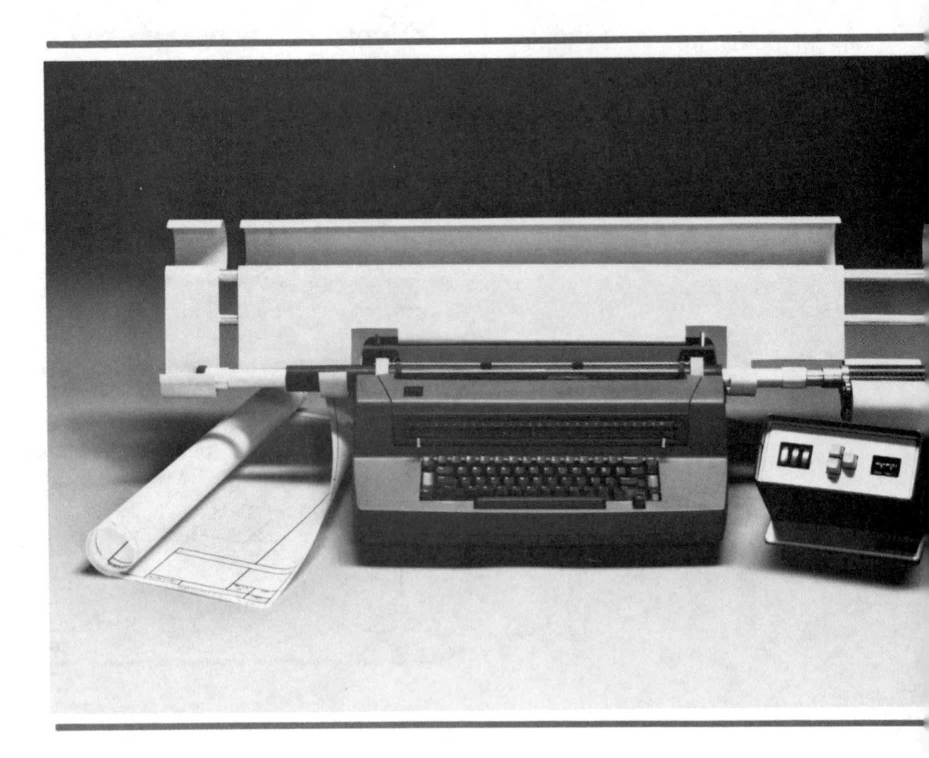

A wide-carriage typewriter.
Courtesy of Diagram Corporation.

convenience. **Clip art** is any picture that can be cut out of any source and used again. Clip art is available in two forms: special clip art books, and in any other form of graphic communication such as magazines, newspapers, and books.

There are companies that produce volume after volume of clip art. There are clip art books containing nothing but people in all shapes, sizes, and positions. There are clip art books containing nothing but animals, cars, trees, and so on.

Clip art can also be taken from other graphic communication forms such as books, magazines, and newspapers. For example, if a person is creating a graphic image for a year-end sale on automobiles, he or she might clip out a picture of a car that appeared in another newspaper advertisement rather than draw one. Using clip art is a popular method for creating images.

Phototypesetting The phototypesetting process is a cross between photography and typing. Typesetting is done on a special machine that contains interchangeable type font negatives, special paper, and a keyboard. A *font negative* is like a broad strip of camera film that contains letters, numbers, and other special characters.

Striking a key on the keyboard is the equivalent of taking a

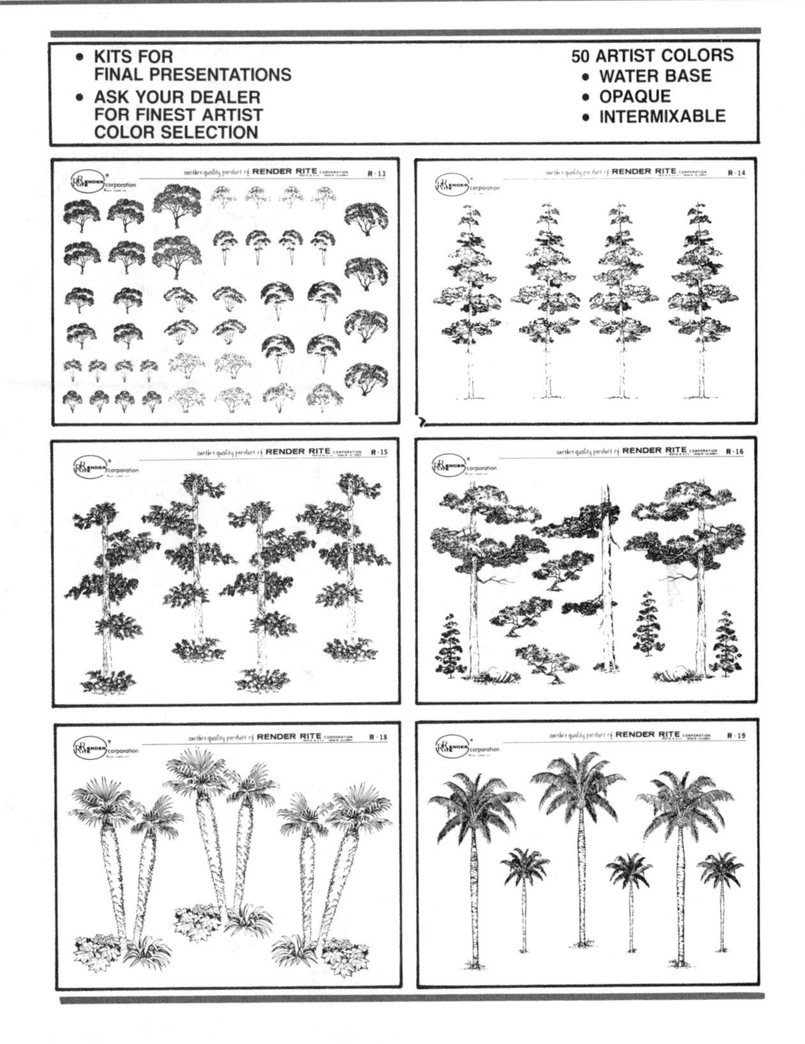

Ruboff sheet used with architectural drawings.
Courtesy of Renderrite.

photograph of that letter, number, or character. The image is created on the special paper as a latent or invisible image. When the special paper is taken from the phototypesetter and run through a developer, the images appear.

Images created on a new phototypesetter are cut out and pasted into position. Phototypesetting is one of the most frequently used image-creation methods in modern graphics shops. Phototypesetting has become an essential process in producing newspapers, books, and magazines. The words you are reading now were produced on a phototypesetter. The printing style was determined by the font used.

Computer Graphics **Computer graphics** is any use of the computer to create graphic images, both static and moving. Of course, in graphic communication, we are interested in producing static

A complete phototypesetting system. *Courtesy of Compugraphic Corporation.*

MCS-8000 digital phototypesetter. *Courtesy of Compugraphic Corporation.*

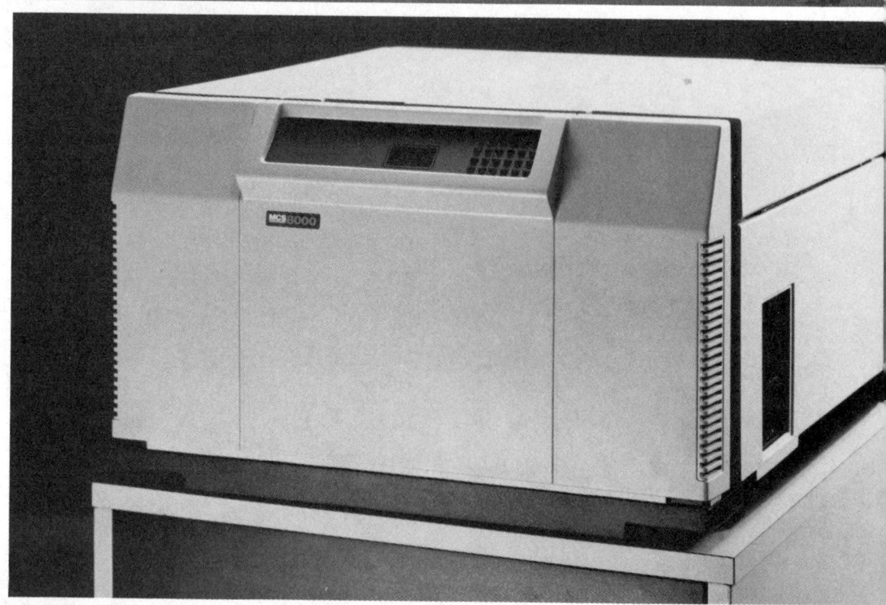

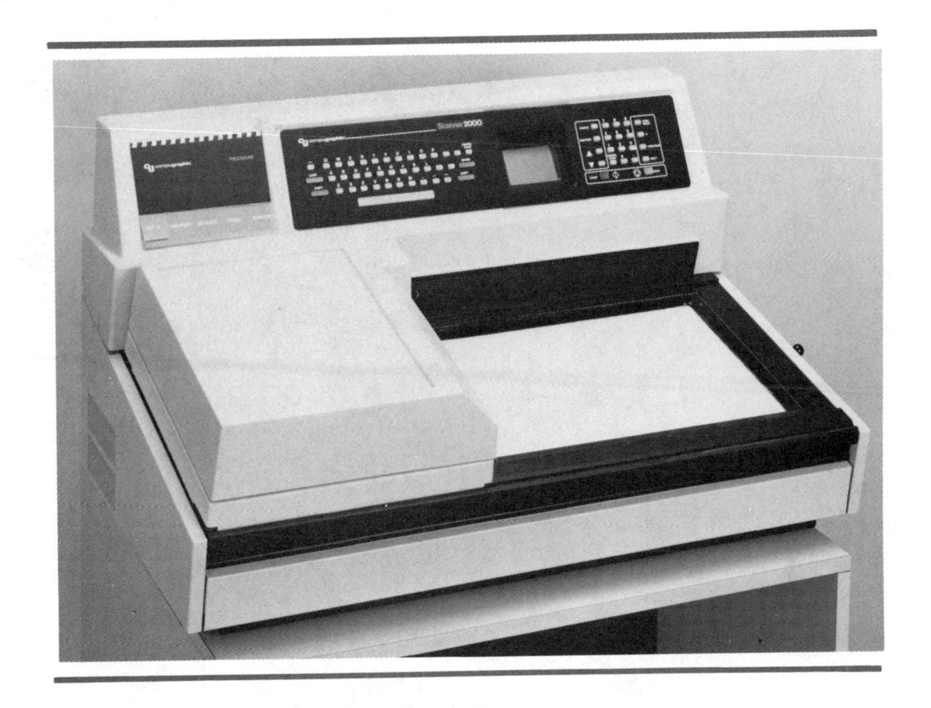

Scanner 2000 laser graphic scanner. *Courtesy of Compugraphic Corporation.*

MCS-8000 digital phototypesetter. *Courtesy of Compugraphic Corporation.*

An example of computer graphics. *Courtesy of auto-trol technology.*

images on a computer. Computer hardware and software are available to allow people to create any type of drawing, from a typed chart to a complete working drawing to cartoon characters. CADD systems have become very popular in architectural engineering and drafting offices.

To create drawings, a computer system must have special software and input/output devices. Graphic software allows users to create the points, lines, planes, circles, and arcs that are the building blocks of all drawings.

The input devices used to create drawings can be the keyboard, digitizer, light pen, puck, or mouse. The *keyboard* allows people to issue drawing commands such as LINE, CIRCLE, and ARC. The computer responds by drawing these things. A *digitizer* is like an electronic sketch pad. A *light pen* and *puck* are devices used to interact with the digitizer just as a pencil is a device that interacts with paper. A *mouse* is a device that allows people to interact directly with the display terminal.

The principal output devices in computer graphics are the display terminal, plotter, and printer. The *display terminal* is the televisionlike device that displays the picture as it is created. A *plotter* is an electromechanical device that, upon command, draws what appears on the display screen. A *printer* is the same device used in word processing, but in computer graphics it can also be used to produce simple drawings and pictures.

Drawings produced by a printer are composed of many small letters, numbers, and characters. Have you ever seen a drawing produced on a printer? If you look closely, you will see that the lines are actually a series of tiny characters, such as X's, O's, Y's, or Z's.

PRODUCTION

When a product has been designed and the images needed have been created, the next step is production. *Production* involves making as many copies of the product as needed. For example, if the product is a newspaper, several thousand copies might be needed. If it is some other product, only three or four copies might be needed. Of the different production processes available for producing multiple copies of a product, the process used depends on the (1) type of product, (2) the expected quality of the product, and (3) the number of copies needed. The most frequently used production processes in graphic communication are

1. Relief printing
2. Screen process printing
3. Intaglio printing
4. Lithographic printing
5. Xerography
6. Duplicating

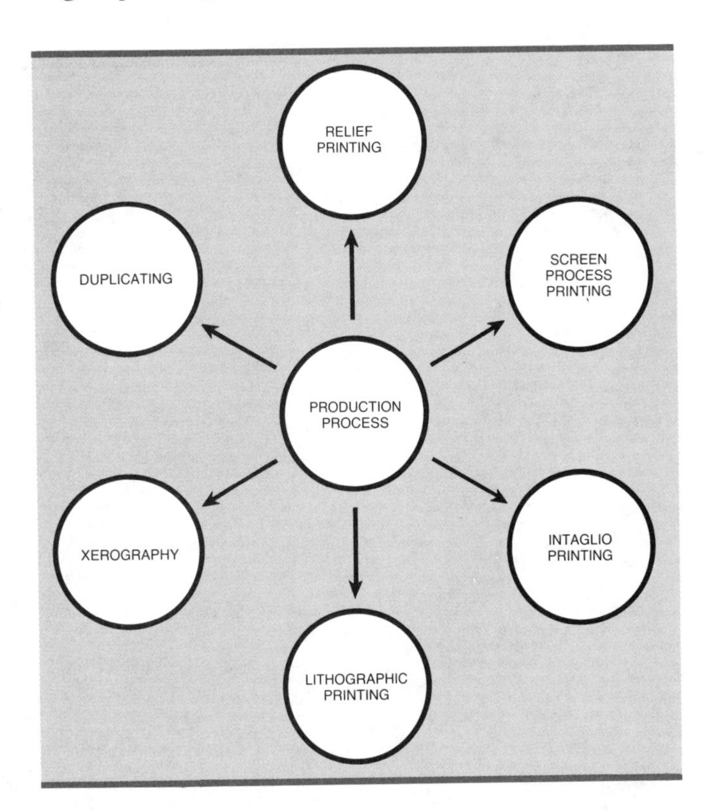

Production processes.

Relief Printing

Relief printing involves pressing inked relief characters against paper. The characters are put together on a carrier called a *plate* to form words and phrases. Each character is a block carrying a raised backwards letter. The letter is backwards so that it will produce a right-reading image when pressed against the paper. The individual characters can be metal and cast in a foundry, carved wood, or linoleum.

Screen Process Printing

Not all printing is done on paper. There are graphic communication messages carried on T-shirts, windows, soda cans, beer cans, and many other things. Since you can't run a T-shirt or a window through a press, there must be a way of printing on them and other nonpaper surfaces such as wood, glass, cloth, plastic, and metal. There is such a process. It is known as screen process printing.

Relief printing.

Screen process printing involves pushing ink through a special stencil attached to a tightly stretched screen. This screen is usually made of cloth and stretched across a wooden frame. The different processes for making the stencil used in screen process printing are the paper stencil, hand-cut film, and photographic stencil methods. The simplest is the paper stencil method. It illustrates how screen process printing is accomplished. The stencil is cut from a water-resistant paper, such as the brown kraft paper that grocery bags are made of. A screen is stretched across a wooden frame and secured. The stencil is attached to the screen. Ink is spread liberally across the stencil with a squeegee or brush. The ink passes through the opening in the stencil, through the screen, and on to the receiving material, the T-shirt, for example.

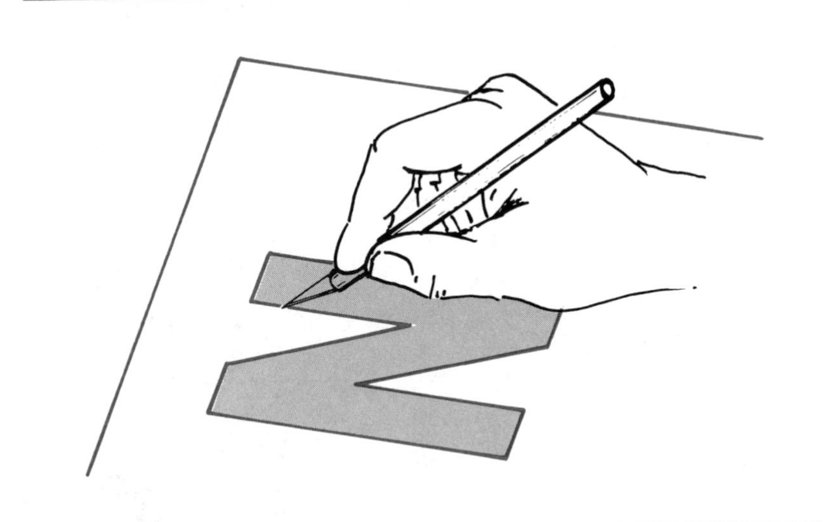
Making a paper stencil.

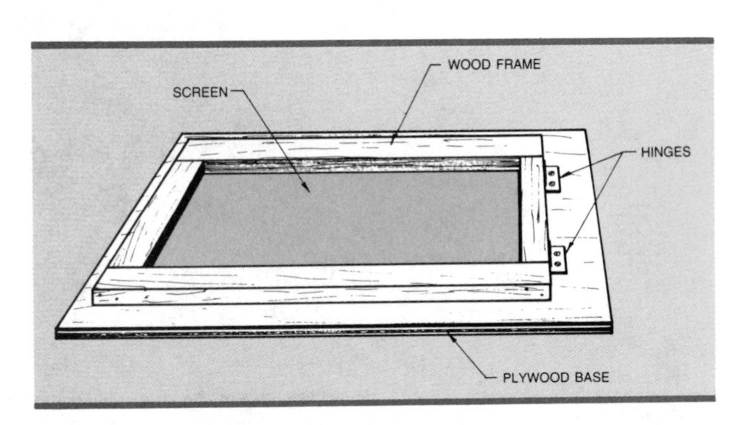

Wooden frame for screen process printing.

Intaglio Printing

Intaglio printing involves etching an image into a sheet of hard plastic or metal called a plate. It is the opposite of relief printing. After the image has been etched, the plate is covered with oil-based ink. Then the ink is wiped away with a cloth, leaving ink only in the grooves etched into the plate. By pressing a sheet of paper on top of the plate, the image can be transferred to the paper. This occurs because the ink remaining in the etched grooves is transferred to the paper. In the same way, a master or a blanket can be made for running copies on a press.

Intaglio printing.

Lithographic Printing

Lithographic printing is usually just called offset printing. It involves printing with a flat-surface printing plate. The term "offset" describes how the image to be printed is transferred from the plate to the paper. An offset press is a machine that has three rollers: the plate roller, the blanket roller, and the impression roller. There are also smaller rollers for applying ink and water.

The master to be copied is rolled around the plate roller and secured. It is a right-reading master. A rubber blanket is rolled around the roller. As these two rollers rotate, the image from the plate is transferred or offset to the blanket.

At this point, the image is reversed—it reads backwards. Then a sheet of paper is fed between the blanket roller and the impression roller. As these rollers rotate, the image is transferred to the paper. At this point, it becomes right-reading. Offset printing is one of the most commonly used methods for producing large quantities of a graphic communication product. It is a fast and relatively inexpensive printing process.

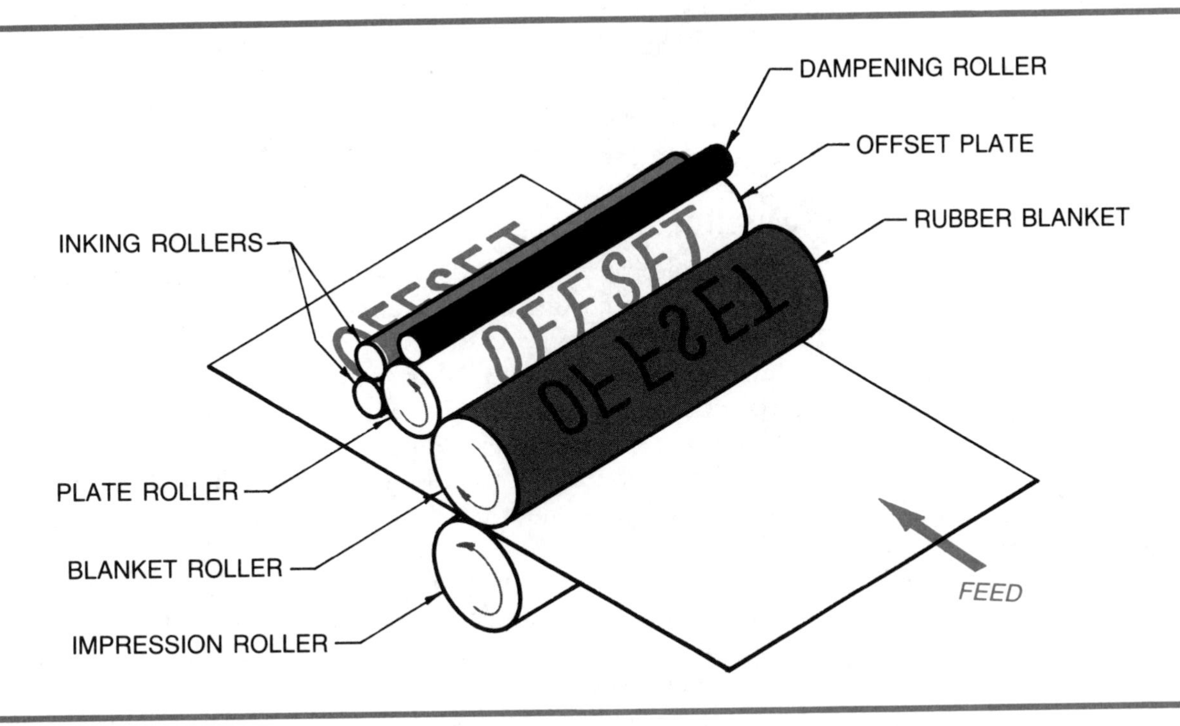

Offset printing.

Xerography

Xerography is a form of production that you may already be familiar with. You may even call machines that copy by xerography "xerox machines." Although this is not correct, it has become a common practice.

The Xerox Corporation does not make xerographic copiers, but IBM, A.B. Dick, 3M, and Savin do.

The term "xerography" actually means dry writing. Unlike the other forms of printing presented so far, xerographic copiers use a dry process—an electrostatic process—to make copies. When a xerographic copy is made, an original copy is placed on the bed of the copying machine. Electrostatic energy then attracts a special, dense, dry powder known as *toner* to the copying page. The toner arranges itself to create an exact duplicate of the original document. Do you have access to a xerographic copier? If there is one in the office of your school, ask the secretary to show you how it works. Ask to see the toner, but don't touch it. Toner is hard to get off of your hands and even harder to get out of your clothes.

Xerographic copying is used in situations when high-quality copies are needed fast. The number of copies made is small and so is the cost. However, if the number is large, offset printing is less expensive. Xerographic printing has become an essential process in most modern offices.

Duplicating

At one time, *spirit* and *mimeograph* duplication were the most widely used forms of making copies in offices and schools. In recent years, these **duplicating** processes have increasingly given way to xerographic copying. However, because they are fast and inexpensive, they are still used a great deal. Spirit and mimeograph duplicating involve preparing a special master and placing it on the drum of a duplicating machine. As the drum rotates, blank paper is fed under it. As each blank sheet feeds under the drum, the image is transferred from the master to it.

Spirit and mimeograph duplicating are used in low-volume copying situations. The master will not hold up for long copying runs. These types of duplicating are popular in schools for copying tests, papers, and handouts. If there is a mimeograph or a spirit master copying machine in your school, ask the secretary to show you how to operate it.

Finishing

After the required amount of a product has been produced, it must be finished. **Finishing** can involve a variety of different operations: for example, trimming, folding, collating, drilling, binding, stapling, and packaging.

Finishing actually encompasses any operation required to get the product ready for the market or the consumer. For example, if the product is the newspaper, finishing involves folding the sheets, trimming, stacking, and packaging the newspapers into bundles for shipping.

If the product is a book, the pages must be cut, trimmed, collated into numerical order, bound, and packaged for shipping. After a product has been finished, it is ready to be distributed to the public.

SOME CAREERS IN GRAPHIC COMMUNICATION TECHNOLOGY

Many careers are associated with graphic communication technology. Like all of the other technology areas, these careers can be grouped into categories that make them easier to learn and understand. Most careers associated with graphic communication technology fall into one of the following categories: professional, technical, skilled worker, and marketing.

Professional Career Fields

Professionals include the scientists, managers, chemists, professors, and teachers who work in fields related to graphic communication technology. Scientists and chemists involved in research to grow stronger, bigger trees faster, to make better paper, and to discover new and better copying processes and materials are involved in graphic communication. Teachers of graphic communication technology subjects from junior high school through college also fall into this category. It includes teachers of English, journalism, writing, graphics and printing, photography, drawing, illustration, and communication arts. Editors, managers, authors, and technical writers are also included here.

Professional careers in graphic communication technology normally require a college degree and often require advanced college degrees.

Technical Career Fields

Design technicians, layout technicians, composers, technical illustrators, and lab technicians are some of the job titles in this category. The education and training required of a technician ranges from one year of specialized training from an area vocational center to two- and three-year associate's degrees from a community college.

Marketing Career Fields

Marketing is an important part of graphic communication technology. Graphic communication products must be marketed so that they reach the desired audience.

Newspapers use marketing personnel to sell advertising and to sell subscriptions to the newspaper. In fact, advertising, not subscriptions, is where a newspaper makes its money. Book publishers use marketing personnel to sell books. Magazine publishers use marketing personnel to sell advertising and subscriptions.

The amount of education required to work in marketing varies a great deal. However, many people who work in marketing have an associate's degree from a community college or a baccalaureate degree; some even have advanced degrees.

Skilled Workers

Skilled workers are used a great deal in graphic communication technology settings. They perform jobs associated with image generation, production, and finishing. Typesetting and printing are skilled fields. Binding, folding, and trimming tasks are also performed by skilled workers.

Skilled workers must have specialized training relating to their jobs. Such training is available in high schools, area vocational schools, trade schools, and community colleges. Such training programs can be from nine months to two years. Some skilled workers in graphic communication technology still enter the field through the apprenticeship training route.

SUMMARY

Graphic communication involves communicating visually through the use of letters, numbers, symbols, characters, pictures, and other visual images. Images used for graphic communication are static. Various forms of graphic communication are books, newspapers, magazines, T-shirts, billboards, and road signs.

Important events in the development of graphic communication were the invention of alphabets, stamping, movable type printing, offset printing, photography, and phototypesetting.

There are four stages in the production of graphic communication products: design, image creation, production, and finishing.

The various production techniques used in graphic communication include relief printing, screen process printing, intaglio printing, lithographic printing, xerography, duplicating, and finishing. Finishing is any task necessary to get the product ready for use on the market.

Some widely practiced careers in graphic communication are scientists, managers, chemists, professors, teachers, design technicians, layout technicians, composers, technical illustrators, lab technicians, marketing personnel, typesetters, printers, and binders.

KEY WORDS

Graphic communication
Hieroglyphics
Stamping
Movable type printing
Lithographic printing
Offset printing
Photography
Phototypesetting
Design
Graphic image
Cold composition
Hot composition
Word processor
Clip art
Computer graphics
Relief printing
Screen process printing
Intaglio printing
Xerography
Duplicating
Finishing

SOME QUESTIONS TO TRY

1. What is graphic communication?
2. Name three different examples of graphic communication.
3. Name and explain three important developments in graphic communication.
4. Briefly explain the four stages in producing graphic communication.

TECHNOLOGY LEARNING ACTIVITIES

1. Carve a stamp out of a soft wood, such as balsa wood, for stamping your first name.
2. Divide the alphabet among your classmates and carve movable type blocks for each letter out of a soft wood such as balsa. Use the movable blocks to make a sign for your room such as FIRE EXIT.
3. Set up a screen process printing station for printing on T-shirts. You will need to make or obtain the following:
   a. A stencil for the word or message
   b. A wooden frame with clamps
   c. A screen
   d. Oil-based ink
   e. A squeegee
   f. T-shirts
4. Etch an intaglio printing plate of your school's logo, and use it to print the logo on T-shirts and posters.
5. Make a timeline of important developments in graphic communication.

CHAPTER 10

ELECTRONIC COMMUNICATION

OBJECTIVES

After completing this chapter, you should be able to

- define electronic communication.
- list some of the significant developments in electronic communication.
- explain how people communicate electronically.
- list some of the careers associated with electronic communication.

Communication is one of the oldest technology areas. Since the dawn of civilization, there has been a need to communicate. There are many different ways to communicate. We communicate when we talk. We communicate when we smile or frown. We can communicate with hand signals, smoke signals, and light signals.

Any communication system has three components:

- The **sender**
- The **transmitter**
- The **receiver**

For there to be communication, a message and/or an image must be sent, transmitted, and received. The most modern form of communication is electronic communication.

WHAT IS ELECTRONIC COMMUNICATION?

Electronic communication is communication that is aided by electronic devices, instruments, or systems. This means that electronics aids in the sending, transmitting, and receiving of messages and/or images.

Many electronic communications systems are available. The most common are radio, telephone, and citizens band (CB) radio. We'll look at these and others in depth later. For now, let's see how electronic communication came about.

Components of a communication system.

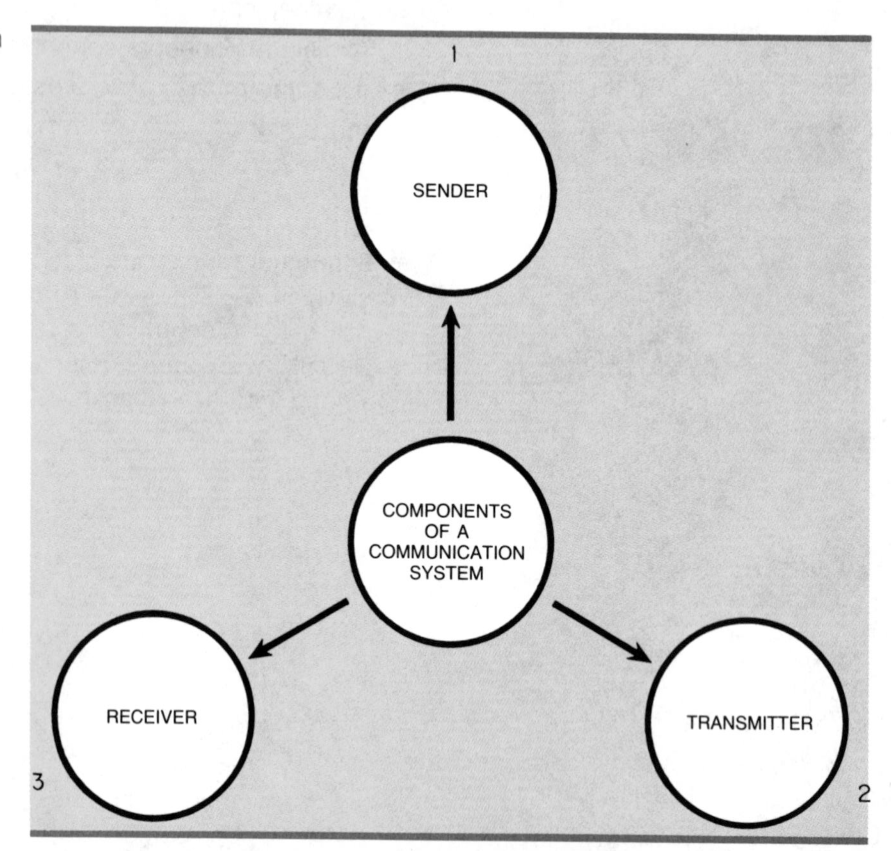

DEVELOPMENT OF ELECTRONIC COMMUNICATION

People had been communicating through various means for thousands of years. Even before they had developed languages, people communicated using grunts, facial expressions, and hand signs.

Common languages made communication much simpler. But they did not solve the problem of communicating across distance. To do this, people relayed messages by runners, signal flags, and even smoke signals. Each method had drawbacks and limitations. These limitations eventually led to electronic communication.

Some of the most important names in the development of electronic communication were Morse, Bell, Marconi, Alexanderson, and Zworykin. Joseph Henry took one of the first major steps in electronic communication. He developed the earliest version of the **telegraph** in 1831. Samuel Morse developed the language of the telegraph, a series of dots and dashes that came to be known as **Morse code**. In addition, he encouraged others to improve on the telegraph equipment developed by Joseph Henry. Consequently, in 1844, Morse was able to demonstrate a vastly improved telegraph system.

Alexander Graham Bell invented the **telephone**, which he patented in 1876. The telephone was much more convenient than the telegraph because it actually transmitted the human voice instead of a complicated code that had to be interpreted by experts.

Guglielmo Marconi conducted experiments that eventually led to **radio**—sending messages without wires. In 1901, he successfully sent a message across the Atlantic Ocean, using improved telegraph

Wireless telephones are a modern form of electronic communication. *Courtesy of Robert C. Atwood.*

GTD-5EAX digital central office switching system. *Courtesy of GTE Communication Systems.*

Conducting a test on the GTD-5EAX digital switching system. *Courtesy of GTE Communication Systems.*

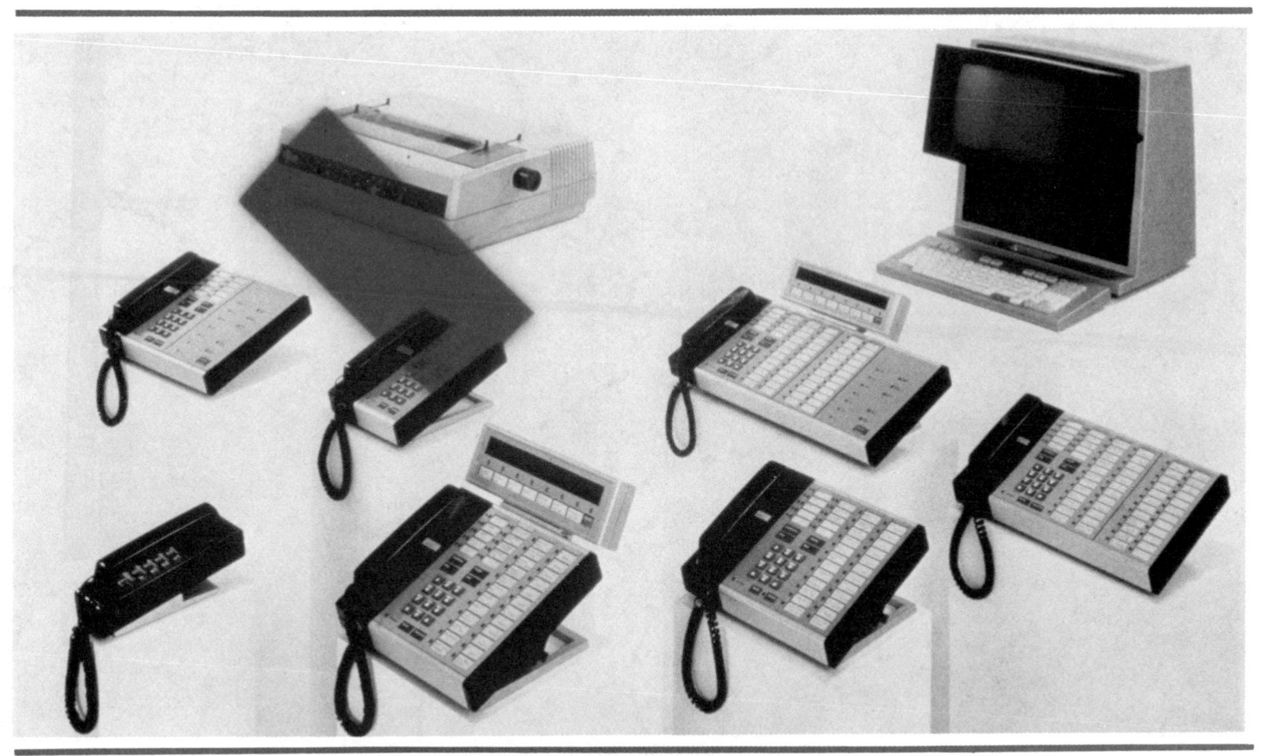

System 75 communications devices. *Courtesy of AT&T.*

System 85 communications system integrates simultaneous voice and data communications. *Courtesy of AT&T.*

Major developments in electronic communication.

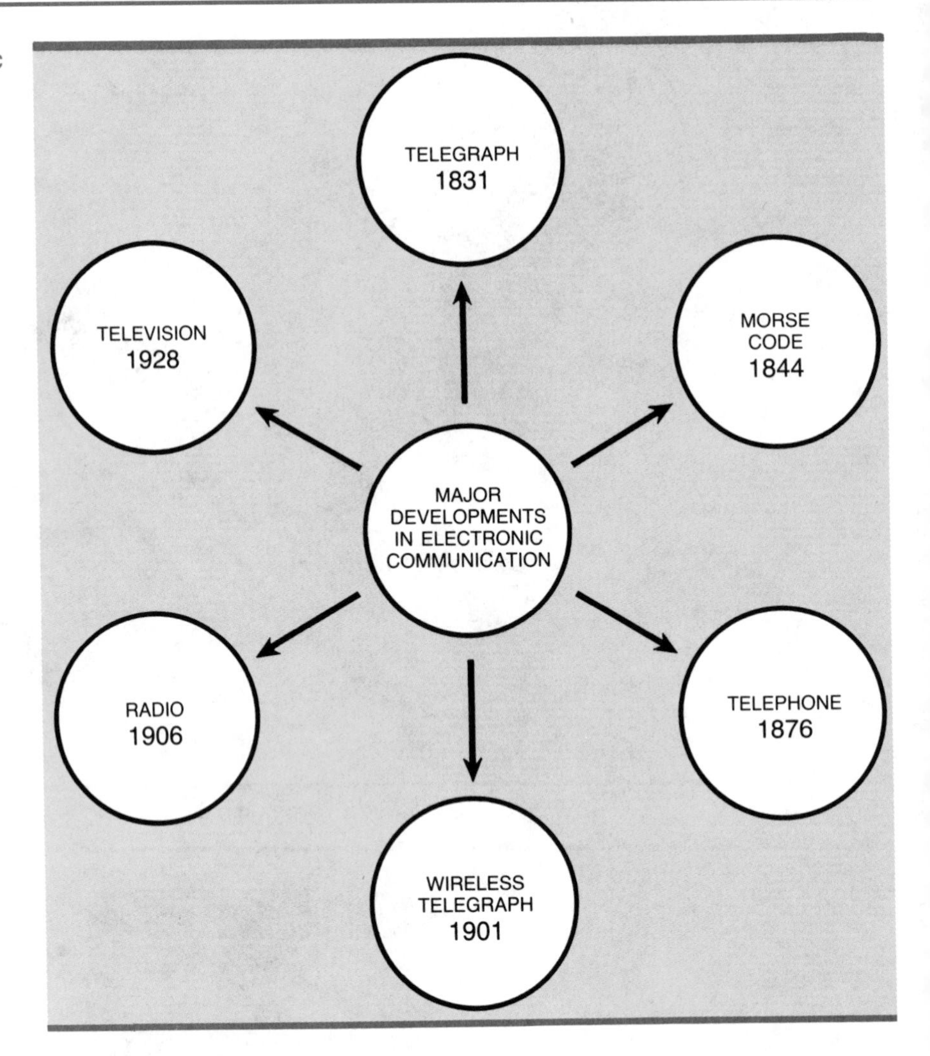

equipment without wires. This was the first use of the **wireless telegraph**.

The next step was radio. Radio involved transmitting the human voice without the use of wires. Ernst Alexanderson developed an electrical alternator that would produce **radio waves**. The first demonstration of his device in 1906 was the first radio broadcast of voice and music in the United States.

Alexanderson's alternator was a large cumbersome device that involved rotating parts at high speeds. The invention of the vacuum tube provided a much smaller, more efficient way to produce radio waves. This invention gave radio the boost it needed to become a major form of electronic communication.

The wire telegraph method.

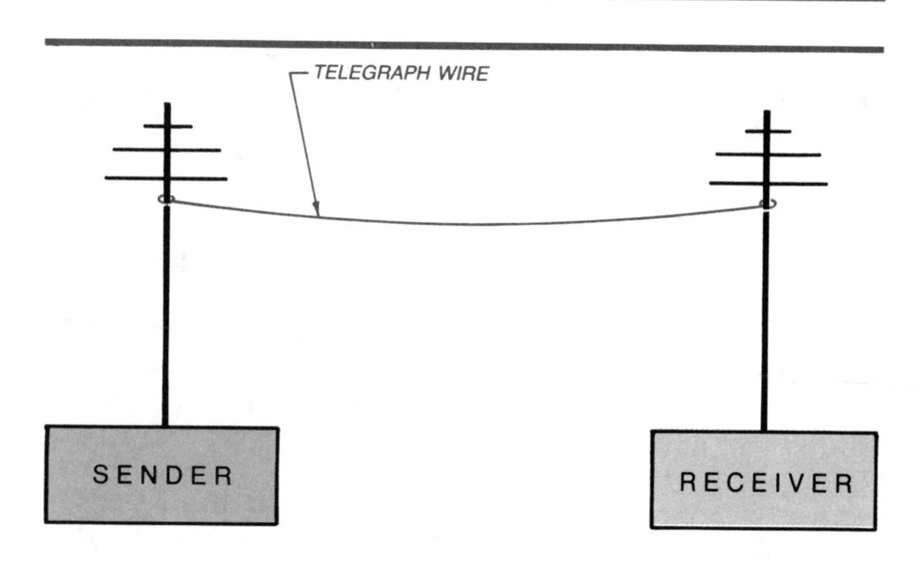

The wireless telegraph method.

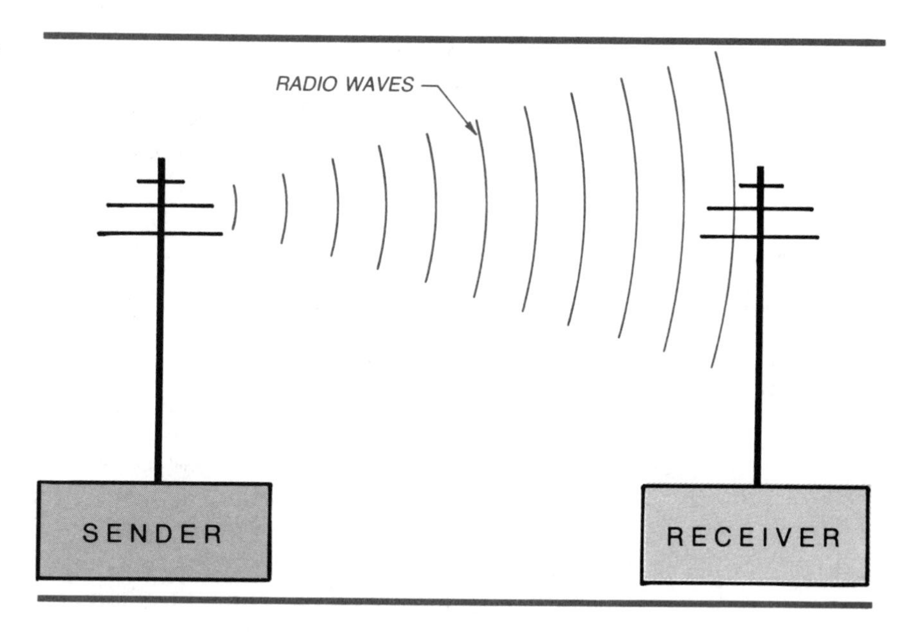

Next came television, which people thought of as radio with a picture. The addition of a picture came about through the pioneering efforts of Vladimir Zworykin. In 1928, he perfected the **cathode ray tube**, or picture tube, which allowed television to develop as a major form of electronic communication.

COMMUNICATING ELECTRONICALLY

The means by which we send, transmit, and receive messages electronically have changed over the years. The first electronic messages were sent as electrical impulses (dots and dashes) over wires to a receiver who decoded them. With the advent, in 1906, of the wireless telegraph, the electrical impulses were replaced by radio waves that travel through the air. But the radio waves were still produced by using a telegraph key. When the key was in the down position, it sent out radio waves. In the up position, it sent out no waves.

By 1906, radio waves could be generated and sent by the large Alexanderson alternator. Through a process known as *modulation*, the actual human voice, as well as music, could be added to the radio waves. This led to a process known as amplitude modulation (**AM**). It works like this. Voice or music are added to the radio waves that are generated. The power or *amplitude* of the radio waves is varied at an audio frequency rate. The radio waves are received. The varied amplitude activitates *diaphragms* that create sound waves.

Amplitude modulation is widely used by radio stations. Have you ever listened to AM radio? What is your favorite AM radio station? The radio waves that carry the voice or music are no longer generated by an alternator. The alternator was replaced by the vacuum tube, which has been replaced by transistors.

Another development that affected how we communicate electronically was frequency modulation (**FM**). You are probably familiar with FM radio. In recent years, it has become more popular than AM radio. In FM broadcasting, only the frequency is varied. The amplitudes of the radio waves that carry the voices and music stay constant all of the time. In practical terms, this makes the sound of FM radio clearer and easier to understand than AM. There are two reasons for this: (1) It cuts down on interference from power lines and lightning. (2) FM lends itself better to transmitting in stereo.

FM and AM **broadcasting** are the two main ways of communicating electronically today. Television combines both AM and FM. In the United States, the television picture is transmitted by AM, and the sound is transmitted by FM.

ELECTRONIC COMMUNICATION SYSTEMS AND DEVICES

People have created various systems and devices for communicating across distance. The most important systems are AM radio, FM radio, and television. Electronic devices are also used in video and music. Video and music are not communication in the strictest sense of the word, but they are important enough to examine in this section.

AM Radio

Most AM radio is broadcast on frequencies between 540 and 1600 kilohertz. AM radio is transmitted as follows:

1. A radio wave is generated by an **oscillator**.
2. The power of the radio wave is increased by an **amplifier**.
3. The microphone signal is amplified by an audio amplifier.
4. A **modulator** amplifies the signal further and varies the power output of the radio frequency (RF) amplifier.

There are four classifications of AM radio stations. Class I stations are the best. They operate on 50,000 watts of power, which is the most an AM station is allowed. Class II stations operate within a range of 10,000 to 50,000 watts of power. At night, they have to reduce their power or, in some cases, go off the air.

Class III stations operate on 5,000 watts of power during the day and reduce their power to 1,000 watts at night. Class IV stations are the smallest. They operate on 1,000 or fewer watts during the day and reduce power at night to as low as 250 watts.

The amount of broadcasting power is important: the more power,

AM radio transmission.

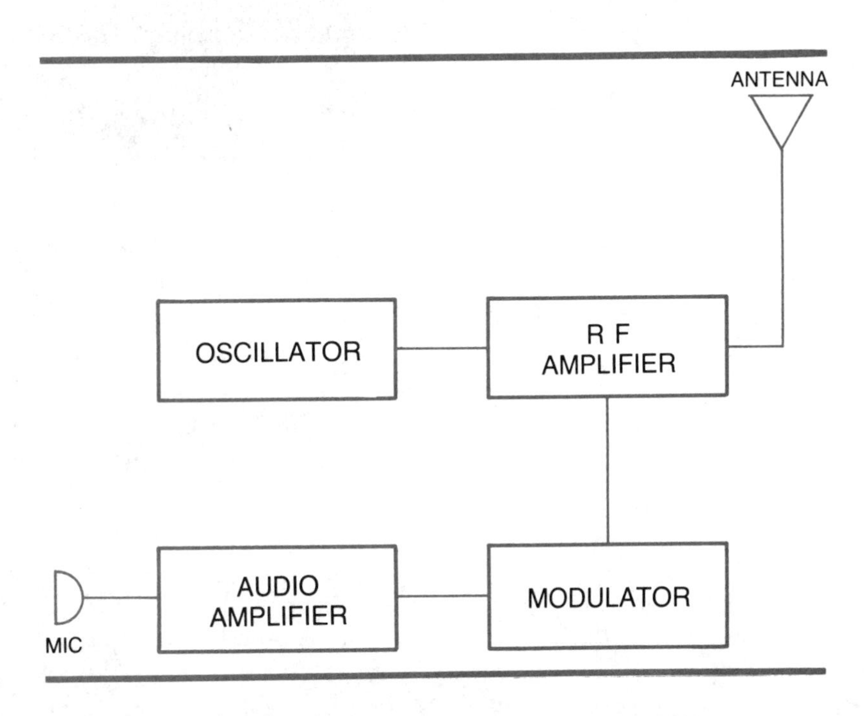

Portable radio/cassette tape systems. *Photo by Deborah M. Goetsch.*

the larger the broadcast area. The larger the broadcast area, the more money the radio station can earn by selling commercial time.

Have you ever been listening to AM radio and heard a station from a distant city? If so, it was probably a Class I or Class II station. Class III stations cover regional areas, usually consisting of one or more cities or towns. Class IV stations are strictly local and serve small cities or towns. Right now, there are over 4,000 AM radio stations in the United States.

FM Radio

FM radio is broadcast on frequencies between 88 and 108 megahertz. FM radio is transmitted as follows:

FM radio transmission.

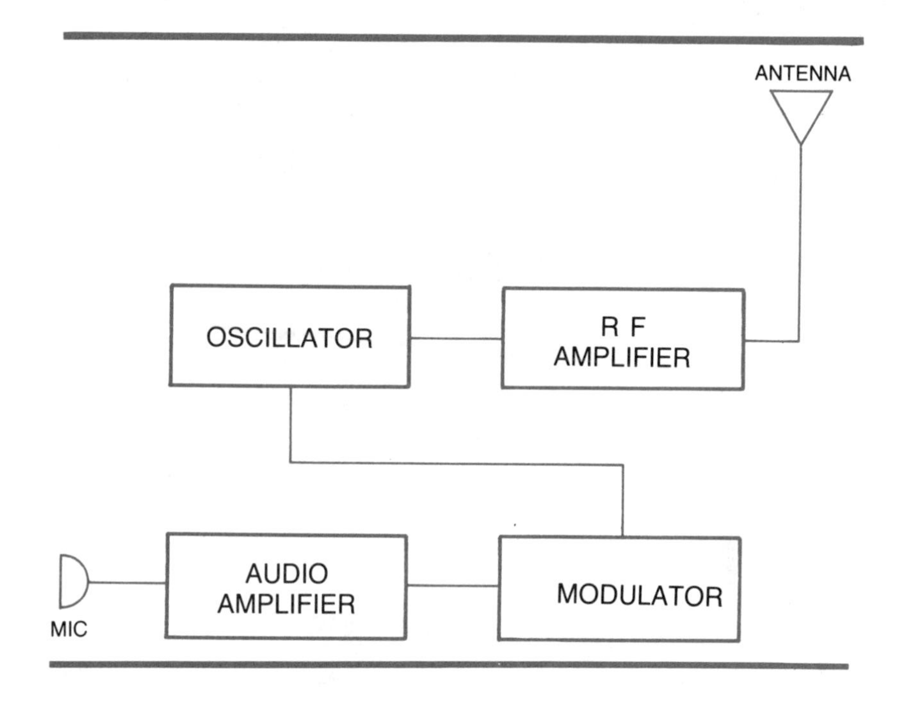

Combination FM-AM cassette tape player systems. *Photo by Deborah M. Goetsch.*

1. A radio wave is generated by an oscillator.
2. The power of the radio wave is increased by an amplifier.
3. The microphone signal is increased by an audio amplifier.
4. A modulator varies the oscillator frequency at a voice or music rate.

Because of the popularity of stereo and the inability of AM radio to transmit stereo, FM radio has grown steadily in popularity. FM radio is now more popular than AM. Which do you prefer?

There are three classifications of FM stations. Class A stations are the smallest. They serve small towns and are limited to 3,000 watts of power. Class B stations can use up to 50,000 kilowatts of power and, therefore, serve a larger area. Class C stations are the most powerful. They can use up to 100,000 watts of power and serve a very large area.

There are over 4,000 FM radio stations in the United States. Do you have a favorite FM station? What is the number of the station?

Television

Television is easily the most popular form of electronic communication. Every day millions of people watch television. There is the traditional network television and the more recent cable television. Experiments with television began as far back as the late 1800s, but television didn't begin to achieve wide-scale acceptance until the 1950s.

Television combines both AM and FM transmissions. The picture we see on television is transmitted by AM. The sound we hear on television is transmitted by FM. Now there are over 900 television stations in the United States and over 100 million television sets receiving broadcasts from them.

In the 1970s, cable television began to grow rapidly. This was due mostly to the development and launching of communications satellites. These satellites orbit the earth in space. They receive television transmissions from TV stations and relay them to receivers in a wide area. To receive these relayed transmissions, you must have a receiving device. The satellite dish is the most popular device for receiving satellite transmissions. Consequently, satellite dish antennas are popping up everywhere. People are installing them on rooftops and in their yards. At the same time, more satellites are being placed in orbit above the earth—the more satellites that relay programming, the more stations you can receive. Do you or does someone you know have a satellite dish?

Before cable television, satellites, and satellite dishes, people could only receive three or four stations. Now, if you have a satellite dish,

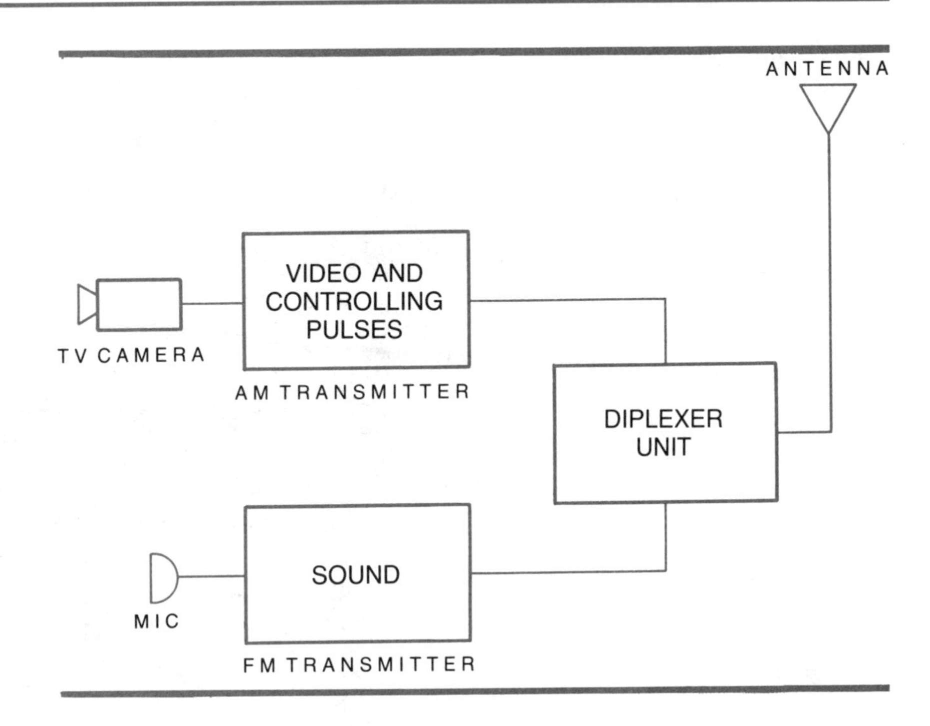

Television transmission.

Televisions are available in all shapes and sizes. *Photo by Deborah M. Goetsch.*

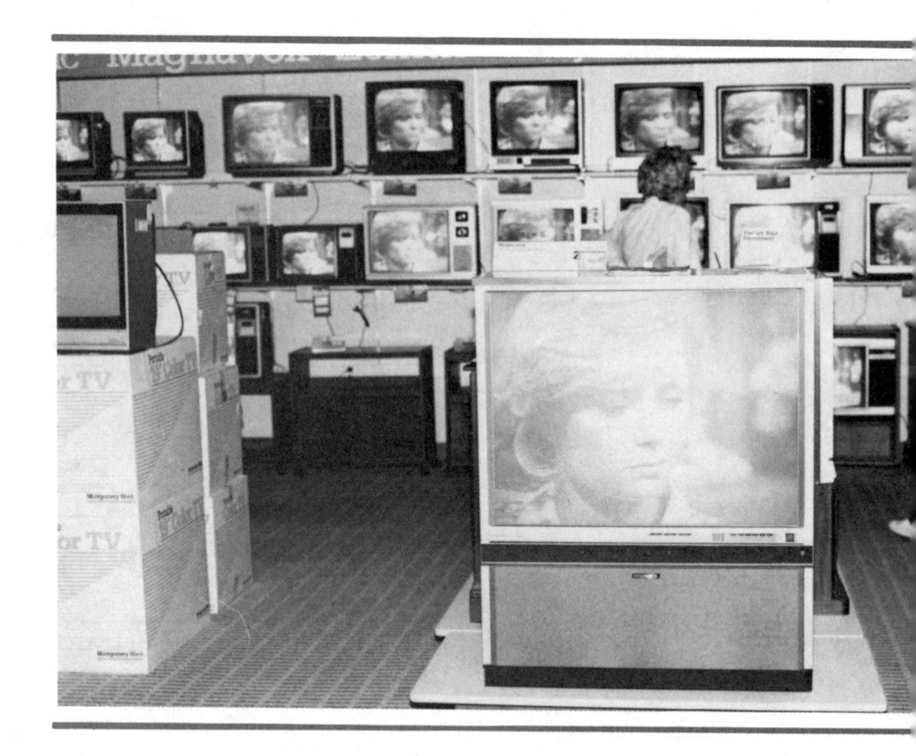

Large-screen televisions are becoming more popular. *Photo by Deborah M. Goetsch.*

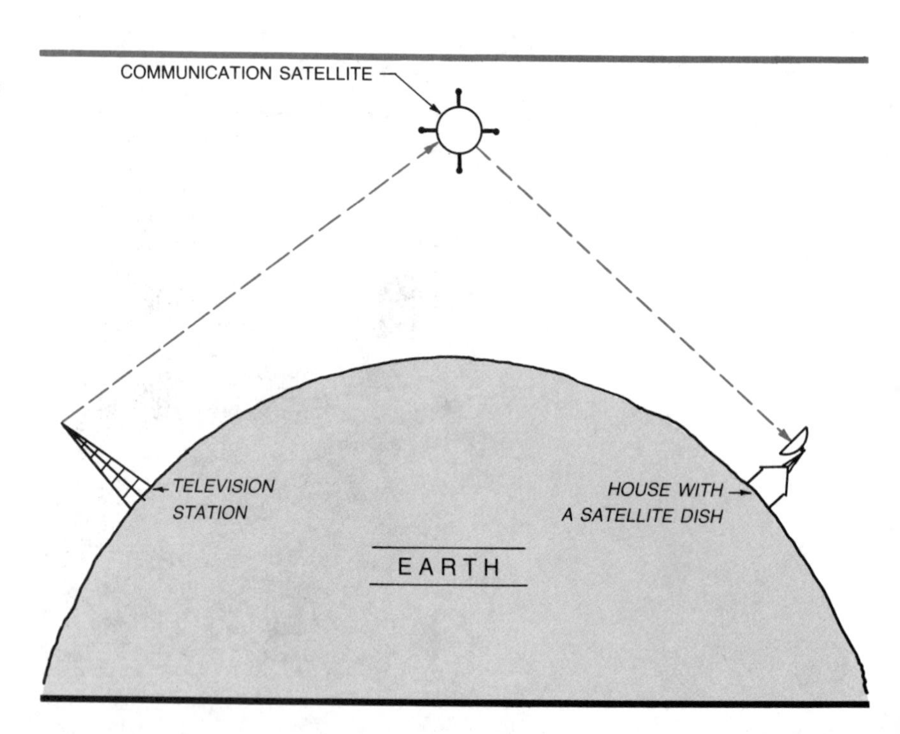

Communications satellite relay of a television broadcast.

Special dishes for receiving satellite-relayed television broadcasts. *Photo by Deborah M. Goetsch.*

you can receive over 100 stations. Some television transmitting stations have begun to scramble the programming they send out via satellite so that receivers who have their own satellite dishes are not able to receive those programs without paying a fee.

RELATED ELECTRONIC DEVICES

You should be familiar with a number of devices that are closely related to the major forms of electronic communication. The communication and entertainment aspects of television have been enhanced by video tape devices, video disk devices, and musical devices.

Video Tape Devices

VCRs tie into a television set and record programming on magnetic tape. With a VCR, a person can tape a program while watching another, or tape a program while away from the house. This enhances both the communication and entertainment potential of television. With a VCR, people don't have to miss any program they want to see.

Video players let people play programs they have recorded and programs that can be bought or rented. Many people have begun to buy or rent movies and videos instead of standing in line at theaters. Video cameras record activities and events on special tapes that fit on video players. These tapes come in several formats: BETA, ½-inch, ¾-inch, and 8-millimeter.

Most people use video cameras in the same way as they used to use home movie cameras. But the small size and portability of video

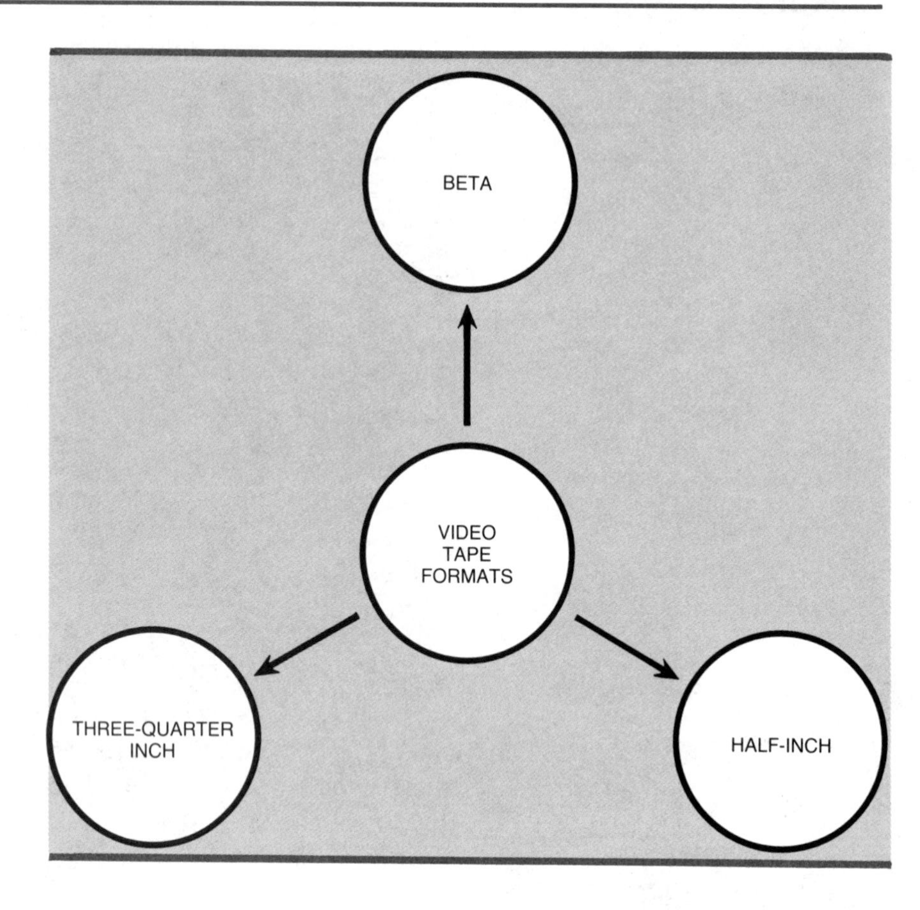

Video tape formats.

cameras has made them popular with network and local news teams also. Using video cameras, newspeople can go almost anywhere and record events. Then the tape can be hurried back to the television station and be used immediately. This is the reason why you can see on the news at night what happened across the world that morning.

At home, video cameras can record special events such as a party, a graduation ceremony, or Christmas morning. The tape can be taken out of the camera, placed in the player, and viewed immediately.

Video Disk Devices

Another approach to video recording and playback is the **video disk**. A video disk is like a record album that contains sound and video. Using a video disk player, people can play movies and other types of programs that have been recorded on a disk.

There are two ways that programming can be recorded and played back on video disks. One is the *electromagnetic* method. The other is

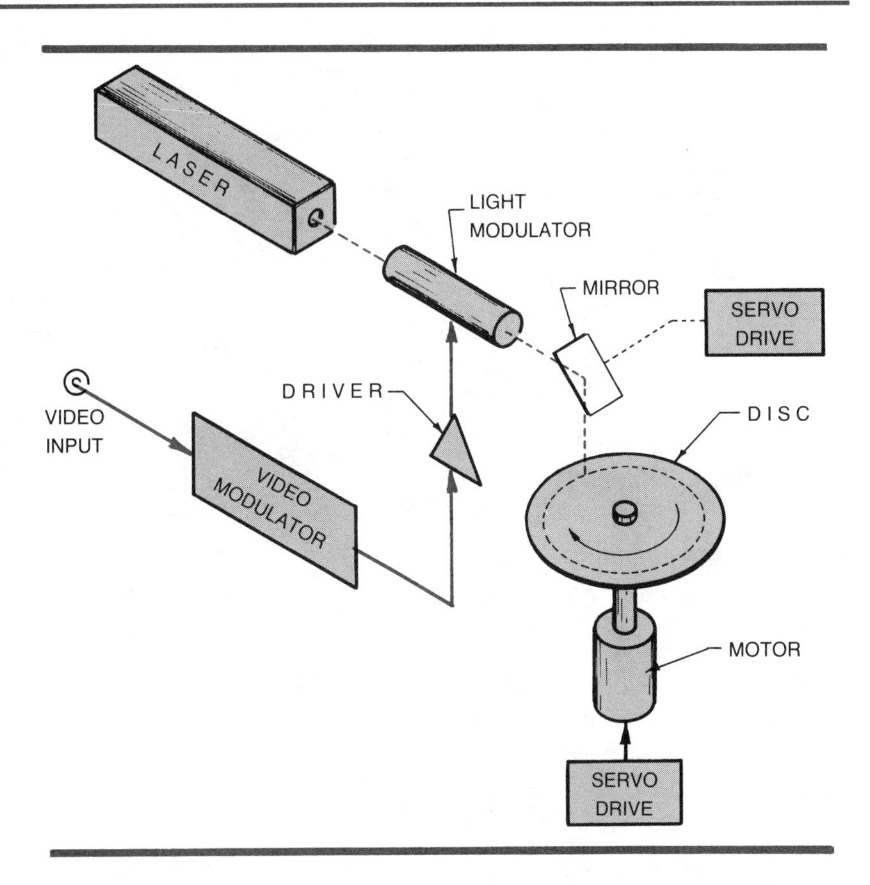

Video disk recording system.

the *laser* method. The most frequently used method is the laser method.

In recording a video disk, technicians use a mirror to direct a laser beam onto a blank disk. The beam contains both the audio and video for the program being recorded. As the disk rotates, the laser beam cuts into it and implants the audio and video information in a spiraling groove, just like those that you see on a record album. The playback process is just the reverse. The laser beam follows the spiraling grooves, picks up the audio and video information contained in the grooves, and displays them on a screen.

Musical Devices

Music is a form of communication. Some sociologists think it is the only universal form of communication. The music-making capabilities of some instruments have been enhanced by electronics.

Steel-stringed instruments are an example of musical instruments that have been enhanced by electronics. An electronic guitar is a good

Video disc playback system.

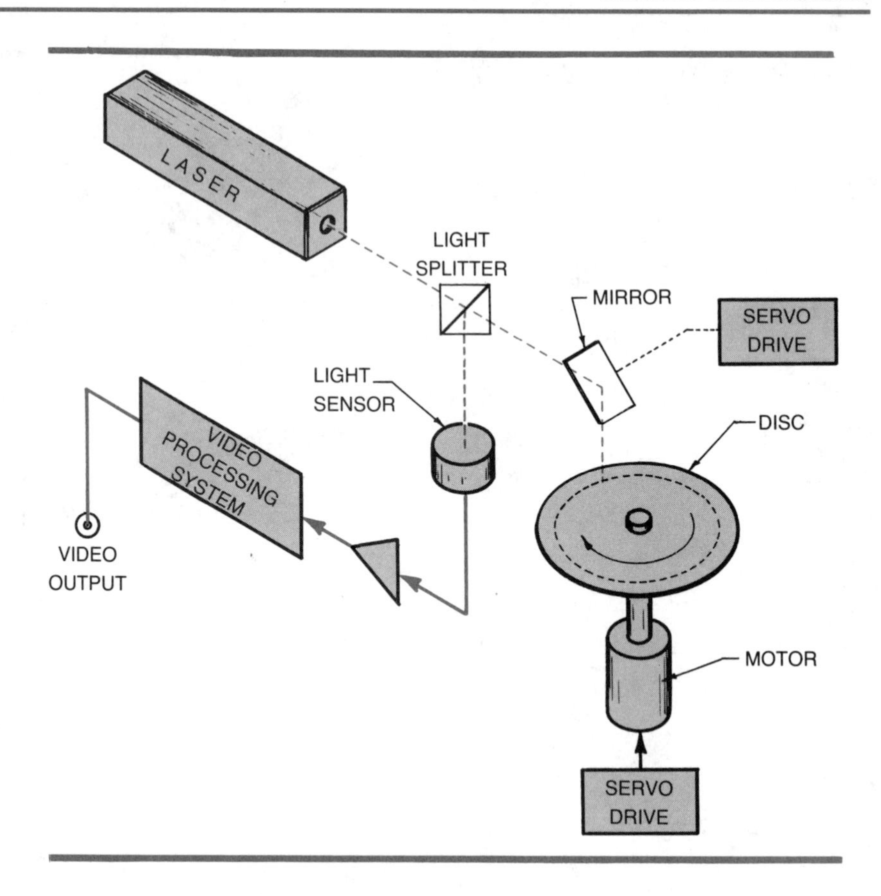

example. An electronic guitar has electromagnetic pickup devices mounted under the strings to amplify the sound. These devices can be used with any steel-stringed instrument.

Electronic organs are popular musical instruments. Audio oscillators allow electronic organs to mimic the sounds of a variety of orchestral instruments. Electronic pianos are also popular, but the best example of electronics and music is the **synthesizer**.

A synthesizer is not a musical instrument at all. It is a collection of electronic devices that, when used together, can make music. The electronic devices that make up a synthesizer are audio oscillators, voltage-controlled audio oscillators, white-noise generators, variable-speed tape recorders, audio filters, keyboards, and control panels. You don't play a synthesizer. You use it to synthesize (combine) scattered pieces of music from various sources into one finished piece. Synthesizers are very popular in recording studios that specialize in rock music, in motion picture studios, and for making television and radio commercials.

DX7 digital synthesizer. *Courtesy of Yamaha Combo Products.*

An electric guitar. *Courtesy of Steinburger Sound.*

Steinburger electric guitar. *Courtesy of Steinburger Sound.*

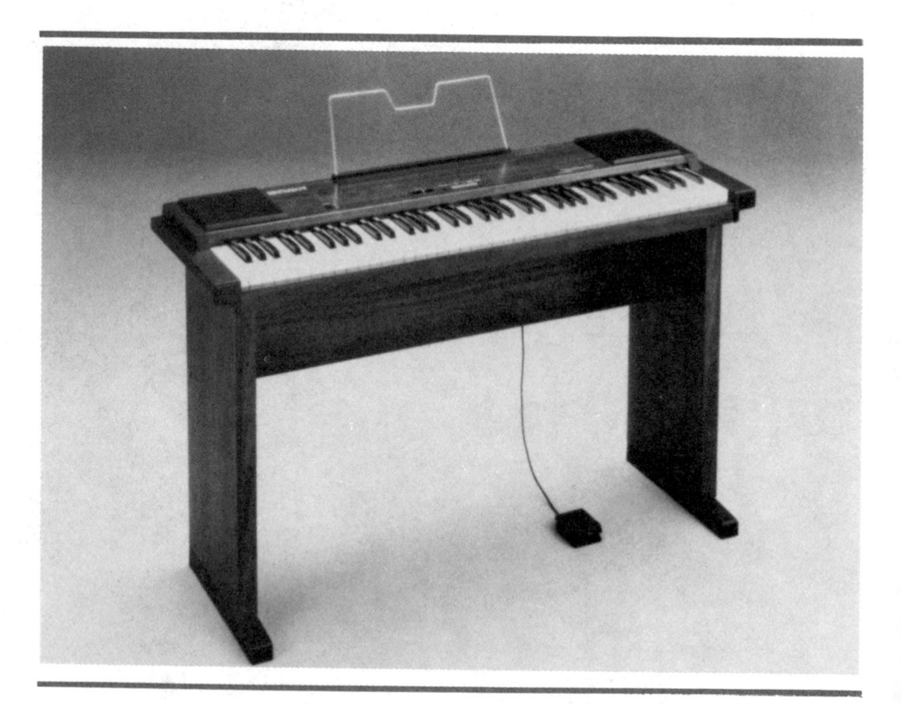

Portable keyboard. *Courtesy of Roland.*

JX-8P digital synthesizer and keyboard. *Courtesy of Roland.*

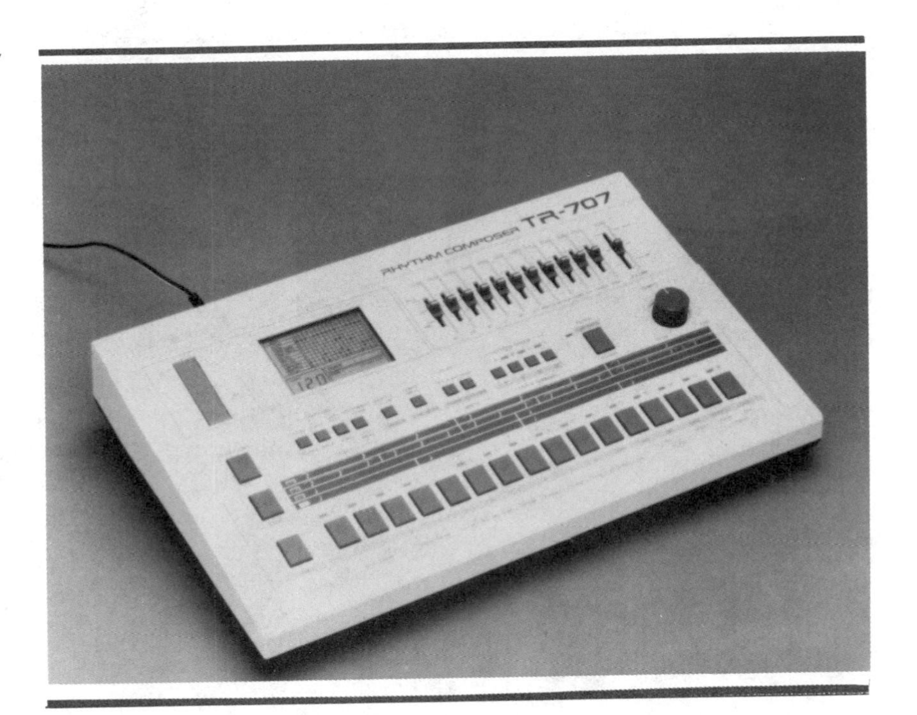

TR-707 rhythm composer. *Courtesy of Roland.*

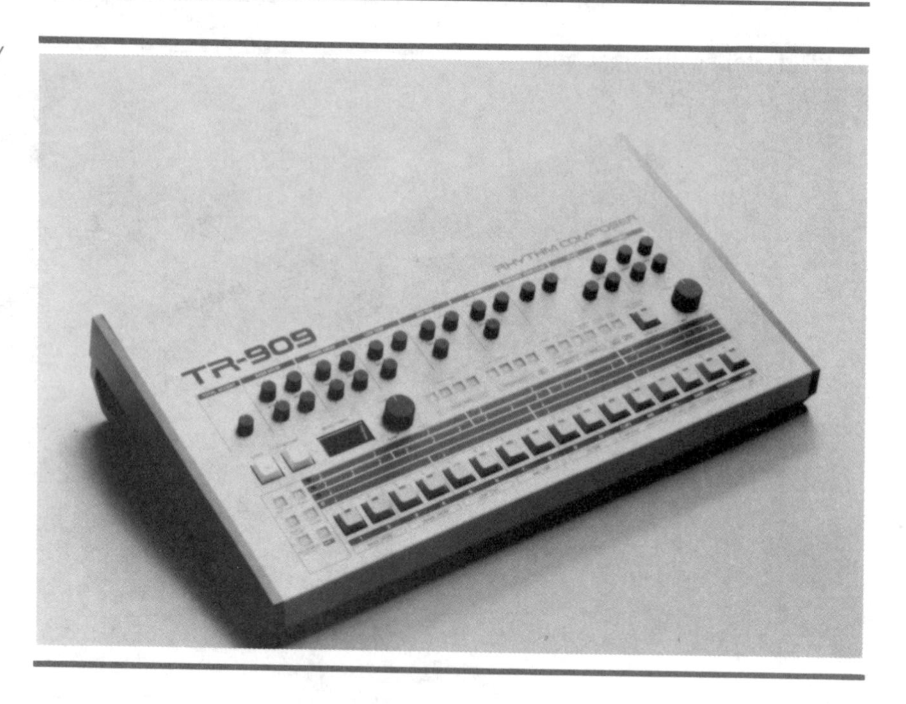

TR-909 rhythm composer. *Courtesy of Roland.*

SOME CAREERS IN ELECTRONIC COMMUNICATION

There are many different careers in electronic communication technology. Most fall into four categories: marketing, engineering, technical, and programming. Marketing personnel sell time to companies to use commercials to sell their products. Selling air time is how radio and television make most of their money. Most marketing personnel for television and radio have four-year college degrees in marketing or two-year associate degrees in marketing from a community college.

Programming personnel are responsible for the planning and broadcasting of music, news, sports, weather, and everything else that is broadcast. Programming jobs include producers, directors, announcers, disk jockeys, and anchorpersons. The education and training required for programming jobs vary greatly from on-the-job training to college degrees.

Technicians are very important in communication technology. They maintain, repair, and operate the electronic equipment and devices used in sending, transmitting, and receiving messages. Technicians require specialized training from a vocational school or a community college. Technicians work for engineers. Electronic communication engineers usually have a college education. In addition, in some cases, they must be licensed by the Federal Communications Commission (FCC). Licensed engineers are responsible for ensuring that radio and television stations comply with all applicable FCC rules and regulations.

SUMMARY

Any communication system has a sender, transmitter, and receiver. The most modern form of communication is electronic communication. Some of the most important developments in electronic communication have been the telegraph, telephone, wireless telegraph, radio, and television.

The two principal methods of broadcasting radio waves are amplitude modulation (AM) and frequency modulation (FM). Radio stations are either AM or FM stations. FM radio signals are clearer than AM signals.

Television broadcasting involves both AM and FM. With television, the picture is AM and the sound is FM. Some recent technological developments that relate to electronic communication are the video cassette recorder, video cameras, video players, video disk devices, and assorted musical instruments.

KEY WORDS

Electronic communication
Sender
Transmitter
Receiver
Telegraph
Wireless telegraph
Morse code
Telephone
Radio
FM
AM
Cathode ray tube
Radio waves
Broadcasting
Oscillator
Amplifier
Modulator
VCR
Video disk
Synthesizer
Television

SOME QUESTIONS TO TRY

1. Name the three components of a communication system.
2. Define electronic communication.
3. Name three important developments in electronic communication.
4. Explain the difference between FM and AM radio.
5. Name several musical instruments enhanced by electronics.

TECHNOLOGY LEARNING ACTIVITIES

1. Go to a hobby store or a radio store and obtain the components for making a crystal radio. Make a crystal radio as a class project.
2. Learn how to transmit your name in Morse code.
3. Make a poster that traces the most important events in the development of radio.
4. Make a poster that traces the most important events in the development of television.
5. Using a video camera, make a short movie involving class members, and show it to the class by using a VCR and player.
6. Make a timeline of important developments in electronic communication.

SAFETY
2594
CASE

PART FOUR

THE BIOTECHNOLOGIES

CHAPTER 11

AGRICULTURAL TECHNOLOGY

OBJECTIVES

After completing this chapter, you should be able to

- define agricultural technology.
- list some of the significant developments in agricultural technology.
- list the most common categories of products of agricultural technology.
- explain how the various products of agricultural technology are produced.
- list some of the careers associated with agricultural technology.

You have learned that the three most basic needs of people are food, clothing, and shelter. Agricultural technology is how people produce most of the food they need. (Hunting is another way.) Agricultural technology also produces many of the materials used to make clothing and shelter. For example, the raw materials for cotton, wool, and leather fabrics are produced by agricultural technology. Trees, the raw materials needed for lumber, plywood, and particle board, and paper for graphic communication, are also produced by agricultural technology.

Trees used to make lumber are a product of agricultural technology. *Courtesy of Weyerhauser Corp.*

Planting seedlings is an agricultural technology process.
Courtesy of Weyerhauser Corp.

DEVELOPMENT OF AGRICULTURAL TECHNOLOGY

The earliest people obtained most of their food by hunting and fishing. This method is called *food collection*. However, the need for surplus food eventually led to the development of **food production** or **agriculture**. After all, how would you like to have to hunt, kill, and clean an animal every time you want something to eat?

One of the most important early developments in agricultural technology was the **domestication of animals**. Domesticated animals could be used to clear away heavy trees, stumps, and boulders. They could be used to carry water, pull a plow, and transport harvested crops. They could also be raised to provide milk, meat, and offspring.

Another important early development was **irrigation**. Before people learned to irrigate, agriculture was limited to land paralleling riverbanks or areas that had plenty of rainfall. Irrigation opened up new land into farming that otherwise would have been waste land.

Early irrigation was done by carrying containers from a water source to the land to be irrigated. Even with domesticated animals, this

Some important developments in agricultural technology.

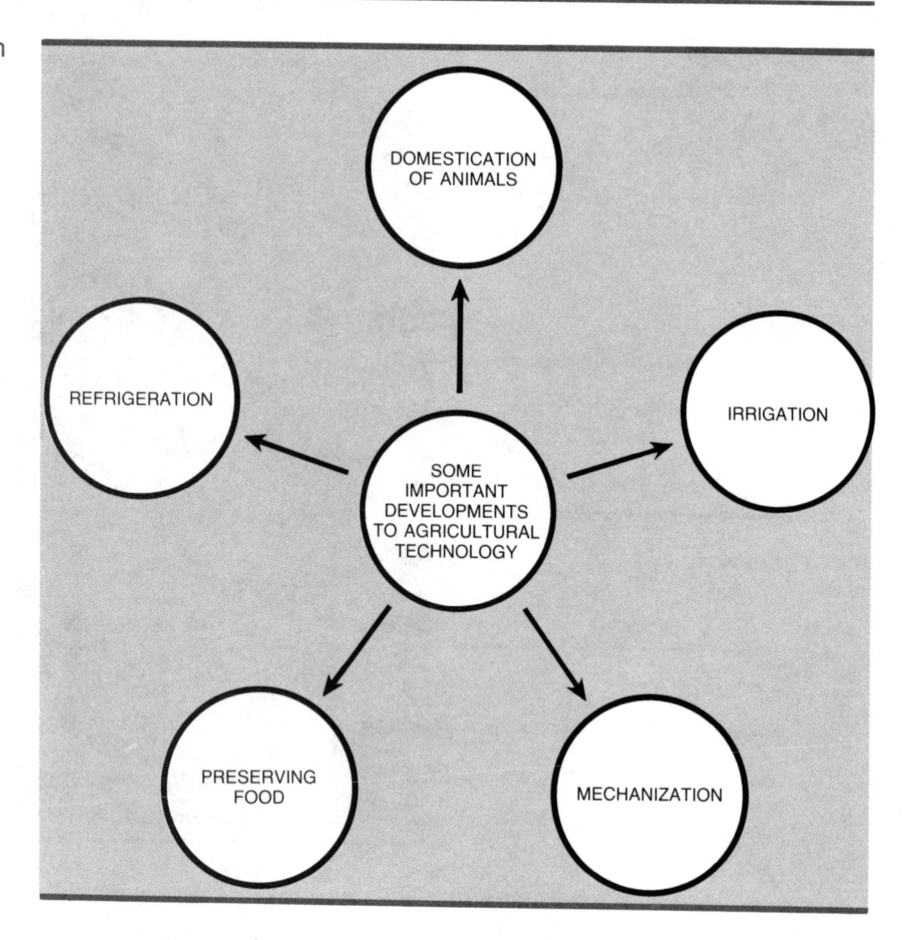

process was slow and ineffective. However, when windmills and waterwheels to pump water and iron pipe to carry it were invented, irrigation became an important agricultural process.

The mechanization of farming was another important development. Once the crude early tools of farming gave way to the metal plow, weed reaper, and seed reaper, agricultural production improved rapidly. A team of horses pulling one of Cyrus McCormick's weed reapers could harvest more wheat in an hour than several people could harvest in a day by hand. Such developments led to the mechanization and automation of certain farming processes.

Food preservation was another major step in the development of agricultural technology. In earliest times, food had to be consumed near the place where it was grown and soon after it was harvested, because there was no way to keep it from spoiling. It wasn't until about 1850 that people began to develop ways of preserving food.

Early preservation methods involved adding sugar or salt, **dehydration**, drying, and canning. Do you know anyone who does

A modern form of irrigation. *Photo by Deborah M. Goetsch.*

The tractor is one of the most important developments in agricultural technology. *Photo by Deborah M. Goetsch.*

Modern planters like this make planting a more productive process. *Photo by Deborah M. Goetsch.*

Cultivating the soil with mechanical disks like this has made planting and cultivating products much easier. *Photo by Deborah M. Goetsch.*

canning? It is still a widely used way of preserving food. Think of all the canned food products in a grocery store. A final development that was important to agricultural technology was **refrigeration**. People had learned how to extend the life of food by using ice years before refrigeration was invented.

It was only natural, then, that the use of ice would lead to the development of refrigeration, which is still a mainstay in agricultural technology.

PRODUCTS OF AGRICULTURAL TECHNOLOGY

There are hundreds of products of agricultural technology. A list would fill a whole book. For convenience, the products of agricultural technology can be placed in several categories: grains, sugar plants, root crops, legumes, tree crops, vine crops, truck crops, animals, cotton, and trees.

Grains

Although they are not usually eaten in their natural form, **grains** are among the most important agricultural products. Important food staples such as breads and cereals are made from grains.

The most widely produced grains are rice, wheat, corn, barley, oats, and rye. Wheat, barley, and rye are ground into flour and used for making bread and other baking products or cereal. Rice and corn are both eaten as vegetables and ground into flour for baking. You have probably eaten corn on the cob, creamed corn, canned corn, cornbread, and maybe even hush puppies made from cornmeal. But did you know that most corn that is grown is fed to cattle?

Do you like cereal? What is your favorite kind? Is it made from oats, wheat, barley, rice, rye, or corn? Usually the name will tell you. There are several different brands of corn flakes, oatmeal, shredded wheat, puffed rice, and so on. The next time you are in the grocery store, examine the different brands of cereal and identify the types of grains used to make each brand.

Sugar Plants

Sugar has developed a bad reputation in the past decade, because too much of it makes people gain weight. However, even though people should restrict their intake of it, **sugar** is still an important product of agricultural technology. In spite of many sugar substitutes that have been developed, hundreds of thousands of tons of sugar are still used each year in this country.

Sugar is used in all of your favorite baked goods, such as doughnuts, cakes, pies, and cookies. It is used in processing and

Products of agricultural technology.

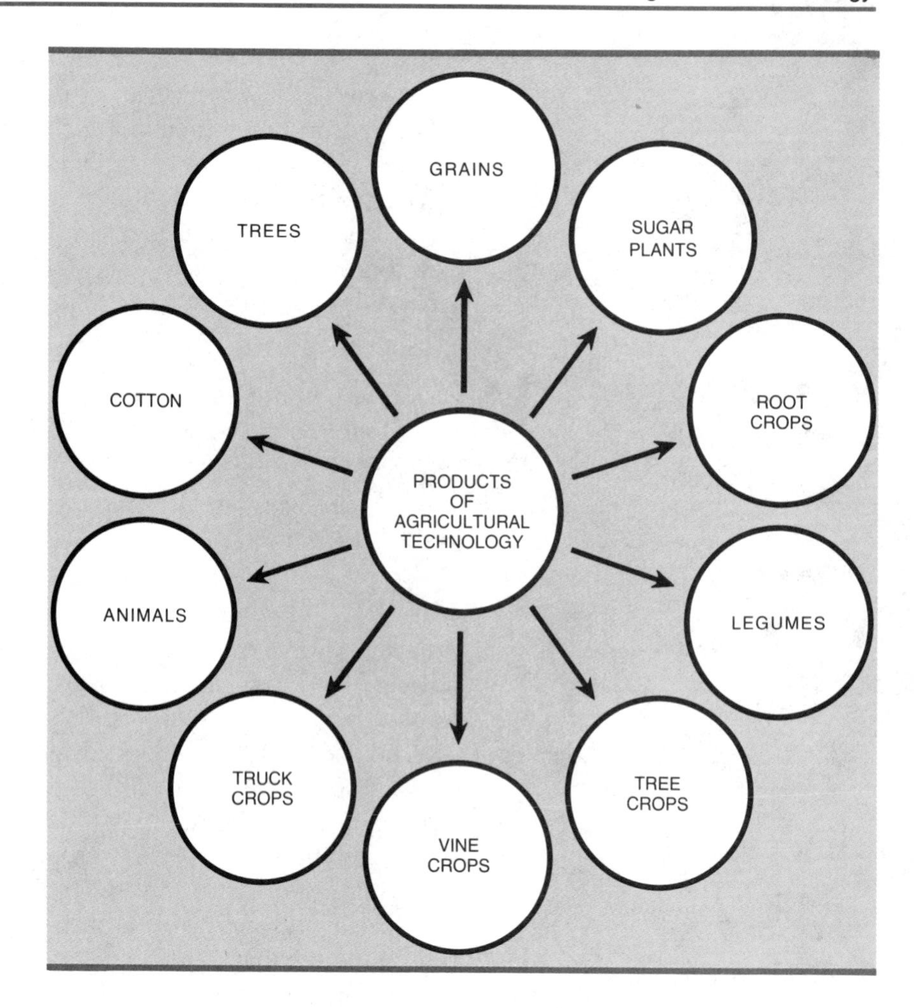

preserving canned and jarred foods. It is used in regular soft drinks and in alcoholic beverages.

The two sources of sugar are sugarcane and sugar beets. Sugarcane grows in stalks that look much like corn stalks without the ears of corn. Sugar beets are root plants. The beet grows underground and sprouts a leafy plant that grows above the ground. Sugarcane and sugar beets are squeezed to collect their sweet juice. This juice is then processed into various forms of sugar. The waste from the stalks and beets can be used as animal food. It is particularly good as cattle food.

Root Crops

Root crops are crops in which the root of the plant is the source of the food. You just learned that a sugar beet is a root crop. In some crops—turnips, for example—people eat both the root and the leafy part that grows above the ground. Carrots, potatoes, sweet potatoes, and onions are all root crops.

Legumes

Legumes are an important product of agricultural technology. They are also a useful tool for farmers. Peanuts, beans, soybeans, and peas are all legumes, and we all eat many of them. But legumes also serve another purpose. They help farmers restore the soil.

Most crops rob the soil of important nutrients needed to raise other crops. But legumes actually put an important nutrient back into the soil. That nutrient is nitrogen. **Nitrogen** acts as a natural fertilizer. Many crops cannot grow without it.

Legumes absorb nitrogen gas from the air through their leaves and combine it with bacteria that exist in their roots. The nitrogen then seeps back into the soil, restoring that which has been used by other crops harvested earlier.

It is common practice to alternate legume crops with other crops every other year. In this way, a farmer can grow a valuable legume crop while restoring the soil depleted by the previous year's crop of grain, sugar plants, or some other nonlegume.

Alfalfa is a legume crop planted to restore nitrogen to the soil.
Photo by Deborah M. Goetsch.

Tree Crops

Tree crops are among the most popular of agricultural technology products. Think of some of the many things that grow on trees: apples, oranges, grapefruit, peaches, pears, coconuts, pecans, walnuts, lemons, tangerines, and bananas are all tree crops.

The climate is important in raising tree crops. Apples require a cool, mild climate. Citrus fruits, such as oranges, grapefruit, and tangerines, require a warmer climate. Most of our citrus fruits are grown in Florida, California, and Arizona. Unexpected cold temperatures can ruin an entire citrus crop overnight.

Peaches require a warm, humid climate. Southeastern states like Georgia are well suited to growing peaches. Bananas need an even more tropical climate. Most bananas eaten in this country are grown in the Caribbean Basin in South American countries.

Vine Crops

The principal **vine crops** are grapes and some types of berries. Unless they are supported, vines will grow and creep along the ground. When they creep along the ground, the amount of fruit they will bear decreases substantially. Consequently, growing vine crops involves building structures and devices to support vines off the ground.

Grapes and berries have three principal uses: (1) They are packaged and sold to be consumed in their picked form as fruit. (2) They are used in making candy, jelly, and juice. (3) They are used in making wine.

Truck Crops

Truck crops are vegetables that were traditionally harvested and taken to market in trucks. Although they may now be shipped by trucks, ships, trains, or even airplanes, they are still called truck crops.

Common truck crops include tomatoes, lettuce, cabbage, squash, broccoli, watermelons, radishes, avocados, cucumbers, and strawberries. Because of advances in refrigeration, truck crops can be shipped long distances without spoiling. Strawberries grown in central Florida can be enjoyed by people in Alaska. Avocados grown in California can be shipped the day they are picked to restaurants in New York and New England.

Animals

Animals raised for food, eggs, milk, wool, and fur are also products of agricultural technology. Such animals include livestock, poultry, and even fish.

Livestock Animals grouped under the heading "livestock" include cattle, sheep, goats, and pigs. Beef cattle are raised for their meat. Dairy cattle are raised for their milk. Although they are not specifically raised for it, cattle also supply leather. In addition, hooves, hides, and horns are used to make glue.

Sheep are raised for their wool and their meat. Americans eat very little sheep meat, called mutton; we prefer beef instead. Wool is a valuable product in clothing production. For example, woolen sweaters, gloves, hats and mittens are popular in colder climates.

Goats are raised for their milk and their meat. Goats' milk is even more nutritious than cows' milk. In America, goats' milk is not as popular as cows' milk, but it is very popular in many Middle Eastern and African countries. In this country, doctors often prescribe goat milk for people who are allergic to cow milk.

Some cattle are raised for milk and dairy products. *Photo by Deborah M. Goetsch.*

Some cattle are raised for meat and by-products such as leather. *Photo by Deborah M. Goetsch.*

The hooves, horns, and hides of cattle can be used for making glue and other by-products. *Photo by Sue Knight Jennings.*

HIGH TECHNOLOGY DOWN ON THE FARM

At one time most Americans were farmers; now only a very small percentage of us make our living from agriculture. Modern farm machinery, new farming techniques and breakthroughs in biotechnology have all combined to make agriculture more efficient. (1) Large combines harvest large fields in a short amount of time. (2) Automatic milking machines are now used in many dairy farms. (3) Irrigation systems bring water to thirsty crops, and help keep productivity high even in dry regions of the world. (4) Farmers use computers to check the latest prices their products will bring, and to keep track of expenses.

SOLVING WASTE DISPOSAL PROBLEMS

Systems of technology can often produce undesirable outputs, such as pollution and garbage. Technology can also be used to deal with these problems. (1) A modern treatment plant insures that sewage can be kept from polluting rivers or lakes. (2) These discarded beverage cans will be recycled into usable aluminum; this process requires far less energy than extracting aluminum from bauxite ore, the main source of this important metal. (3) Garbage can be put to good use in plants such as this one, which generate energy by burning refuse. Smoke from the plant is carefully treated to prevent air pollution.

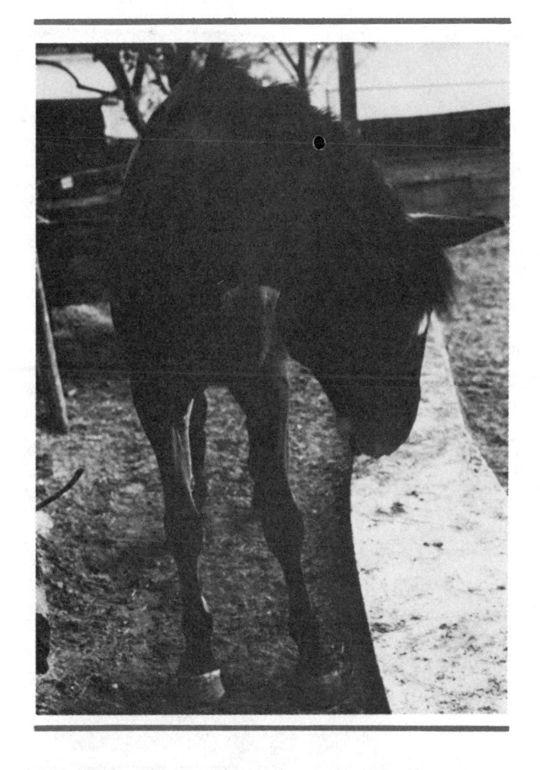

Raising horses is a big agricultural business. *Photo by Sue Knight Jennings.*

Chickens and other poultry animals are products of agricultural technology. *Photo by Sue Knight Jennings.*

Pigs are a valuable agricultural product. *Photo by Sue Knight Jennings.*

Pigs are a valuable animal. They are a good source of meat (bacon, ham, sausage, and pork chops) and they also help to eliminate garbage. Pigs will eat almost any food that humans throw away. Therefore, they give farms an opportunity to turn waste into a useful product.

Poultry Poultry animals are primarily chickens and turkeys. Chickens are raised for their meat and eggs. Turkeys are raised primarily for their meat. Chicken meat sells widely year-round. Turkeys sell best at holiday times, especially Thanksgiving and Christmas. However, in years to come, turkey is expected to become a more year-round product, as health-minded people continue to look for inexpensive alternatives to red meat (that is, beef).

Fish If you go deep-sea fishing and catch a fish to eat, that is food collection, not food production. However, fish can be a product of food production. Fish can be grown and harvested like any other agricultural product.

In fish farming, small fish are put in controlled environments such as ponds. In this way, they can eat, grow, and reproduce without the threat of predators. Fish are an inexpensive and nutritious source of food. When raised in a controlled environment, they grow and reproduce much faster than in a natural element.

Cotton

Cotton has a long, illustrious, and controversial history. It has been a product of agricultural technology for thousands of years. It has become strongly associated with the American Civil War because it was one of the principal crops of the Old South. As such, it was harvested or picked by slaves on those farms that owned slaves.

Most people associate cotton only with clothing. Cotton is important to the textile industry, but it is also a food product. Cotton seeds can be used to make margarine, cooking oil, and animal food.

Until the late 1700s, cotton was a labor-intensive product. This term means many people were needed to plant, harvest, and process cotton. This tended to limit the amount of farm land used to grow cotton, because the more people needed it, the more expensive it was to grow. For those farms that did not have free labor (slaves), cotton could cost more to grow than it would bring on the market when sold. However, just before the turn of the century, an important technological development began to change this situation.

At that time, Eli Whitney invented the cotton gin and revolutionized cotton processing. After cotton has been picked, the seeds must be separated from the cotton. Until the cotton gin was invented, this had to be done by hand, and it was a slow, difficult process. The cotton gin separated the seeds mechanically and, therefore, could do the job much faster. One cotton gin could do the work of hundreds of people. Cotton farming is now almost completely mechanized: planting seeds, irrigating fields, picking cotton, separating seeds, and baling cotton are all done by machines.

Trees

Most people don't think of growing **trees** as farming, but it is. The construction and communication industry use so many trees each year for lumber and paper that this precious natural resource would be depleted if it were not replaced. Tree farming replaces trees. It takes a long time for a tree to grow large enough to be used for lumber. For this reason, tree farming has become so important. Through advances in agricultural technology in such areas as fertilization, soil treatment, and crop management, farmers are able to make trees grow large much faster.

A typical tree farm will have sections containing trees in every stage of growth, from seedlings that have just been planted, to mature trees

Planting a tree is a form of agricultural technology. *Courtesy of Robert C. Atwood.*

that are ready to be cut. The main principle behind tree farming is to plant a new tree every time a mature tree is cut down, and, through technology, make the trees grow as large as possible as fast as possible.

PRODUCING AGRICULTURAL PRODUCTS

In the preceding section, you learned about the different groups of products produced through agricultural technology. In this section, you will learn about the main processes used to produce these products and some of the tools associated with each process. The main processes you will learn about are

1. Preparing the soil
2. Planting
3. Cultivating
4. Irrigating
5. Harvesting
6. Storing
7. Shipping
8. Processing
9. Breeding

Preparing the Soil

Preparing the soil means getting it ready for planting. If you want to plant a garden in your back yard, what do you have to do? You would

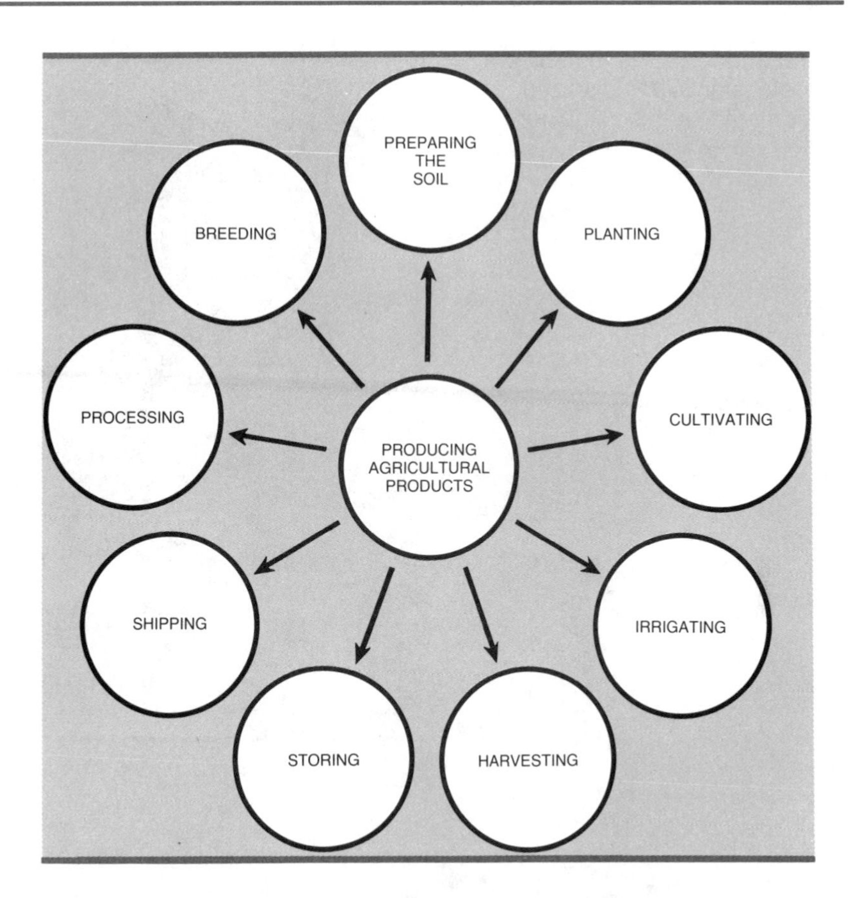

Producing agricultural products.

Tractors are an invaluable aid in preparing the soil. *Photo by Deborah M. Goetsch.*

Tilling and fertilizing the soil are made easier by mechanized equipment like this combine.
Photo by Deborah M. Goetsch.

mark off the area to be gardened. You would clear away the grass, roots, rocks, and other obstacles. You would chop the soil up to make sure it is soft, loose, and fine. Then you would add fertilizer. Now you are ready to plant.

The same tasks are performed on a larger scale on farms. Preparing the soil involves (1) clearing, (2) leveling, (3) tilling, and (4) fertilizing.

Clearing consists of removing trees, grass, underbrush, stumps, and rocks. Farmers use tractors and plows for clearing land. *Leveling* involves grading away high spots and filling in low spots on relatively flat land. On steep hills, it involves terracing. *Terracing* means cutting wide steps into the side of a hill to provide a series of flat, level surfaces for planting.

Tilling involves plowing the soil to break it up and loosen it. *Fertilizing* is done to replace soil nutrients that have been depleted.

Planting

Planting involves placing seeds or small plants in the ground. Some seeds are planted in rows, and others are cast out randomly. Crops

Special mechanized seed spreaders like this make planting crops easier. *Photo by Deborah M. Goetsch.*

such as corn, beans, onions, cotton, melons, and many others are planted in rows. Special seed drillers, which make a series of holes in the ground and then drop seeds into them, are used for row planting.

Crops that are scattered randomly include alfalfa and wheat. The seeds for these drops can be cast by hand or by spreaders. In larger-scale planting operations, seeds can even be scattered from airplanes or helicopters.

Cultivating

Cultivating crops means nurturing them so they will grow. Cultivation protects growing crops from weeds, insects, and other potential sources of damage. Weeds can be controlled in different ways. They can be pulled by hand, turned under by hoeing, or killed using chemicals. Chemicals especially made to kill weeds are called *herbicides*.

Insects are controlled by applying special chemicals called *insecticides*. Insecticides can be applied by special sprayers pulled by tractors. But in large-scale operations, they are often applied by crop dusters. *Crop dusters* are special airplanes that have been adapted

Cultivating products is made easier by modern tractors. *Photo by Deborah M. Goetsch.*

Mechanized disking equipment like this can be pulled by tractors to cultivate crops and turn the soil. *Photo by Deborah M. Goetsch.*

for spraying insecticides. Some are single-wing and some are double-wing (called biplanes). In either case, they are especially maneuverable. They are able to accelerate and climb very rapidly and turn sharply. They are filled with tanks, fuselage, and sprayers along the wings. Have you ever seen a crop duster spraying a field?

Weeds and insects are the major causes of crop damage, but they are not the only ones. Cultivating crops can also mean protecting them from the elements such as cold weather. Some crops will die if the weather becomes too cold. Citrus crops are a good example. If the temperature suddenly drops below freezing (32 degrees Fahrenheit or 0 degrees Celsius), citrus crops will die. Citrus farmers have developed special methods to save their crops when the temperature suddenly falls. If the forecast is for a mild freeze, farmers light *smudge pots* and place them under the trees. The flames from these smudge pots keep the temperature of the trees above freezing. If the forecast is for a hard freeze, the smudge pots cannot produce enough heat to keep the fruit from freezing. When this happens, citrus farmers use another cultivation technique. They sprinkle the trees with a continuous spray of water. The water freezes and encases each piece of fruit in a coat of ice. This sounds like the last thing you would want to do, but it works. Fruit is only in danger if it drops below freezing. When a piece of fruit is wrapped in ice, its temperature stabilizes at 32 degrees, or exactly freezing. The sprayers used for this job are the same ones used to irrigate the fields in warmer weather.

Irrigating

Do you ever water the grass or plants at your house? If so, you are irrigating them. **Irrigation** involves providing the crops with water that is not provided by rain. Some areas receive plenty of rainfall and don't need to irrigate. But in dryer areas, farming cannot take place without irrigation.

Irrigation involves pumping water from a source, such as a well, a spring, or a river, to where it is needed. Then it is applied to the crops. It is carried from the source by pipes. At this point, it can be sprinkled on the crops or dripped on the individual plants. Another irrigation method involves simply flooding the field and letting the water run between the rows of crops, sinking in as it runs. Irrigation helps the crops mature and readies them for harvesting.

Harvesting

Harvesting is collecting the mature crops. Harvesting can range from hand picking to driving large combines. The method used depends on the type of crops and the size of the field being harvested.

A modern form of irrigation. *Photo by Deborah M. Goetsch.*

Large modern combines make harvesting much easier. *Photo by Deborah M. Goetsch.*

Grain crops and cotton are harvested mechanically by large machines called *combines*. A combine combines a reaper and a thresher into one large unit that is like a tractor. The reaper mows down the wheat, the thresher separates the wheet seeds from the straw and chaff. Some combines even package the seeds and bundle the straw. Combines and other harvesting machines and equipment are so expensive that many farmers can't buy them, so they must rent them during the harvesting season. Others contract with harvesting companies. These companies own the equipment and travel from farm to farm for a price.

Most other crops, such as truck crops and tree crops, are harvested by hand. A lot of people are needed. Farmers that grow crops that must be harvested by hand depend on migrant workers. Migrant workers are people who travel around the country harvesting crops for a living. Different types of crops are ready for harvesting at different times of the year. Consequently, a family of migrant workers may harvest citrus fruit in South Florida one month and peanuts in Georgia the next. They might harvest truck crops in Southern

The harvesting of grain crops is made much easier by mechanized equipment like this.
Photo by Deborah M. Goetsch.

California one month and tree crops in Oregon the next. After the crops are harvested, they must be stored.

Storing

Storing of food crops is necessary because of a need for surplus. The primary purpose of food production is to have surplus. A surplus means more than you need at the time. Surplus crops must be stored until they are needed.

Different types of crops must be stored in different ways. Grain

Once grain has been harvested, it must be stored until it is time for it to be processed. *Photo by Deborah M. Goetsch.*

crops are stored in grain elevators, which are large structures specially designed for bringing crops in and shipping them out.

Dairy products and meat must be refrigerated. Therefore, large refrigerator and freezer facilities are needed. Crops such as cotton are packaged in bales and stored in barns and warehouses. When they are needed, the crops are taken from storage and shipped either to market or to processing plants.

Shipping

Shipping of crops is a big business. Harvested crops must be transported to storage. Stored crops must be transported to processing plants. Processed crops must be transported to market. Transporting crops is called shipping.

Crops can be shipped by land, air, and water, but most are shipped by land on trucks and trains. There are special refrigeration trailers for trucks and refrigeration cars for trains. There are special compartments in some ships, barges, and train cars for shipping grain crops. Have you ever seen a tank truck that ships milk?

Trucking is one of the principal ways of shipping agricultural products. *Courtesy of Fruehauf Division of Fruehauf Corporation.*

There are many different types of trucks used to ship agricultural products. *Courtesy of Peterbuilt, Inc.*

Aluminum refrigerated van trailers make shipping of products cross country possible. *Courtesy of Fruehauf Division of Fruehauf Corporation.*

Lightweight pneumatic trailers make shipping of milk and other liquid products cross country possible. *Courtesy of Fruehauf Division of Fruehauf Corporation.*

Numerous types of aircraft are used for shipping crops that must get from one location to another quickly. The military often uses aircraft for shipping food.

Processing

Food **processing** is one of the most important of all agricultural technology processes. Few crops produced through agricultural technology are sold in their natural state. Grains become bread and cereal products. Sugar cane and sugar beets become granulated sugar. Raw milk becomes whole milk, skim milk, condensed milk, cheese, butter, cottage cheese, and other products.

The conversion of harvested crops into food that is ready to consume is called processing. Processing is not just one process. It is a number of different processes that vary according to the crop being processed. Processing grain involves grinding and packaging. Process-

Grain crops must be processed before they can be used. *Photo by Deborah M. Goetsch.*

ing can also mean canning, flash freezing, boxing, cooking, and adding preservatives.

Breeding

Breeding is a good example of agricultural technology at work. Through breeding, agricultural technology has produced better strains of crops and animals. Breeding is done in a variety of ways.

One of the simplest ways of breeding better crops is to use the seeds of the biggest, healthiest plants of one crop to plant the next crop. The same principle applies to animals. By breeding the biggest and best of the herd, bigger cattle can be produced.

One strain of cattle has certain good characteristics, and another strain has other good characteristics. Through breeding, they can produce offspring with the good characteristics of both. Through breeding, agricultural technology has produced bigger, higher-quality strains of many different crops and animals.

Grain elevators store various grain products until it is time for them to be processed. *Photo by Deborah M. Goetsch.*

Breeding of cattle and other farm animals is an important business in agricultural technology. *Photo by Deborah M. Goetsch.*

SOME CAREERS IN AGRICULTURAL TECHNOLOGY

Agricultural technology has become increasingly mechanized and automated, so fewer people are needed. Today, only 3% of the work force is involved in agricultural technology. Most careers associated with agriculture fall into one of four groups: professionals, skilled workers, operatives, and unskilled workers.

Professional Careers

Professional-level careers in agricultural technology include scientist, chemist, and top-level managers. Scientists and chemists are involved in research to produce better fertilizers, better food preservatives, soil additives, crops, animals, and means of growing all types of crops faster and better. Scientists and chemists must have advanced college degrees.

Managers run the large corporate farms, research companies, and food processing firms associated with agricultural technology. They have such titles as president, vice president, administrator, and manager. Most top-level managers have college degrees.

Skilled Workers

Skilled workers is the largest category of agricultural technology careers. Farmers are skilled workers. They must be skilled at all of the various tasks required to prepare the soil, plant, cultivate, irrigate, harvest, store, and breed crops. Farm equipment mechanics and butchers are skilled workers. Farm equipment mechanics maintain the different types of equipment and machines used on the modern farm. They might be called on to repair a milking machine, a large combine, or a diesel tractor, for example.

Butchers are an important part of meat processing. They transform slaughtered cattle and pigs into steaks, hams, roasts, sausage, bacon, hamburger, and other cuts of meat and meat products.

Farmers can get their training on the job, or in formal training programs offered in high schools, area vocational centers, and community colleges. Farm equipment mechanics and butchers must have specialized training that can be received from an area vocational center or community college.

Operatives

Operatives are the people who operate the equipment, tractors, and trucks used in preparing the soil, in planting, cultivating, harvesting, storing and shipping crops. Operatives drive combines, forklifts, tractors, and trucks of different sizes. Operatives can receive their training from area vocational centers or on the job.

Unskilled Workers

Unskilled workers in agricultural technology are principally the people who harvest the crops that must be harvested by hand. Most harvesting personnel are migrant workers. Migrant workers learn on the job the very basic skills required of them. The only skill associated with harvesting crops is learning to harvest them rapidly. In some cases, migrant workers are paid according to how much they pick.

SUMMARY

Agricultural technology is the way people produce most of the food they need. Some important early developments in agricultural technology were the domestication of animals, irrigation, mechanization, preservation of food, and refrigeration. The products of agricultural technology fall into several categories: grains, sugar plants, root crops, legumes, tree crops, vine crops, truck crops, animals, cotton, and trees.

There are nine processes used in producing the products of agricultural technology: preparing the soil, planting, cultivating, irrigating, harvesting, storing, shipping, processing, and breeding.

Some of the most common careers associated with agricultural technology are scientists, chemists, managers, farmers, farm equipment mechanics, butchers, equipment operators, truck drivers, and harvesters.

KEY WORDS

Agriculture
Domestication of animals
Irrigation
Food production
Food preservation
Dehydration
Refrigeration
Grains
Sugar
Root crops
Legumes
Tree crops
Vine crops
Truck crops
Animals
Cotton
Trees
Nitrogen
Preparing the soil
Planting
Cultivating
Irrigation
Harvesting
Storing
Shipping
Processing
Breeding

SOME QUESTIONS TO TRY

1. Name three important early developments in agricultural technology.
2. Name three grain crops.
3. Name two sugar crops.
4. Name two root crops.
5. Name two legume crops.
6. Name two tree crops.
7. Name a vine crop.
8. Name two truck crops.
9. Name two types of livestock.
10. What animals are considered poultry?
11. Explain cultivating.

TECHNOLOGY LEARNING ACTIVITIES

1. As a school project or as a home project, grow a truck crop, root crop, or vine crop. Perform all of the steps from planting through harvesting and keep records of what was done in each step.
2. Go to a grocery store and make an appointment with the manager of the vegetable department. Find out where each of the various types of vegetables sold comes from.
3. Do one of the following activities.
   a. Obtain some unprocessed cotton and separate the seeds by hand.
   b. Obtain some wheat and separate the seeds from the chaff and the straw.
4. Take a field trip to a farm and witness all of its operations.
5. Take a field trip to a processing plant and watch the various steps in processing food.
6. Bring in a processed, packaged food product (can of beans, loaf of bread, box of cereal) and trace its development for the class from planting to its current state.
7. Make a timeline of important developments in agriculture.

CHAPTER 12

HEALTH CARE AND MEDICAL TECHNOLOGY

OBJECTIVES

After completing this chapter, you should be able to

- define health care technology.
- explain the various ways that health can be measured.
- list the most common sources of health problems.
- explain how good health is maintained.
- define medical technology.
- list the most common sources of medical care.
- list some of the careers associated with health care and medical technology.

Some of the greatest advances in history have been in the areas of health care and medical technology. The tools, resources, and processes of these two overlapping technology areas have changed considerably over the years, and now these areas represent some of the most sophisticated technology in the world.

Computers are an important part of modern medical technology.
Photo by Deborah M. Goetsch.

Microelectronic diagnostic equipment is common in most modern health care facilities.
Photo by Deborah M. Goetsch.

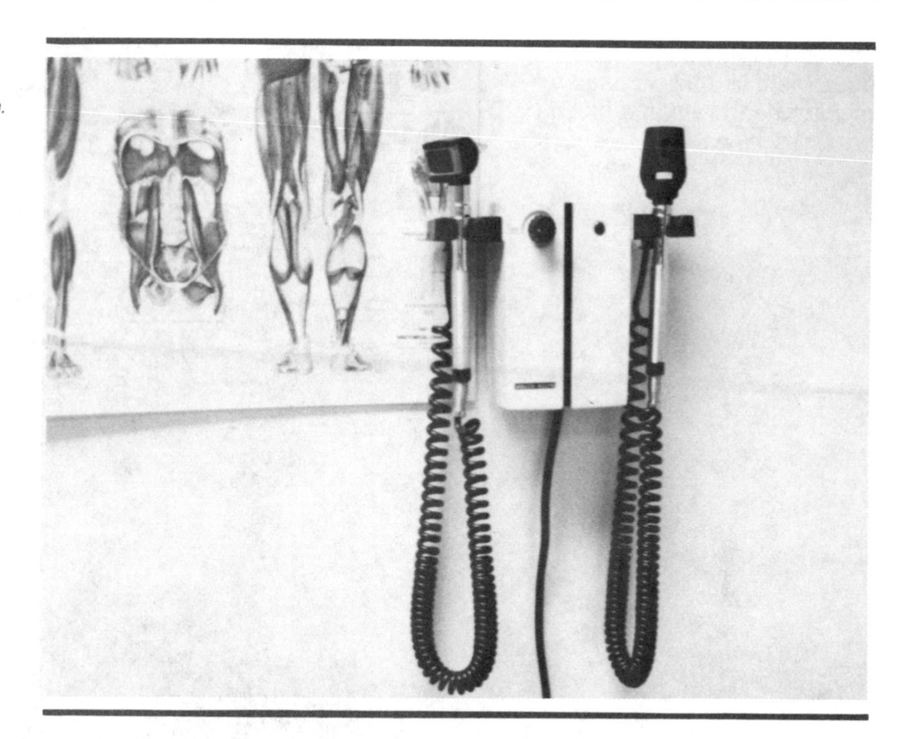

Even the most basic of health care and medical tools are now electronic. *Photo by Deborah M. Goetsch.*

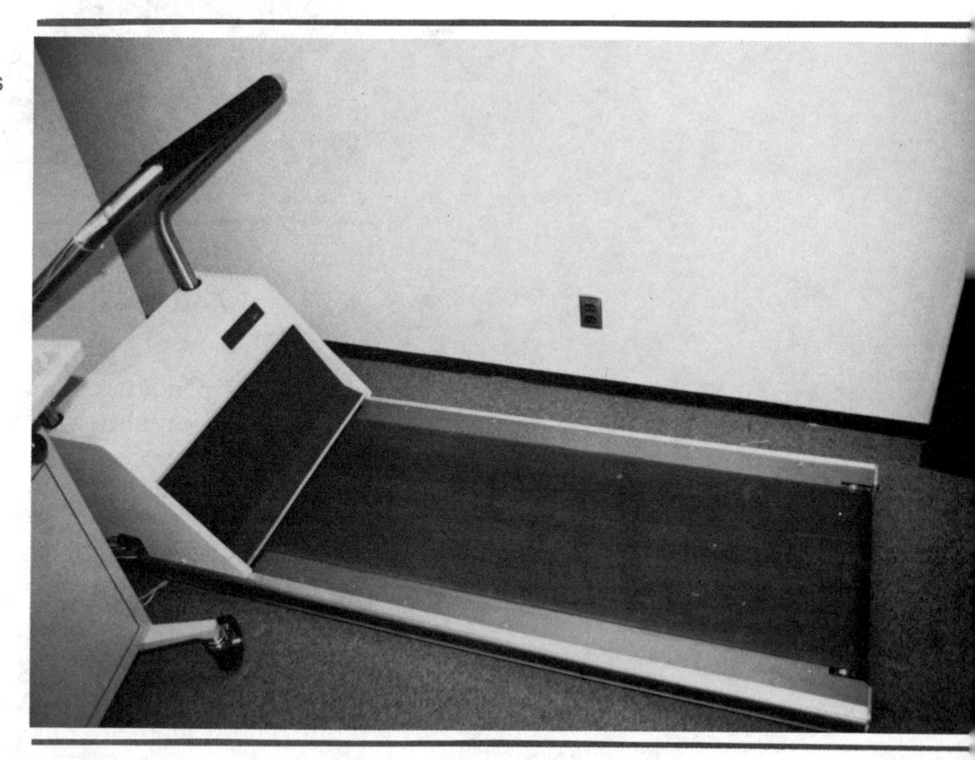

Treadmills are used for diagnosing and measuring the health of people under conditions of physical stress. *Photo by Deborah M. Goetsch.*

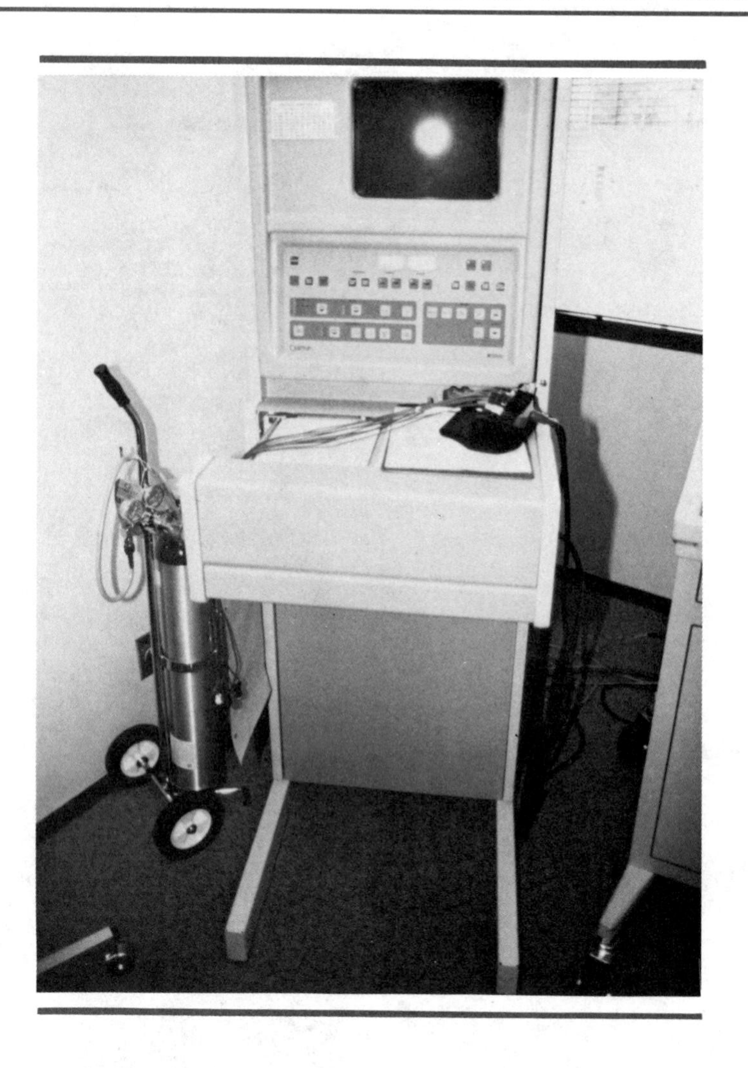

Microelectronic biomedical equipment like this is used to monitor heart rate and blood pressure. *Photo by Deborah M. Goetsch.*

WHAT IS HEALTH CARE TECHNOLOGY?

Health is a state or condition of the body or mind. **Physical health** concerns the body. **Mental health** concerns the mind. Often the state of one can affect the state of the other. A body or mind in good condition is healthy. A body or mind in poor condition is unhealthy.

Sickness or **illness** are states of being in which a certain amount of health has been lost. The loss could be temporary or curable, or it could be permanent, progressive, or terminal. A *terminal* illness is one in which the sick person is not expected to get well.

Health care technology is technology that is concerned with helping people maintain a healthy condition. Just as there are different degrees of sickness, there are different degrees of health. For exam-

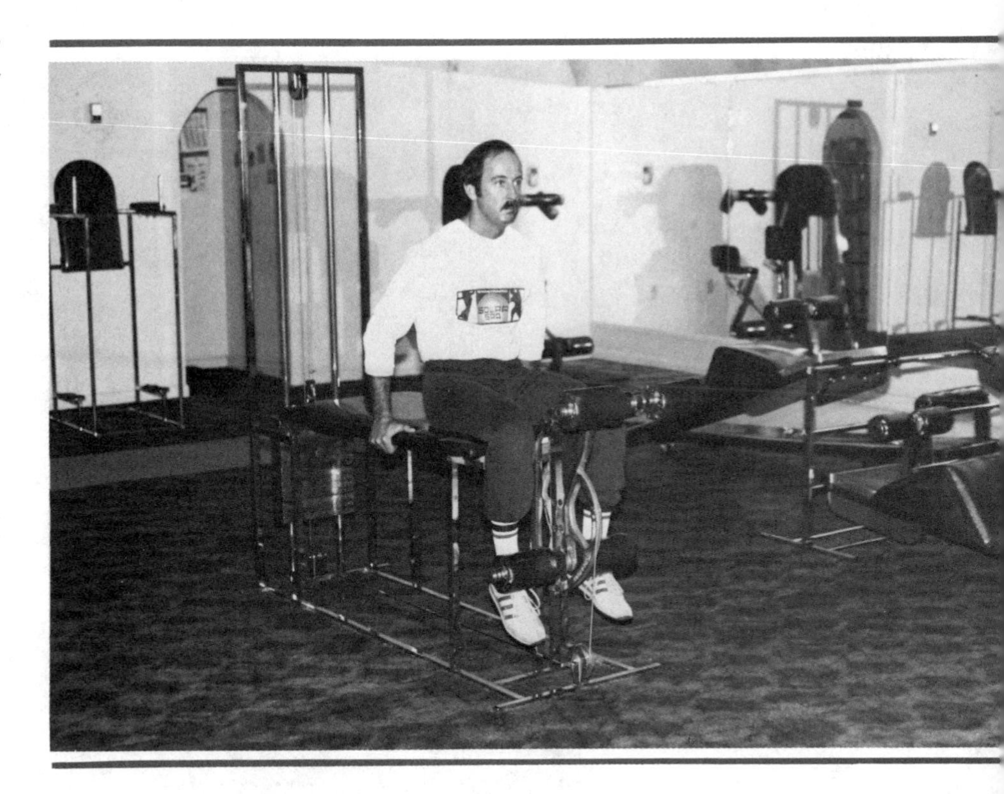

Athletes must work hard to stay in top physical condition. *Photo by Deborah M. Goetsch.*

ple, a well-conditioned athlete would normally be healthier than a healthy nonathlete.

On the other hand, an athlete might not be healthier than anyone else. It depends on several factors. These factors can be measured. Part of health care is knowing what these factors are, how to measure them, how to monitor them, and how to change them for the better.

MEASURING HEALTH

The human body is like the weather. It has certain things about it that can be measured to tell how it is doing. With the weather, we can measure the temperature, wind chill factor, wind strength, humidity, and barometric pressure. These things tell us if we are going to have good weather or bad weather.

With the human body, we measure temperature, blood pressure, pulse rate, respiration, and other things such as the fat-to-lean ratio of the body and blood composition. These measurements tell us how we are doing healthwise.

What is considered normal for each of these measurements varies from person to person. However, health care professionals have established acceptable ranges for people of different age groups for

Personal profile chart for recording a person's measurable health factors.

PERSONAL PROFILE

NAME ______________________ AGE ______

SEX ____________ SSN ____________

MEASUREMENT	ACTUAL	RANGE
PULSE RATE		
TEMPERATURE		
BLOOD PRESSURE		
RESPIRATION		
FAT-TO-LEAN RATIO		
BLOOD COMPOSITION		
a. CHOLESTEROL		
b. TRIGLYCERIDES		
c. SODIUM		
d. PROTEIN		
e. CALCIUM		
f. IRON		

each of these factors. If a person's blood pressure or pulse rate falls outside of the established range for his or her age group, it could mean that he or she is having a health problem or is going to have one. You should be familiar with the body's **vital signs**: temperature, pulse, blood pressure, and respiration. A vital sign is one of the first things to check to see if a person is healthy.

Temperature

Although there can be slight variations, the normal temperature for humans is considered to be 98.6 degrees Fahrenheit or 37 degrees Celsius. When a person gets sick, the temperature may rise. This happens because the body is working to fight the illness. This work

Measures of health.

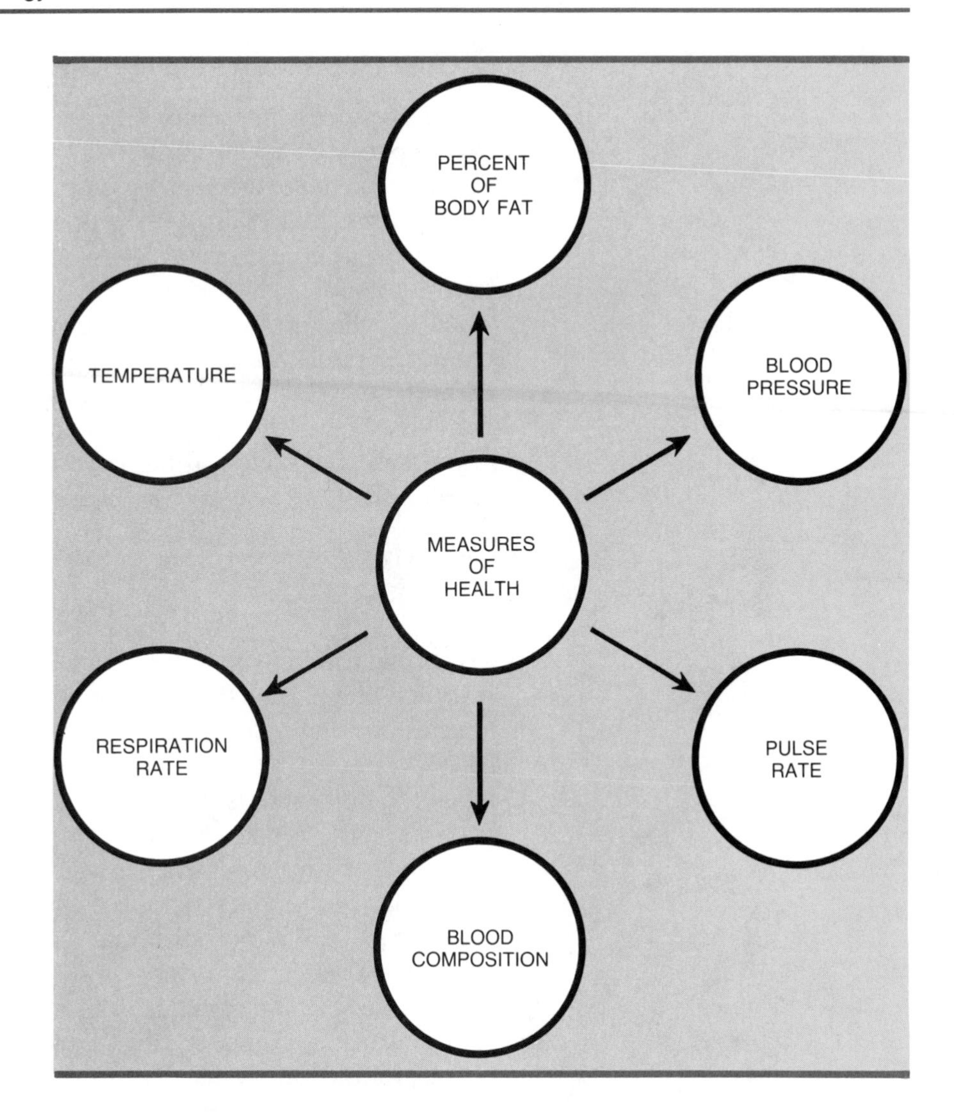

can cause the body to overheat. If a person's temperature rises, it is an indication that something might be wrong.

A person's **temperature** is checked with a **thermometer**. A thermometer is a long, thin, mercury-filled glass tube with a scale on one side. Higher temperatures cause the mercury to rise in the glass tube. The temperature determines how high the mercury rises. The temperature is read on the scale printed on the side. Newer thermometers are electronic with digital readouts. They give direct numerical readouts of the temperature. Have you ever seen a digital thermometer? Have you ever read the scale on a regular thermometer?

Blood Pressure

Blood pressure is important. High blood pressure can lead to strokes. **Blood pressure** is the amount of pressure exerted by the blood as it travels through the arteries. Blood pressure is measured by wrapping a blood pressure cuff around the upper arm and pumping it full of air. A gauge attached to the cuff measures the pressure at the highest and lowest points.

The highest point is when the heart has just beat and a rush of blood is surging through. The lowest point is right after that surge and before the next one. The blood pressure is recorded with two numbers, such as 120 over 70. The top number is the highest amount of pressure: the bottom is the lowest. Do you know what your blood pressure is? Have you ever had your blood pressure checked?

Pulse Rate

The **pulse rate** tells how many times per minute the heart beats. The heart is a pump. Each time it beats, it sends blood surging through the arteries carrying badly needed **oxygen** to all parts of the body. What a person's pulse rate should be depends on several factors, including when the pulse rate is taken.

If the pulse is taken when a person is sitting down and resting, it is normally between 60 and 72 beats per minute. A slightly higher or lower pulse rate does not necessarily mean something is wrong. However, a very rapid pulse rate while resting could indicate that a problem exists.

While engaged in physical activity such as walking, running, bike riding, dancing, skiing, skating, or any other strenuous activity, it is normal for the pulse rate to be over 150 beats per minute. Health care professionals are interested in how fast the pulse climbs as well as how rapidly it returns to normal after an activity. As a general rule, a strong, healthy heart beats fewer times per minute than a weak heart. It takes longer to elevate the pulse rate of a strong heart and less time for it to return to normal after physical activity. Well-conditioned athletes, such as those who run marathons, may have resting pulse rates as low as 30 to 40 beats per minute.

Do you know how to check your pulse? One way is to press the index and middle fingers of one hand against the inside of the wrist near the thumb. Press gently and wait. You should be able to feel a pulse. To save time, count the number of beats you feel in 6 seconds and multiply by 10, or count the number of beats you feel in 10 seconds and multiply by 6. What is your pulse rate?

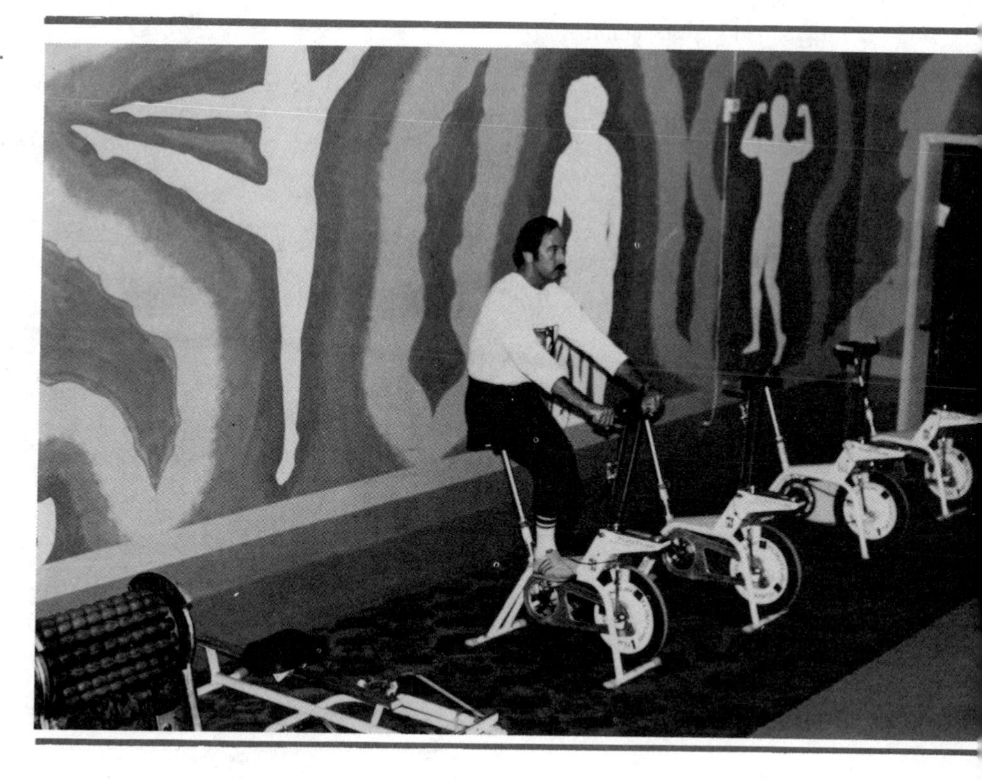

During strenuous physical activity, the pulse rate increases.
Photo by Deborah M. Goetsch.

Respiration

Respiration is the rate of breathing. It is measured by how many breaths you take per minute. Have you ever noticed that you breathe faster when you run or play hard? The harder your body works, the more oxygen it needs. The more oxygen you need to take in, the faster you have to breathe. Just like a car that is going faster or climbing a hill needs more gas, a body that is working harder needs more oxygen.

Health care professionals or technicians measure the respiration rate by listening to the lungs through a **stethoscope**. Have you ever seen a stethoscope? It is that device that sometimes hangs around the doctor's neck. He puts the forked ends into his ears and presses the disk at the other end against a person's chest or back when giving a physical examination.

Other Health Factors That Can Be Measured

Body Fat Health care professionals also check such things as the percent of the body that is fat and the composition of the blood. A

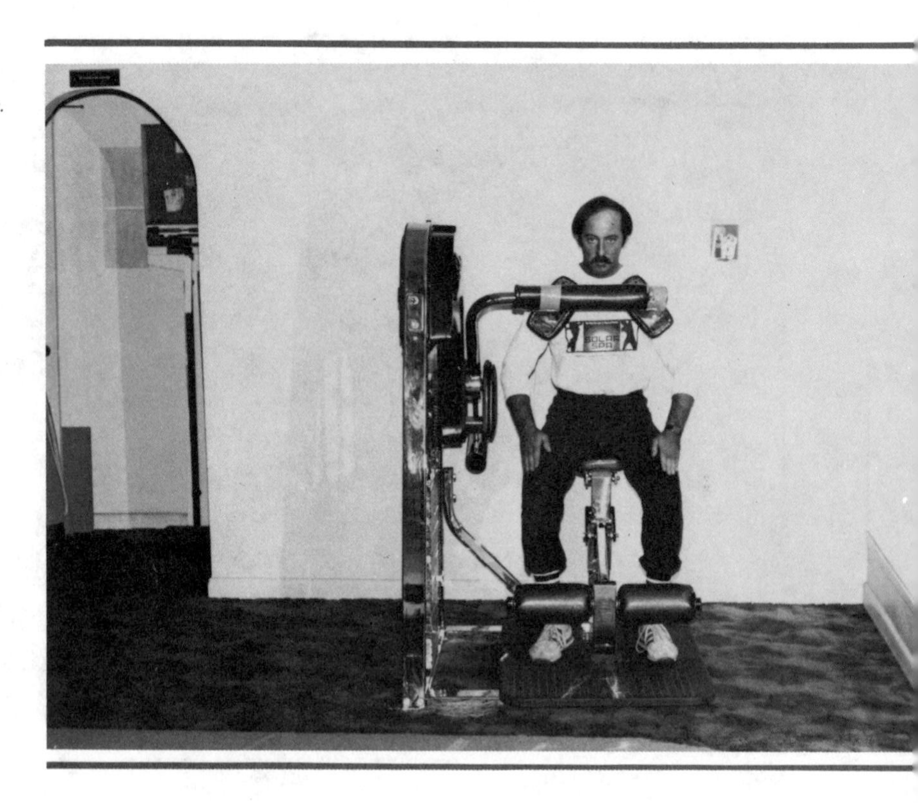

During strenuous physical activity, the respiration rate increases. *Photo by Deborah M. Goetsch.*

healthy person will tend to have a lower percentage of body fat. For men, 15 to 18% body fat is considered normal, depending on the age of the individual. However, less than 15% is better. For women, 18 to 22% is usually considered normal. A higher than normal fat content can lead to heart and other health problems.

Many people keep a close watch on their weight. Doctors have charts that tell people how much they should weigh according to their height. You should be careful of such charts, though. They can be deceptive and inaccurate. What a person weighs is not as important as how much of his or her total weight is fat.

You cannot measure your own body fat. A **fat-to-lean ratio** requires special knowledge and equipment. However, it is a good idea to have it measured periodically. This is especially true for people over 30 years of age. Fat content of the body can be checked in different ways. One way is to immerse a person in water and record his or her weight while the body is submerged. Then the weight of the person while submerged in water can be compared to the actual weight. The submerged weight and the actual weight can give an indication of the percent of body fat.

Body fat can also be checked by using calipers to measure the thickness of the layers of fat on the body at different critical points. Then by using the measurement in special formulas, doctors can calculate the fat content. More recently, computers and electronic equipment have been developed that can do the job easily. By placing electrodes at different places on the body, computers can send electric impulses into the body, measure the resistance of the body to the passage of those impulses, and thereby measure the amount of body fat.

Blood Composition **Blood composition** is also an important indicator of health. By analyzing your blood, health care professionals can learn a great deal about your health. Such an analysis will tell if you are getting the right amount of important nutrients such as **iron**, **calcium**, and protein. It will also tell if you are getting too much of potentially dangerous things such as sodium, cholesterol, or triglycerides.

Sodium comes from salt. Too much of it can be dangerous, especially to people with high blood pressure. **Cholesterol** is animal fat. Too much of the wrong kind of cholesterol can lead to clogged arteries and health problems. **Triglycerides** come from eating sweets. Too many triglycerides can lead to clogged arteries and heart problems.

SOURCES OF HEALTH PROBLEMS

You now know some ways to measure the level of health of the human body. Another important part of health care is knowing some of the causes of health problems or poor health. These things cause the measurements discussed in the previous section to be bad.

Two of the best ways to stay healthy are to avoid, as best you can, those things that could lead to health problems and to prepare or treat those that cannot be avoided. Most sources of health problems fall into the following categories:

1. Diseases
2. Accidents
3. Technological problems
4. Psychological problems
5. Genetic problems

Diseases

Certain **germs**, **bacteria**, **parasites**, and other organisms can cause sickness in people. All such organisms are very small. They can only be

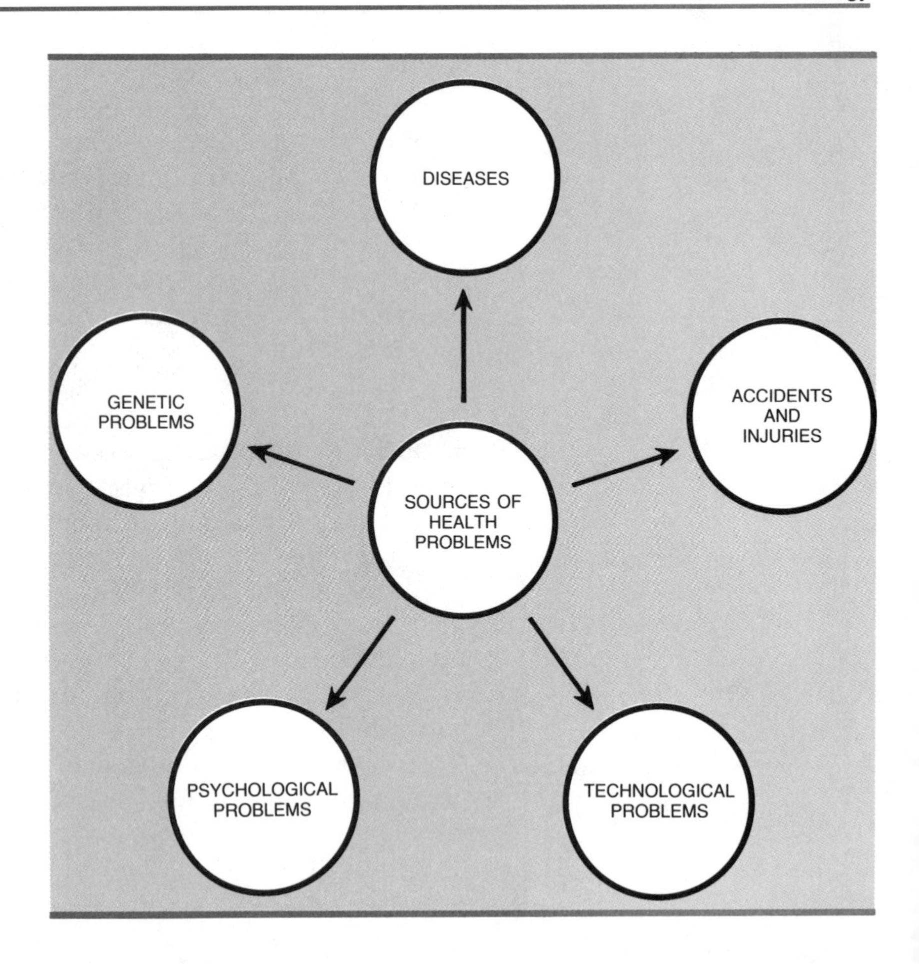

Sources of health problems.

seen under a microscope. Consequently, they are called **microscopic organisms**. There are many different types of **diseases**; some are serious and some are not. Measles, mumps, chicken pox, diphtheria, whooping cough, rubella, scarlet fever, cancer, and even the common cold are all diseases caused by a variety of germs.

People avoid germs and the diseases they cause in different ways. *Prevention* is the most important method for avoiding diseases. Prevention involves practicing good personal hygiene, good community sanitation, and avoiding contact with people who have a communicable disease. A **communicable disease** is one that can be passed along from one person to another. Not all diseases are communicable.

Immunization is another method for avoiding disease. **Immunization** makes the body immune (that is, gives the body resistance) to the germs that cause certain diseases. It is usually done by introducing a small number of the germs into the body to give it a chance to build defenses against the germ. If the number is small enough, the body

SPACE STATIONS

(1) This concept picture shows a large space station in orbit around the earth. Such stations will serve as "jumping off" points for longer voyages into space. Factories and laboratories will also be built into them, and will take advantage of the weightless conditions present in space. Space shuttles will be used to carry supplies and people from the earth to the space station.

MINES ON THE MOON

Someday in the not too distant future space explorers will begin to extract usable minerals and chemicals from the crust of the moon. (1) This concept illustration shows a space vehicle ready to take off, bound for a distant space station with a cargo of metals and other minerals. Living quarters for the miners are in the structures to the right. The earth is seen rising above the horizon. (2) Large shovels dig away on the lunar surface. In the yellow tanks to the right material is heated to release hydrogen, oxygen, and water. Other valuable compounds will also be taken from the ore.

Even the simple act of applying a Band-Aid is a form of health care. *Courtesy of Robert C. Atwood.*

will be able to overcome the germs. In the process, it builds up its defenses against the germ. If the germ later gets into the body, it can do no harm. Immunizations are usually given by injection (needles). You probably had to be immunized (sometimes called vaccinated) before you entered the first grade.

When prevention and immunization are not practiced, or when they don't work, medical treatment is the only other option. Medical treatment is discussed later.

Accidents

Injuries that result from **accidents** are another cause of health problems. It is easy to have an accident that causes an injury. You have probably had plenty of them already. Broken bones, scrapes, cuts, and bruises are all a part of growing up. Have you ever broken a bone or gotten a cut that required stitches?

Not all injuries are the result of accidents. Every year people are purposely injured in robberies, rapes, murders, and suicides. Many people are also injured in transportation-related accidents. Such problems are discussed in the next section.

The most effective way to avoid accidents and injuries is to be careful. Think before you act. Every year several people receive serious neck and spinal cord injuries by diving into shallow water. If they had looked before they leaped, they could have prevented their injuries.

All vehicles must be driven safely to avoid dangerous accidents. Superglide FXR and Lowrider Sport Edition FXRS. *Courtesy of Harley Davidson.*

Super Glide® FXR

Low Rider™ Sport Edition FXRS

Professional truck drivers are noted for being safe drivers and obeying the rules of the road. All drivers should follow their good example. *Courtesy of Peterbuilt.*

Technological Problems

As a rule, technology has been good for people, but it does have another side. The best example of **technological problems** is the automobile. It has been one of the most important technological developments in history. Yet over 50,000 people are killed in automobile accidents each year. Almost 2 million are injured in automobile accidents.

Airplane accidents happen less frequently than automobile accidents. Still, people are killed or injured each year in aircraft disasters. Even NASA's space program, which many consider the technological marvel of the century, has been involved in accidental deaths. On January 29, 1986, seven astronauts were killed when NASA's space shuttle exploded shortly after taking off from the Cape Canaveral, Florida, launch pad. Nineteen years earlier, three astronauts had been killed in a launch pad explosion.

In spite of rigorous testing, new medicines and drugs are occasionally released that turn out to be dangerous. An infamous example is the drug thalidomide. This drug was given to pregnant women in the early 1960s as a sedative. Later it was learned that it caused severe birth defects in over 10,000 children worldwide.

Another example is nuclear energy. Nuclear energy was developed as an alternative to fossil fuels, such as coal and oil, for producing electricity. It works well and is widely used. However, radioactive waste from nuclear generating plants now poses a serious health hazard. Leaks at nuclear plants have also threatened the lives of people who live near them.

Chemicals have been important products of technology. They have improved the quality of life for people all over the world. However, they can also be dangerous. Derailed train cars and overturned trucks carrying chemicals have caused a number of deaths. Potential leaks at chemical-producing plants pose a serious health problem. In 1985, a leak in a chemical plant silently but quickly killed thousands of people as it spread through a small town in India.

Hundreds of buildings, homes, ships, and airplanes are currently being inspected to determine if they contain asbestos. At one time, asbestos was popular material for insulation and fire protection. However, in recent years it has been connected with lung cancer.

Other health hazards resulting from technology include exhaust fumes from automobiles and industrial smokestacks, cigarette and cigar smoke, and certain sugar substitutes. Technological advances in the areas of agricultural, manufacturing, communication, construction, and health care technology have clearly improved the quality of life for people, but the same advances also have the potential for unplanned dangerous side effects.

Psychological Problems

Psychological problems can also lead to health problems, both mental and physical. Family, school, work, and personal problems can lead people to seek relief through alcohol and drugs. When abused, these substances can lead to a **chemical dependency** that can endanger a person's physical and mental well-being.

The pressure of life can lead to health problems, even without drug involvement. Prolonged worrying and stress can lead to internal disorders such as ulcers, and cause a person's blood pressure and pulse rate to rise dangerously. Probably the most common social problem affecting the health of people is poor eating habits. Poor eating habits can lead to a variety of health problems. The most common offenses are eating too much and eating the wrong types of food.

In many cases, obesity and heart problems are the result of eating too much of the wrong types of food for a long time. Every person needs the right number of calories every day. When a person takes in more calories than needed, the calories are stored as fat. Too much fat can lead to serious health problems, such as heart attacks and strokes. Foods that are high in cholesterol and sugar pose the greatest threat.

Other health problems can be worsened by a poor diet. Eating properly gives the body the right amounts of vitamins, minerals, proteins, carbohydrates, and other nutrients it needs to stay healthy. When people fill up on junk food, they take in calories but no nourishment. Proper eating means eating the right foods in the right amounts.

Genetic Problems

Genetic problems are inborn. This means that the problem exists when a baby is born and that it was passed along through the genes of the mother or father or both. Genetic problems usually result in birth defects that can be physical or mental or both.

Modern advances in medical technology have allowed doctors to do a better job of determining, or detecting, genetic problems in parents so that they can be warned before having children. Technology has also improved the care and treatment available to children born with birth defects.

MAINTAINING HEALTH

Health maintenance has become an important and popular part of health care technology. Health maintenance has become so popular that it is now a major industry. Health clubs and spas can be found in most communities. Stores that cater specifically to health maintenance are also popular. They specialize in such health maintenance products as running shoes, aerobic dancing shoes, tennis shoes, weights, jogging suits, racket ball and tennis items, stopwatches, scales, athletic clothing, and numerous other health maintenance items.

Maintaining health is a four-step process:

1. Practice prevention.
2. Get regular checkups.
3. Eat the proper diet.
4. Exercise regularly.

People should be familiar with each step in the process, and they should practice each step.

The four steps to maintaining health.

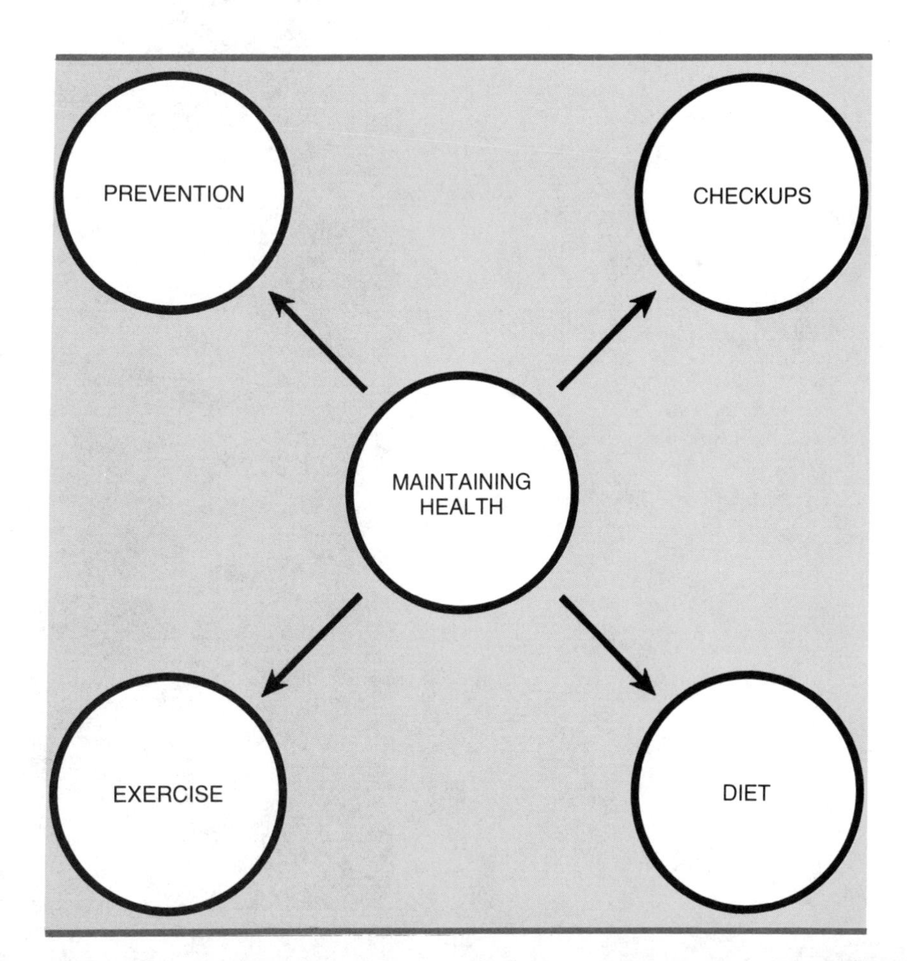

Microelectronic stationary bicycles have become a popular form of exercise. *Photo by Deborah M. Goetsch.*

The athletic shoe industry has become a major one. There are special shoes for tennis, basketball, running, walking, and aerobic dancing. *Photo by Deborah M. Goetsch.*

Practice Prevention

Prevention means doing everything possible to promote good health while avoiding poor health. In other words, it means avoiding or overcoming those potential threats to health that were presented earlier. People can prevent diseases by making sure that they have had all vaccinations and inoculations recommended by medical science. They can also practice good personal hygiene and avoid contact with people who have communicable diseases.

People can prevent accidents by exercising caution. Be careful! If you run along a road, wear bright clothing and watch for cars. If you dive, check the depth of the water first. Say *NO* to drugs and alcohol. Never ride in a car with someone who has been drinking, and always wear a seat belt. Eat properly and get plenty of exercise and rest.

People have less control over technological and genetic problems. But by practicing prevention in the areas they can control, people can lower their overall risks of suffering health problems.

Get Checkups

Annual checkups are a good way to maintain health. Many potential health problems can be stopped before they become serious if they are detected early enough. Checkups can detect these problems. In a typical checkup, the doctor will check all vital signs, make x-rays, and ask questions. Often the patient is the most important person in helping detect problems. By responding truthfully to the doctor's questions, patients can help detect signs of pending trouble.

One of the most important features of annual checkups is that they develop a written medical history of the patient that can then be monitored. If a patient gains a great deal of weight or vital signs change significantly, it will be readily apparent in the next checkup.

Eat a Proper Diet

A person's diet is the total of everything that person eats and drinks. What we eat and how much we eat are important. The human body is like a machine that needs just the right amount of fuel of the right mixture to run properly. Food is the fuel of the human body. The amount of food needed varies from person to person according to age and activity level. Generally speaking, younger people who are still growing need more fuel than older people whose bodies have stopped growing. Active people need more fuel than people who are not physically active.

The proper mixture of food does not change much regardless of age although it does change some. The essential elements of food are

1. Protein
2. Carbohydrates
3. Fats
4. Vitamins
5. Minerals
6. Roughage

All people need these essential elements in just the right amounts. Getting them is one of the best ways to maintain good health.

Protein Protein is needed for building and repairing muscle tissue. For this reason, weightlifters eat high-protein foods. Protein comes primarily from meat. Beef, chicken, fish, pork, and mutton are all good sources of protein. Certain plant foods, such as peanuts, beans, and

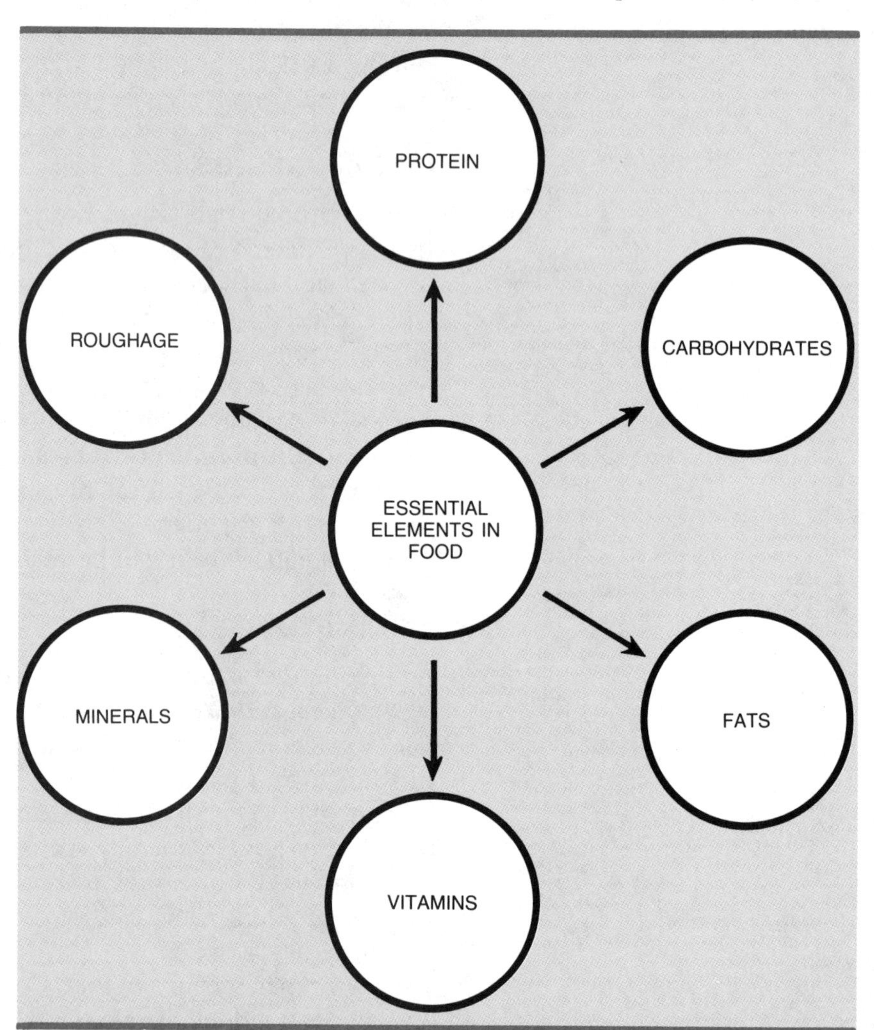

The essential elements in food.

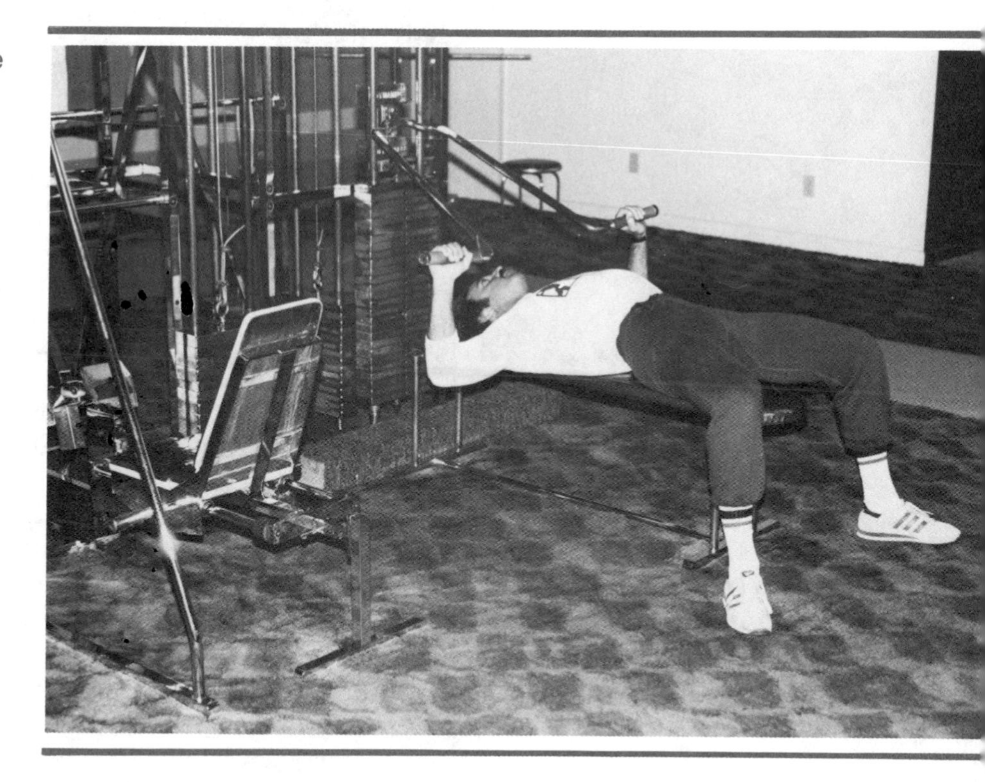

Protein is very important to weight lifters. It helps rebuild the muscles they tear down while lifting weights. *Photo by Deborah M. Goetsch.*

rice, are also sources of protein. All dairy products, such as milk, cheese, and yogurt, are also good sources of protein.

The problem with getting protein from meat and dairy products is that you also get fat—sometimes more than you need. This problem can be overcome by eating only the most lean cuts of red meat, relying more on white meat (chicken, turkey, and fish), relying more on plant sources (rice and beans), and eating dairy products low in fat.

Carbohydrates **Carbohydrates** provide energy. They are the real fuel in food. Carbohydrates come from sugar and starch. You know what sugar is. Starchy foods are potatoes, bread, and some white vegetables.

It is important to take in only the amount of carbohydrates needed by the body. This need, as we stated earlier, depends on age and activity levels of an individual. When it takes in more carbohydrates than it uses, the body stores them away as fat. Too much fat can lead to health problems.

Fats **Fats** are the most potentially unhealthy element in food. Fats are very high in calories. A *calorie* is a measure of how much heat can be produced by burning the fuel provided by food. The human body needs small amounts of fat, but it is easy to get too much fat. Fat is found in meats, particularly pork and beef, and in butter, margarine, whole milk, cream, cheese, and cooking oil. Fat is also found in peanuts, cotton seeds, and soybeans.

Fat from meats and dairy products is known as animal fat. Fat from plants is known as vegetable fat. Of the two, vegetable fat is less of a health problem because animal fat is high in cholesterol, a substance that clogs the arteries and leads to health problems.

To make sure you don't take in too much fat, (1) avoid foods that are high in fat, such as fatty cuts of beef and pork; (2) avoid foods that are prepared with fat, such as fried chicken and french fries; and (3) substitute vegetable fat for animal fat if fats must be used. For example, use margarine and vegetable oil instead of butter and cooking oil made from animal fat.

Vitamins and Minerals **Vitamins** are catalysts. They themselves aren't used. They allow other essential elements to be used. Protein, carbohydrates, and fats cannot be properly used by the body unless the right vitamins and minerals are available in the body in the right amounts.

There is a recommended daily allowance (RDA) of all vitamins. The best way to get all the vitamins needed is to eat a balanced diet. For those people who don't, vitamins can be purchased in tablet form and taken as directed on the label.

Minerals are also important. The minerals needed by the human body are some of the same minerals available in the earth, such as iron and zinc. The most important minerals to the human body are iron and calcium. Iron is needed in the blood. Calcium is needed in the bones. Green leafy vegetables, such as spinach, collards, and turnip greens are good sources of iron. Milk and dairy products are the best sources of calcium. Minerals can also be purchased in tablet form.

Roughage **Roughage** is an important element for digestion. Good digestion is essential for good health. Roughage can be obtained from high-fiber foods such as the hard fruits and vegetables (for example, apples, pears, carrots, lettuce, broccoli, and cauliflower). Grain foods, like oatmeal and bran, are also good sources of roughage. Scientists believe that the proper amount of roughage in the diet can be important in preventing certain types of internal cancer.

Exercise Regularly

Regular exercise is also important in maintaining health. The heart, lungs, arteries, and muscles are kept in good shape through regular

Skateboarding is a form of aerobic exercise. *Courtesy of Robert C. Atwood.*

Bicycling is a form of aerobic exercise. *Courtesy of Robert C. Atwood.*

exercise. The heart, lungs, and arteries are known as the body's **cardiovascular system**.

Exercises designed to keep the cardiovascular system in shape are called **aerobic exercises**. Aerobic exercises include running, brisk walking, jogging, swimming, and dancing. For the exercise to produce any benefit to the cardiovascular system, it must drive the pulse rate to

Baseball and softball can provide vigorous exercise. *Courtesy of Robert C. Atwood.*

Organized team sports such as soccer are popular ways to get exercise. *Courtesy of Robert C. Atwood.*

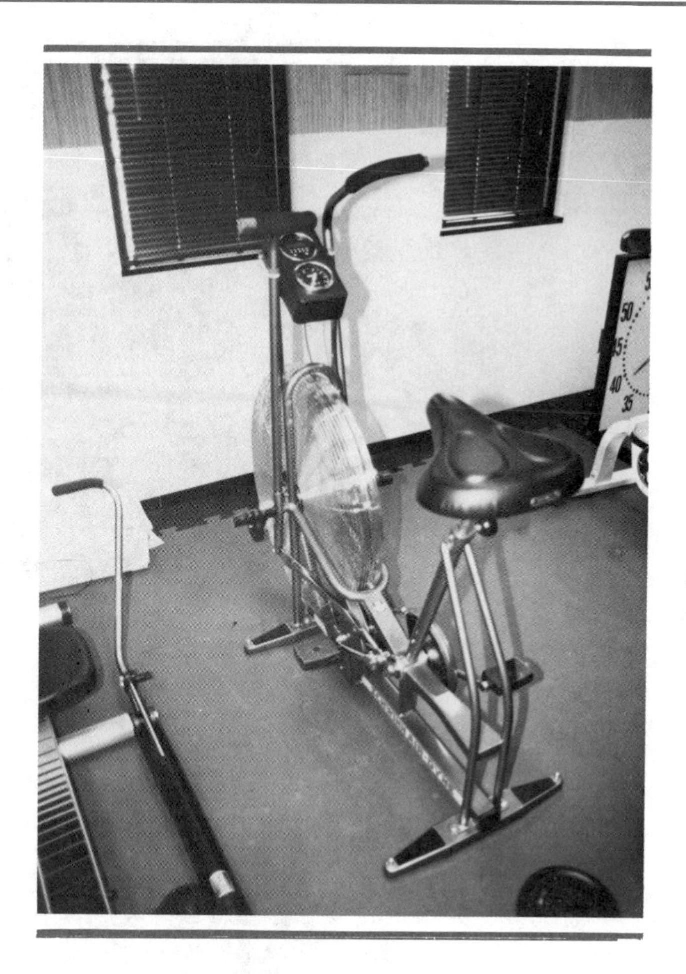

Riding a stationary bicycle is a form of aerobic exercise. *Photo by Deborah M. Goetsch.*

70% of its maximum level and keep it there continuously for at least 20 minutes. You should exercise at least three times per week.

Exercises designed to build and tone muscles are known as isometric or resistance exercises. In **isometric exercises**, the muscles exert force against opposite forces, such as heavy weights or immovable objects. The most popular type of isometric exercise is weight lifting.

Using weights, people can systematically exercise all of the muscles in as little as 30 to 45 minutes. Isometric exercises should be done three times per week. It is important to have a day of rest between isometric workouts to give the muscles time to grow and rebuild themselves. Many people alternate aerobic and isometric workouts. Three of each type of workout per week on an alternating basis is a good exercise plan.

Weight training is a form of isometric exercise. *Photo by Deborah M. Goetsch.*

Many types of home weight-training stations can be found in almost any department store. *Photo by Deborah M. Goetsch.*

WHAT IS MEDICAL TECHNOLOGY?

When health maintenance breaks down for some reason, or when it is not practiced, medical technology must take over. Medical technology is used to diagnose, treat, and rehabilitate people with health problems.

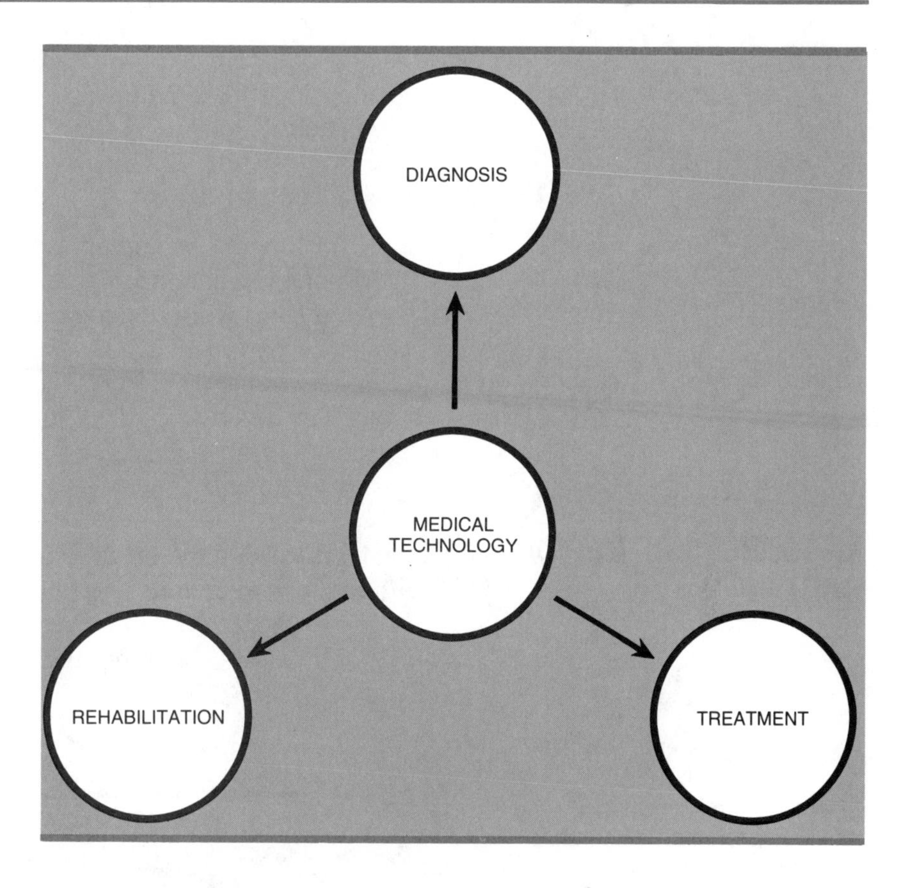

Medical technology.

Diagnosis

Diagnosis is the way doctors figure out what is wrong with patients when the problem is not obvious. This can be compared to an automobile mechanic figuring out what is wrong with a car that isn't running properly. Doctors use three methods of diagnosis: (1) measuring and monitoring the vital signs, (2) conducting **chemical analysis**, and (3) conducting **physical examinations**. You already know about vital signs.

Analyzing the blood, solid wastes, and liquid waste of the body is a good way to identify the causes of health problems. Physical examination techniques include making x-rays, looking into the ears, looking down the throat, or looking into the eyes with special instruments.

Treatment

Treatment means doing whatever is necessary to cure the health problem. Treatment ranges from something as simple as giving a patient an aspirin, to something as drastic as major surgery. Stitching a bad cut, putting a cast on a broken bone, cleaning a scrape, and implanting an artificial heart are all forms of treatment. Have you ever been treated for an injury or illness in a hospital or in a doctor's office?

Rehabilitation

Rehabilitation means getting a person who has been treated for a health problem back to normal. Rehabilitaion might involve extended rest, or it might involve a program of diet and exercise. The type of rehabilitation treatment given depends on the type of health problem a patient has. Rehabilitation of a patient who has had a severe stroke generally involves helping that person learn to walk again, to use his or her hands or arms again, and to talk normally again.

SOURCES OF MEDICAL CARE

Where do people go for diagnosis, treatment, and rehabilitation? There are several sources available. Doctors' offices are a common source of medical care. Clinics are another. The most comprehensive source of medical care is the hospital.

Hospitals are the most common and comprehensive source of medical care. *Photo by Deborah M. Goetsch.*

Community-based immediate-care centers are becoming increasingly popular as inexpensive alternatives for minor medical emergencies. *Photo by Deborah M. Goetsch.*

Hospitals offer a wide range of services and capabilities from emergency care to long-term care. For less serious health problems, emergency care or immediate-care centers are available in some communities. These centers offer the same types of services as emergency rooms in hospitals, but they don't normally have the capability of dealing with critically ill patients or serious, traumatic injuries.

SOME CAREERS IN HEALTH CARE AND MEDICAL TECHNOLOGY

Many careers are associated with health care and medical technology. Most fall into one of three categories: professionals, paraprofessionals, and technicians. There is a certain amount of controversy surrounding several career fields as to whether they are categorized as professional, paraprofessional, or technical.

Professional Career Fields

Professional careers in health care and medical technology include scientists, chemists, physicians, psychiatrists, psychologists, dentists, counselors, managers, and teachers. In addition, some people place nurses who have four-year college degrees or more in this category. Scientists and chemists are involved with developing new and better drugs and analyzing body chemistry as part of the diagnosis step in medical treatment.

Physicians and dentists are the doctors—general practitioners, specialists, and surgeons who treat patients. Psychiatrists and psychologists treat mental health problems. Counselors assist psychiatrists and psychologists. There are teachers at all levels who teach the formal courses that health care and medical professionals must take to become qualified and stay qualified.

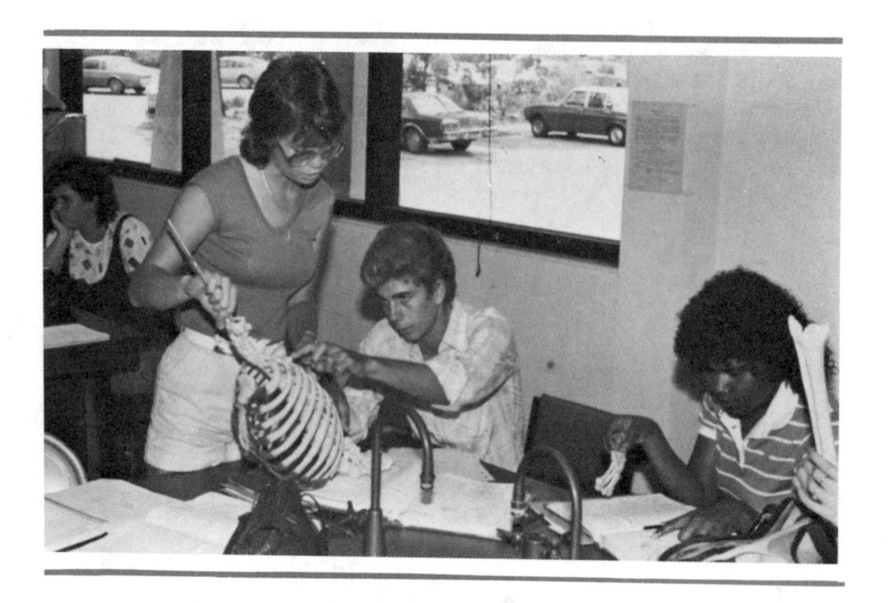

Health care students study human anatomy. *Courtesy of Robert C. Atwood.*

Managers are the people who run health care and medical technology facilities such as hospitals, wellness centers, clinics, and immediate care centers. Professionals in health care and medical technology typically have advanced college degrees and spend from 4 to 12 years in college.

Paraprofessionals

Paraprofessionals assist physicians in diagnosing, treating, and rehabilitating patients. They are highly trained and skilled, but they can perform only specified tasks under the supervision of a professional. Paraprofessionals include physical therapists, paramedics, registered nurses with less than a four-year college degree, and medical assistants. Most paraprofessionals have two or three-year associate degrees from community colleges or advanced training from an area vocational center or technical school.

Technicians

Many different types of technicians are associated with health care and medical technology. Laboratory technicians assist scientists, chemists, and physicians. Nurses and dental assistants are classified as technicians. Biomedical technicians maintain and repair all of the sophisticated electronic equipment that has become essential to health care and medical technology. Technicians must have specialized training,

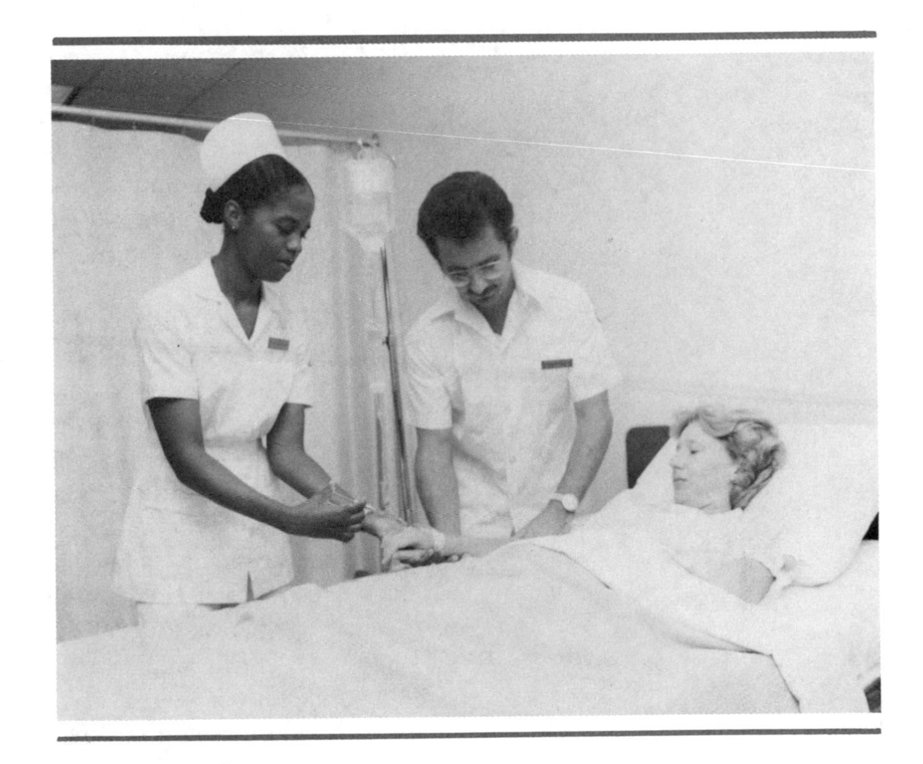

Nursing is an important health care occupation. *Courtesy of Robert C. Atwood.*

usually ranging from one to three years in a community college, technical school, or private training program sponsored by hospitals.

SUMMARY

Health is a state or condition of the body or mind. Physical health concerns the body. Mental health concerns the mind. Health care technology is concerned with helping people maintain health. Health can be measured. Temperature, blood pressure, pulse rate, respiration, fat-to-lean ratio, and blood composition measurements tell the qualified person how healthy a patient is.

The blood pressure is an important measure of health. High blood pressure can lead to strokes or heart attacks. Blood pressure is the amount of pressure applied by the blood as it travels through the arteries. The pulse rate tells how many times per minute the heart beats.

Respiration is the rate of breathing. It is measured by how many breaths per minute a person takes.

Blood composition is important. By analyzing a person's blood, health care professionals can learn a great deal about that person's health.

Health problems fall into one of the following categories: diseases, accidents, technological problems, psychological problems, and genetic problems. Diseases are caused by microscopic organisms called germs. The best ways to prevent diseases are to practice good personal hygiene and good community sanitation and to avoid contact with people who have a communicable disease. Another way is immunization—making the body immune to the germs that cause certain diseases.

Injuries that result from accidents are another cause of health problems. Psychological problems can also lead to health problems. Family, work, school, and personal problems can lead people to seek relief through alcohol and drugs, which can lead to a chemical dependency that endangers physical and mental well-being. Genetic problems are inborn—the problem was passed along to the baby through the genes of the mother or father or both.

Health maintenance has become an important and popular part of health care technology. Maintaining one's health is a four-step process: (1) practice prevention; (2) get regular check-ups; (3) eat the proper diet; and (4) exercise regularly.

Regular exercise keeps the cardiovascular system (heart, lungs, and arteries) and muscles in good shape. Exercises for the cardiovascular system are called aerobic exercises. They include running, brisk walking, jogging, swimming, and dancing. Exercises for the muscles are known as isometric exercises. In isometric exercises, the muscles exert force against opposite forces such as heavy weights or immovable objects.

When health maintenance breaks down or is not practiced, medical technology must take over. Medical technology is the diagnosis, treatment, and rehabilitation of people with health problems.

Diagnosis is the way doctors figure what is wrong with patients when a problem is not obvious. Treatment involves doing whatever is necessary to cure the health problem. Rehabilitation means getting a person who has been treated for a health problem back to normal.

Several sources of medical care are available in most communities. Doctors' offices and clinics are common sources of medical care. The most comprehensive source is the hospital.

Some of the most widely practiced career fields associated with health care and medical technology are scientists, chemists, physicians, psychiatrists, psychologists, dentists, counselors, nurses, therapists, paramedics, lab technicians, dental assistants, and biomedical electronics technicians.

KEY WORDS

Health care technology
Physical health
Mental health
Sickness
Illness
Temperature
Thermometer
Oxygen
Stethoscope
Sodium
Cholesterol
Triglycerides
Iron
Calcium
Diseases
Accidents
Technological problems
Psychological problems
Genetic problems
Germs
Bacteria
Parasites
Microscopic organisms
Communicable disease
Immunization
Isometric exercise
Aerobic exercise
Diagnosis
Treatment
Rehabilitation
Chemical analysis
Physical examination
Blood pressure
Pulse rate
Respiration
Fat-to-lean ratio
Blood composition
Vital signs
Carbohydrates
Fats
Vitamins
Roughage
Cardiovascular system
Chemical dependency

SOME QUESTIONS TO TRY

1. What is health care technology?
2. Name three factors to check to measure a person's health.
3. Name four sources of health problems.
4. Name the four steps for maintaining health.
5. What is medical technology?
6. Name two sources of medical care available in most communities.
7. Name five careers in health care and medical technology.

TECHNOLOGY LEARNING ACTIVITIES

1. Develop a health care maintenance card on which you can record the following: name and age, pulse rate, blood pressure, respiration rate, and weight.
2. Arrange for a nurse or a medical technician to visit your class and give demonstrations on how to measure the pulse rate, blood pressure, and respiration rate.
3. Select a buddy from among your classmates. Practice measuring the pulse rate, blood pressure, and respiration rate. Record the results on your health maintenance card.
4. Arrange for a physical education teacher to come to your class and help each student plan an exercise program of both aerobic and isometric exercises. Record all of the exercises and how often you are to do them on the back of your health care maintenance card.
5. In a small notebook, record each day for a week everything you eat. Have a health or biology teacher visit your classroom and analyze your diet and make recommendations for improvement.
6. Practice your exercise program and your new diet for a month. Then recheck your pulse rate, blood pressure, and respiration rate. See if there is any change.
7. Make a timeline of important technological developments in health care and medical technology.

NETED
SYMED
SIM
FSIM
DOC
PICED
"ZERO POSITION"
/ MOVE / TURN / SIZE / RESET / SET / GRID /

PART FIVE

ISSUES AND OUTLOOKS IN TECHNOLOGY

CHAPTER 13

PROBLEMS OF TECHNOLOGY

OBJECTIVES

After completing this chapter, you should be able to

- list some of the common personal and societal problems of technology.
- list some of the common environmental problems of technology.
- explain how technological problems can be prevented.
- explain some of the future careers in technology.

People have improved their lives by technology since the beginning of time. In transportation, people have gone from walking to traveling in space. In communications, people have gone from face-to-face conversation to talking to astronauts in space. In health care and medicine, the average lifespan has increased from 40 years to over 70 years. In construction, people have gone from crude huts to multistory skyscrapers.

Technology has been good, but there are problems that we should know about. Personal, societal, and environmental problems can result from technology.

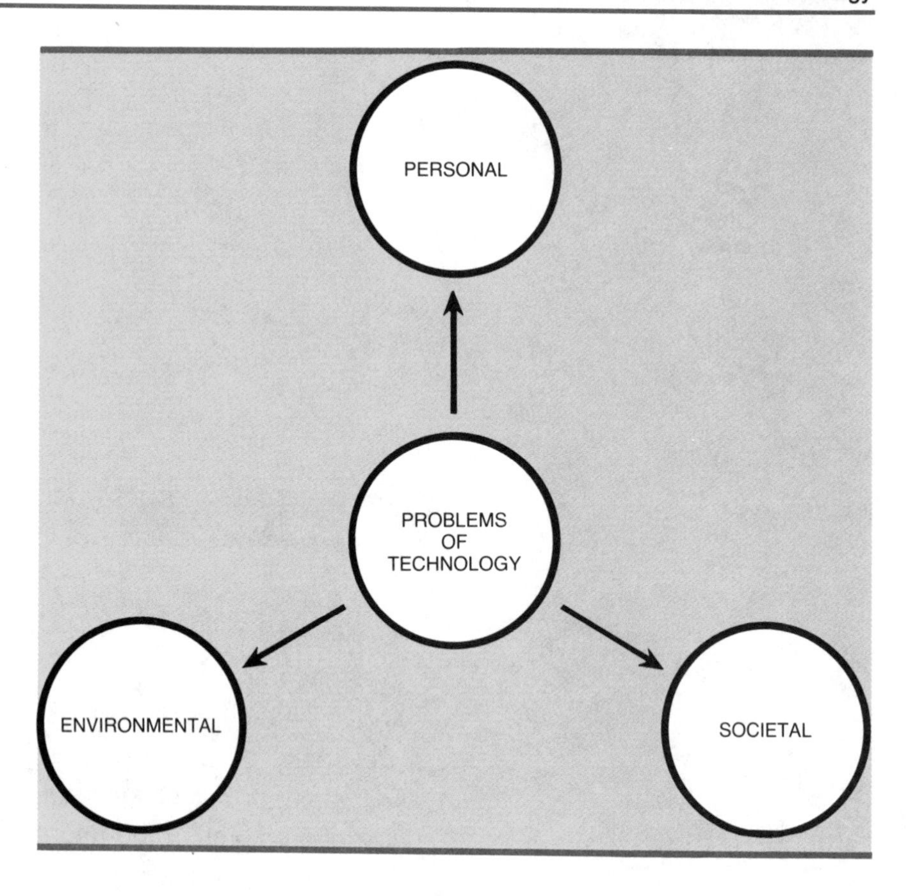

The problems of technology.

PERSONAL AND SOCIETAL PROBLEMS

People create technology to improve their lives, but they are also the victims of the problems that sometimes accompany technological development. These problems fall into one or more of the following categories

1. Overcrowding
2. Development of drugs
3. Accidents
4. War and disarmament
5. The changing nature of work
6. Uprooting

Overcrowding

Construction technology has allowed for the development of huge cities. In the United States, New York, Los Angeles, Detroit, Chicago, Miami, and Atlanta are some of the largest, most heavily populated cities in the world. In these cities, thousands of people live in large apartment buildings and tenements. Thousands of people work in

Checklist of specific problems associated with technology.

CHECKLIST OF SPECIFIC PROBLEMS ASSOCIATED WITH TECHNOLOGY

_____ CRIME AND ANXIETY FROM OVERCROWDING IN CITIES
_____ HEALTH PROBLEMS FROM SEDENTARY LIFE STYLES
_____ RISK OF NUCLEAR WAR
_____ DRUG ABUSE AND ACCIDENTAL DRUG PROBLEMS
_____ TRANSPORTATION RELATED ACCIDENTS
_____ LOSS OF JOBS DUE TO AUTOMATION
_____ LOSS OF JOBS DUE TO MECHANIZATION
_____ UPROOTING OF HOMES TO MAKE WAY FOR EXPANSION
_____ DESTRUCTION OF NATURAL RESOURCES
_____ AIR POLLUTION
_____ WATER POLLUTION
_____ SOIL POLLUTION
_____ LEAKS IN NUCLEAR PLANTS
_____ DISPOSING OF NUCLEAR WASTE
_____ NUCLEAR DESTRUCTION

Buildings like the Water Board Administration Building in Denver, Colorado, are especially designed to prevent the types of problems normally associated with overcrowding. Designs such as this one give people a pleasant, calm atmosphere within which to live and work.
Courtesy of the Prestressed Concrete Institute.

huge high-rise office buildings. And every day, thousands of people crowd into the various forms and means of transportation in these cities.

These large cities are technological wonders on the one hand, but, on the other hand, the crowded living conditions in them can lead to problems. **Overcrowding** of people leads to such problems as anxiety and **depersonalization**. These problems can lead to health problems and antisocial behavior. Incidents of all forms of violent crime tend to be higher in crowded cities than in more sparsely populated areas. Rape, assault, and murder are common occurrences in some large cities. **Sociologists** draw parallels between such crimes and overcrowded conditions. Many people who live in large cities move to quieter areas when they retire.

Drug Problems

Scientists and chemists are constantly seeking new and better drugs to treat disease and sickness. These drugs have worked wonders, but they can be, and have been, abused. The drugs associated with **drug problems** and crime, such as cocaine and heroin, have legitimate medical uses. When abused, they lead to addiction (chemical dependency). Drug addiction is linked to thousands of cases of crime, violent deaths, school dropouts, and accidental deaths each year, not including the many people killed in alcohol-related traffic accidents each year.

Even medical science can accidentally misuse drugs. This has happened occasionally when a drug has been approved for use but later is found to be harmful. One such tragic incidence of this happened in the 1960s with the drug thalidomide. Thalidomide was developed as a sedative for pregnant women. It was supposed to help them avoid morning sickness. And it worked well in this regard. Thousands of women took the drug and later learned that it caused serious birth defects. More than 10,000 thalidomide babies were born throughout the world with missing limbs or tiny flippers instead of arms or legs.

Another example of drug problems caused by medical science was the swine flu vaccination program of the 1970s. Through this program, doctors encouraged people to be vaccinated against swine flu. The medical world was fearful that a swine flu epidemic was about to break out in this country. Many people at the urging of their personal physician or a television commercial took the vaccination. For the most part, it caused no problems. However, in some people, it brought on temporary paralysis. Some people died. These things happened because the swine flu vaccine had not been sufficiently tested to determine all of the possible side effects that it could have.

Accidents

Although transportation technology has improved our lives, thousands of accidental deaths occur each year through land, water, and air travel.

Every year over 50,000 people are killed in automobile **accidents**. Additionally, almost 2 million people are injured in automobile accidents each year. Several hundred more deaths and **injuries** can be attributed to air and water transportation accidents. The year 1985 was one of the worst years in the history of air travel for accidents and injuries. Over 2,000 people were killed in aircraft disasters.

NASA Space Program is a true technological marvel and has traditionally been one of the safest modes of transportation. However, in January of 1986, NASA's space shuttle, Challenger, exploded shortly after takeoff from Cape Canaveral killing all seven people aboard. This tragedy marked the first fatal inflight accident in NASA's history. It underscored for the entire nation the price people sometimes pay for technological progress.

The first automobile was a major step forward in ground transportation. However, automobile accidents now account for over 50,000 deaths per year. *Courtesy of Spizzirri Publishing Company, Inc.*

Modern airliners such as this Boeing 737 contain every possible type of safety feature to cut down on the possibility of accidents. *Courtesy of Boeing Aircraft Corporation.*

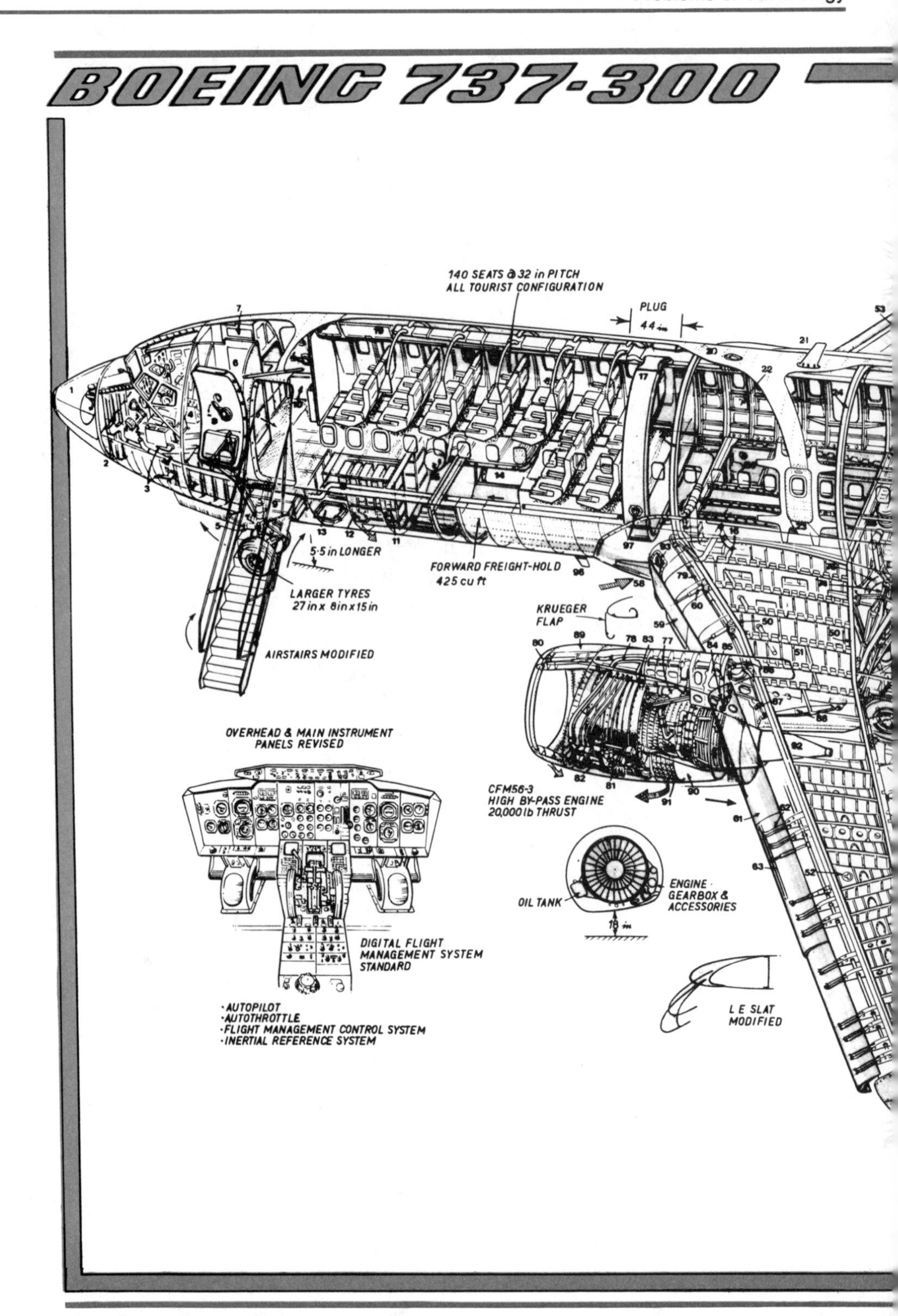

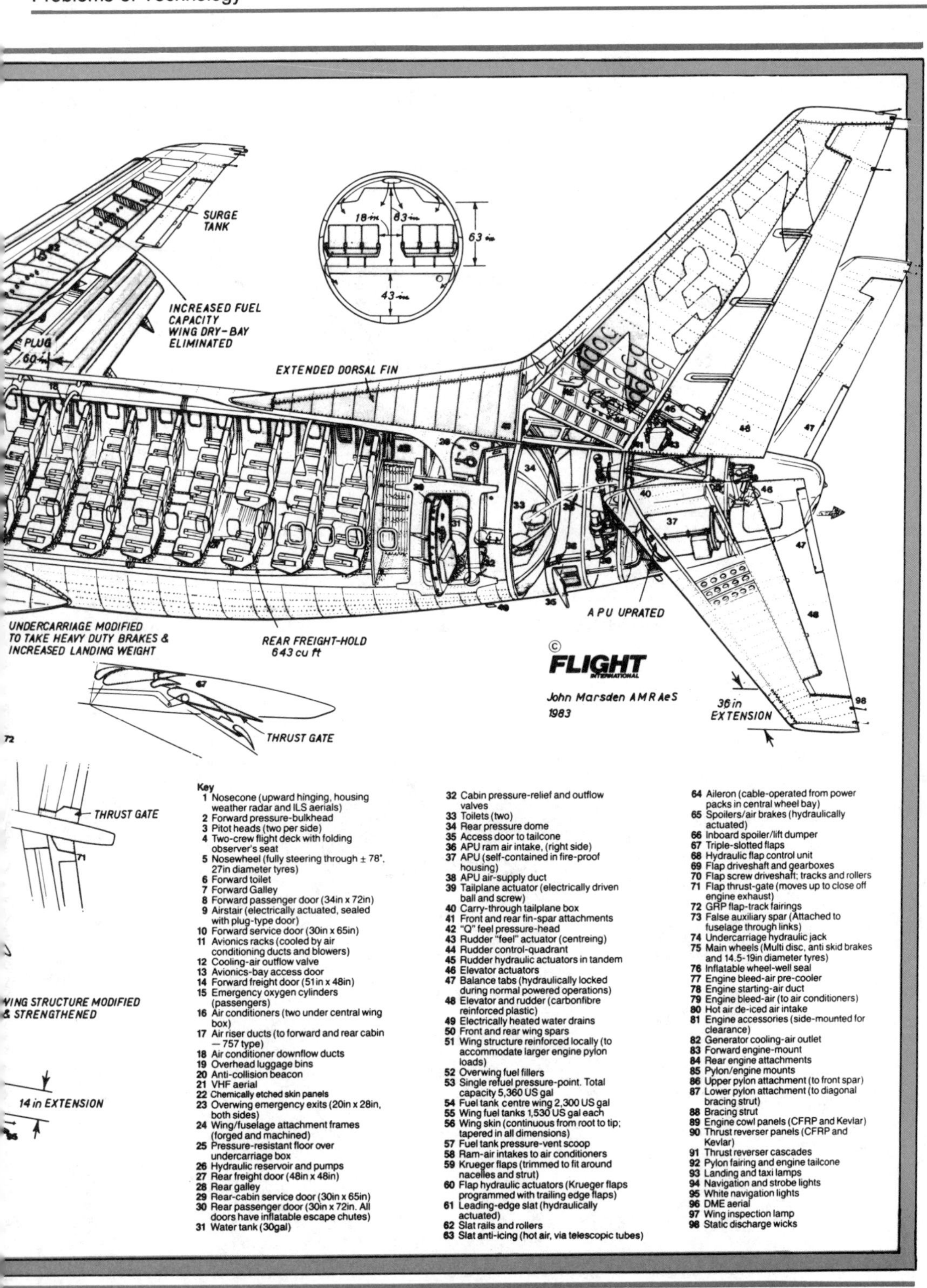
SURGE TANK
18 in
63 in
63 in
43 in
INCREASED FUEL CAPACITY WING DRY-BAY ELIMINATED
PLUG
60 in
EXTENDED DORSAL FIN
APU UPRATED
UNDERCARRIAGE MODIFIED TO TAKE HEAVY DUTY BRAKES & INCREASED LANDING WEIGHT
REAR FREIGHT-HOLD 643 cu ft
© FLIGHT INTERNATIONAL
John Marsden AMRAeS
1983
36 in EXTENSION
THRUST GATE
THRUST GATE
WING STRUCTURE MODIFIED & STRENGTHENED
14 in EXTENSION
Key
1 Nosecone (upward hinging, housing weather radar and ILS aerials)
2 Forward pressure-bulkhead
3 Pitot heads (two per side)
4 Two-crew flight deck with folding observer's seat
5 Nosewheel (fully steering through ± 78°, 27in diameter tyres)
6 Forward toilet
7 Forward Galley
8 Forward passenger door (34in x 72in)
9 Airstair (electrically actuated, sealed with plug-type door)
10 Forward service door (30in x 65in)
11 Avionics racks (cooled by air conditioning ducts and blowers)
12 Cooling-air outflow valve
13 Avionics-bay access door
14 Forward freight door (51in x 48in)
15 Emergency oxygen cylinders (passengers)
16 Air conditioners (two under central wing box)
17 Air riser ducts (to forward and rear cabin — 757 type)
18 Air conditioner downflow ducts
19 Overhead luggage bins
20 Anti-collision beacon
21 VHF aerial
22 Chemically etched skin panels
23 Overwing emergency exits (20in x 28in, both sides)
24 Wing/fuselage attachment frames (forged and machined)
25 Pressure-resistant floor over undercarriage box
26 Hydraulic reservoir and pumps
27 Rear freight door (48in x 48in)
28 Rear galley
29 Rear-cabin service door (30in x 65in)
30 Rear passenger door (30in x 72in. All doors have inflatable escape chutes)
31 Water tank (30gal)
32 Cabin pressure-relief and outflow valves
33 Toilets (two)
34 Rear pressure dome
35 Access door to tailcone
36 APU ram air intake, (right side)
37 APU (self-contained in fire-proof housing)
38 APU air-supply duct
39 Tailplane actuator (electrically driven ball and screw)
40 Carry-through tailplane box
41 Front and rear fin-spar attachments
42 "Q" feel pressure-head
43 Rudder "feel" actuator (centreing)
44 Rudder control-quadrant
45 Rudder hydraulic actuators in tandem
46 Elevator actuators
47 Balance tabs (hydraulically locked during normal powered operations)
48 Elevator and rudder (carbonfibre reinforced plastic)
49 Electrically heated water drains
50 Front and rear wing spars
51 Wing structure reinforced locally (to accommodate larger engine pylon loads)
52 Overwing fuel fillers
53 Single refuel pressure-point. Total capacity 5,360 US gal
54 Fuel tank centre wing 2,300 US gal
55 Wing fuel tanks 1,530 US gal each
56 Wing skin (continuous from root to tip; tapered in all dimensions)
57 Fuel tank pressure-vent scoop
58 Ram-air intakes to air conditioners
59 Krueger flaps (trimmed to fit around nacelles and strut)
60 Flap hydraulic actuators (Krueger flaps programmed with trailing edge flaps)
61 Leading-edge slat (hydraulically actuated)
62 Slat rails and rollers
63 Slat anti-icing (hot air, via telescopic tubes)
64 Aileron (cable-operated from power packs in central wheel bay)
65 Spoilers/air brakes (hydraulically actuated)
66 Inboard spoiler/lift dumper
67 Triple-slotted flaps
68 Hydraulic flap control unit
69 Flap driveshaft and gearboxes
70 Flap screw driveshaft; tracks and rollers
71 Flap thrust-gate (moves up to close off engine exhaust)
72 GRP flap-track fairings
73 False auxiliary spar (Attached to fuselage through links)
74 Undercarriage hydraulic jack
75 Main wheels (Multi disc, anti skid brakes and 14.5-19in diameter tyres)
76 Inflatable wheel-well seal
77 Engine bleed-air pre-cooler
78 Engine starting-air duct
79 Engine bleed-air (to air conditioners)
80 Hot air de-iced air intake
81 Engine accessories (side-mounted for clearance)
82 Generator cooling-air outlet
83 Forward engine-mount
84 Rear engine attachments
85 Pylon/engine mounts
86 Upper pylon attachment (to front spar)
87 Lower pylon attachment (to diagonal bracing strut)
88 Bracing strut
89 Engine cowl panels (CFRP and Kevlar)
90 Thrust reverser panels (CFRP and Kevlar)
91 Thrust reverser cascades
92 Pylon fairing and engine tailcone
93 Landing and taxi lamps
94 Navigation and strobe lights
95 White navigation lights
96 DME aerial
97 Wing inspection lamp
98 Static discharge wicks

War and Disarmament

When people cannot settle their differences peacefully, they sometimes fight. Fighting between countries is called **war**. Unfortunately, war occupies a large part of the history of mankind. Rarely has there been a period in history in which war was not taking place somewhere in the world. Since establishing itself as a country through its War of Independence with Great Britain, the United States has been involved in the War of 1812, the Spanish-American War, World War I, World War II, the Korean War, and the Vietnam War. In each conflict, the technology of war was more effective and more deadly. Today at least two countries, the United States and the Soviet Union, have the military technology to destroy the entire world in minutes. Such a danger has caused the world to begin seeking ways to disarm. But **disarmament** is a slow and difficult process. As battles continue to be fought (Afghanistan, the Middle East, Central America), the need to disarm is critical.

The most serious threat to the world is thermonuclear war. The United States and the Soviet Union both have nuclear weapons in **submarines**, **strategic bombers**, **surface-to-air missiles**, and **intercontinental ballistic missiles**, which are ready to be used in

The potential for worldwide nuclear war is one of the most serious problems associated with technology. *Courtesy of Ken Elliott.*

a moment's notice. Many safeguards are taken to ensure that accidental or unapproved use of nuclear weapons does not take place, but people still fear that it could happen. If it did, the earth could be destroyed.

The Changing Nature of Work

Since the beginning of time, technology has continually changed the nature of work. It has allowed people to do more work in less time with less effort. In terms of productivity, technology has been good. But it has caused some problems.

The changing nature of work has contributed to health problems. The typical diet in this country is one designed to provide plenty of energy for physically active people. It is high in protein, carbohydrates, and fat. This diet was good when people worked hard all day with their hands. However, in most cases, machines now do most of our physical work.

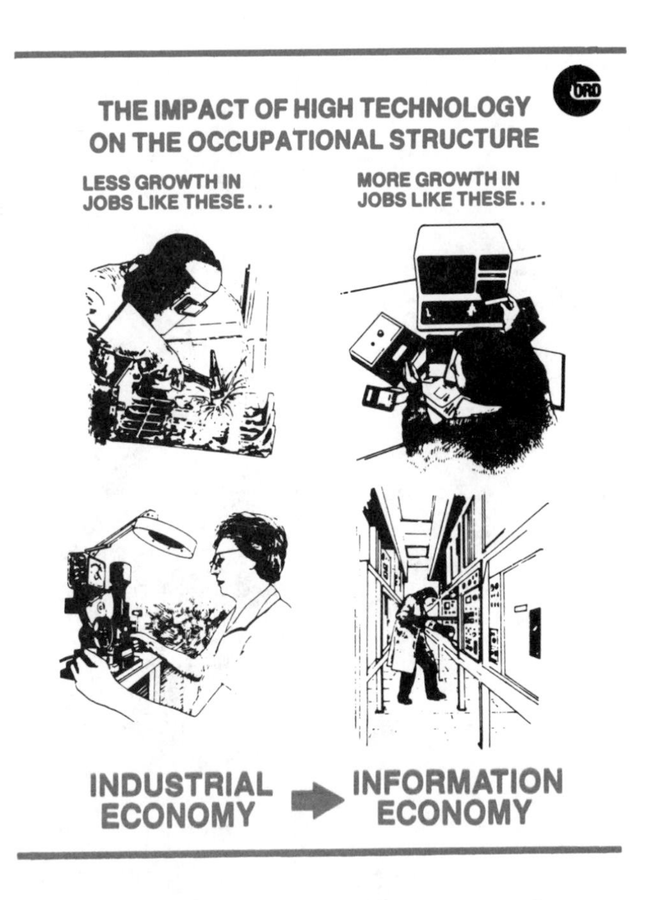

Technology is changing the economy from an industrial-based economy to an information-based economy.
Courtesy of The Center for Occupational Research and Development.

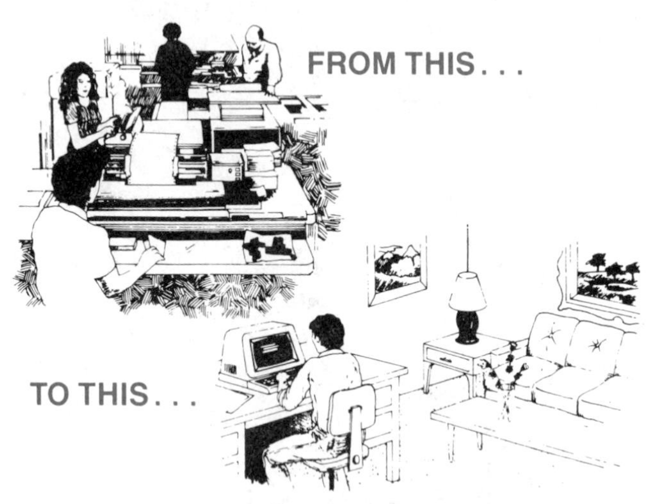

HIGH TECHNOLOGY IS CHANGING
THE WAY WE WORK . . .
•Decentralization of the workplace
FROM THIS . . .
TO THIS . . .
•More people linked to the economy
RURAL CITIZENS
HOMEMAKERS
SENIOR CITIZENS
DISABLED PERSONS

Technology is changing the way people work. *Courtesy of The Center for Occupational Research and Development.*

In many cases, people changed to a less-physical life style, but they did not change their diets. They used to burn up energy from carbohydrates by working hard all day. Now the carbohydrates turn to fat and clog the arteries. Thus, the 1970s was a decade of heart attacks.

One of the concepts behind the changing nature of work has been mechanization. **Mechanization**—using machines to do work previously done by hand or with hand tools—has improved immensely the productivity of people, especially in manufacturing technology and agricultural technology. In agricultural technology, the reaper, thresher, seeding machine, and milking machine have allowed farmers to do much more work in much less time.

Mechanization of farm work and advances in the soil treatment, fertilizers, insecticides, and food processing have made the American farmer the most productive farmer in the world. In fact, farmers have become too productive for their own good.

HIGH TECHNOLOGY SOMETIMES LEADS TO

WORKER DISPLACEMENT

- Some people **LOSE JOBS**.
- Some people **RETRAIN** within the company or in a related field.
- Some people **CHANGE CAREERS** entirely.

Technology sometimes leads to worker displacement. *Courtesy of The Center for Occupational Research and Development.*

By the 1980s, the productivity of the American farmer had reached the point where more food was being produced than the market could use. Such surpluses drove down the price of food until food began to cost some small farmers more to grow than it would sell for. Many farmers borrowed a lot of money to keep from going out of business. When their crops continued to sell for less than it cost to produce them, some farmers could not repay the loans and went bankrupt.

Automation processes through computer-aided manufacturing (CAM) and robotics have changed work arrangements in manufacturing plants. CAM machines and robots are doing work previously done by people. Therefore, people lost jobs. To get new jobs, they had to retrain, in most cases, and sometimes move to other areas of the country. Losing a job and being forced to relocate for another job can be a traumatic experience.

SIX CHANGES ON THE EMPLOYMENT SCENE

- Moving toward an information and service economy
- Traditional semiskilled factory jobs on the wane
- More low-skilled jobs which involve little or no training and low pay
- Two out of three new workers female and/or minority
- In demand--a well trained technical workforce for high-tech industries
- A growing trend--mid-career changes

A summary of six major changes taking place on the employment scene. *Courtesy of The Center for Occupational Research and Development.*

Computers are a major part of automation and manufacturing. *Courtesy of Cincinnati Milacron.*

Uprooting

One of the problems with technological progress is **uprooting** people and interrupting their lives. As cities grow, new and better highways are needed. Often new highways go right through the middle of existing neighborhoods. When this happens, people must give up their homes and property. This practice is called the right of **eminent domain**. It means that the government does have the right to take an individual's property for the good of the public. Of course, the government must pay the fair-market value for any property it takes, but even so, people find it hard to leave homes and neighborhoods where they grew up.

ENVIRONMENTAL PROBLEMS

Some of the most serious technological problems exist in the environment. Most environmental problems can be categorized as follows:

1. Depletion of natural resources
2. Pollution
3. Nuclear-energy-related problems

Depletion of Natural Resources

Advances in technology have caused people to use natural resources faster than they can be replaced in many cases. They need for lumber has led people to cut down trees at a rapid pace, in some cases eliminating large areas of timber.

Fossil fuels such as coal and oil are natural resources. They were plentiful at one time, but with the rapidly increased use of electricity and gasoline, they are being used up so fast that they may run out.

Pollution

Pollution is one of the most serious problems that has resulted from technological advances. The earth is struggling with pollution of air, water, and soil. Burning fossil fuels pollutes the air. Smoke from industrial smokestacks and exhaust from automobiles, aircraft, boats, and ships also pollute the air.

Air Pollution Air pollution is called **smog**. In some large cities, air pollution has reached such high levels that periodic smog alerts must be issued. A smog alert means that the air is dangerous to breathe. Medical scientists believe that air pollution is directly linked to such health problems as emphysema and lung cancer.

Water Pollution Water pollution is also a serious problem. Oil spills have become a major cause of water pollution. When large

JT-15D turbofan jet engine. The jet engine is one of the most important technological advances in history. However, if exhaust from the jet engines is not properly controlled, it can contribute to air pollution. *Courtesy of Pratt & Whitney.*

tankers are damaged at sea, oil sometimes spills into the ocean. Offshore oil rigs also sometimes spill over. Oil spills pollute the water and kill sea life in the surrounding area. If the spill washes ashore, it can cause serious damage to the coastline and upset the ecological balance along the coastline.

Water pollution also occurs when industrial waste, sewage, and trash are poured into rivers, streams, or lakes, or allowed to seep into the underground water table. Dumping heated water into rivers, lakes, and streams pollutes them. This sometimes happens around electric generating plants. These plants need steam to power the turbines that power electric generators. As steam cools down, it becomes water. If this water is dumped into surrounding rivers, lakes, or streams before it cools sufficiently, it can cause the temperature in the water to rise rapidly. Such rapid rises in water temperature can cause the death of fish and other water life. These fish kills are a serious form of water pollution.

Soil Pollution Soil pollution occurs when the ground becomes poisoned from **insecticides**, chemicals, **nuclear waste**, or industrial waste. Chemical and industrial dumping sites are scattered across the country. At one time, a common way of disposing of dangerous chemicals involved sealing them in metal drums and burying the drums. Through time, underground moisture can rust the metal drum. Chemicals will then seep out and poison the ground. If that ground is farmed, or if the area is developed as a housing project, or if the **chemicals** seep into the water table, serious health problems can occur.

Nuclear-Energy-Related Problems

Nuclear energy is an important technological advance. Nuclear-powered ships are more powerful and efficient than diesel-powered ships. Nuclear energy is also an alternative to fossil fuels for creating steam to power steam turbines in electric generating plants. At one time, nuclear energy was looked on as the energy source that would probably replace fossil fuels in generating electricity. Right now, approximately 20 percent of the electricity generated in this country is generated by nuclear energy.

Nuclear energy, however, has more potential for negative side effects than any other product of technology. The potential problems fall into three categories: (1) nuclear destruction, (2) nuclear leaks, and (3) nuclear-waste disposal.

Nuclear materials can be used to produce bombs and missiles that are capable of immense destruction. **Nuclear leaks** pose a potential threat to people who live near nuclear generating plants. Numerous safety precautions are taken to prevent nuclear leaks, but accidental leaks are still possible. A nuclear leak occurs when radioactive materials from the core of a nuclear reactor accidentally spill and contaminate the water or soil around the reactor.

Nuclear-waste disposal is one of the most serious problems associated with nuclear energy. Nuclear waste is radioactive and dangerous, and it stays that way for many years. So it cannot just be dumped into the water or into the ground, and it cannot be burned. So what can we do with it? At this point, there is no good answer to this question. Scientists have been unable to find a completely safe way to dispose of nuclear waste. For now, it is sealed in thick concrete containers and buried. Until a fail-safe method of disposing of nuclear waste is found, nuclear energy must be limited and closely monitored.

SOLVING THE PROBLEMS OF TECHNOLOGY

People create technology to improve the quality of life. When the unplanned side effects of technology create problems, people must solve them. People are working to solve the problems of technology

PRATT & WHITNEY PW2037

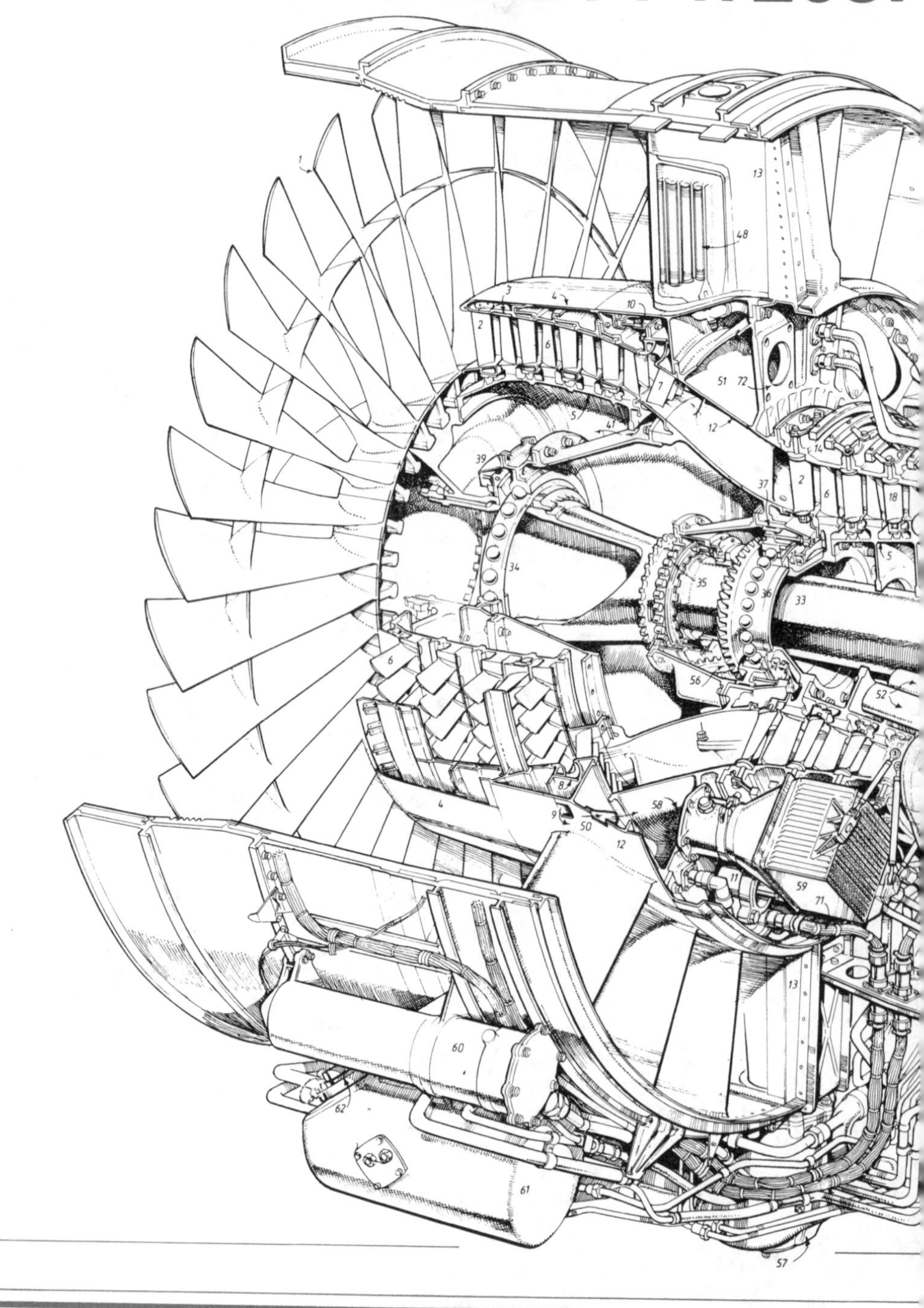

Jet engines such as this Pratt & Whitney PW2037 are especially designed to not pollute the air. *Courtesy of Pratt & Whitney.*

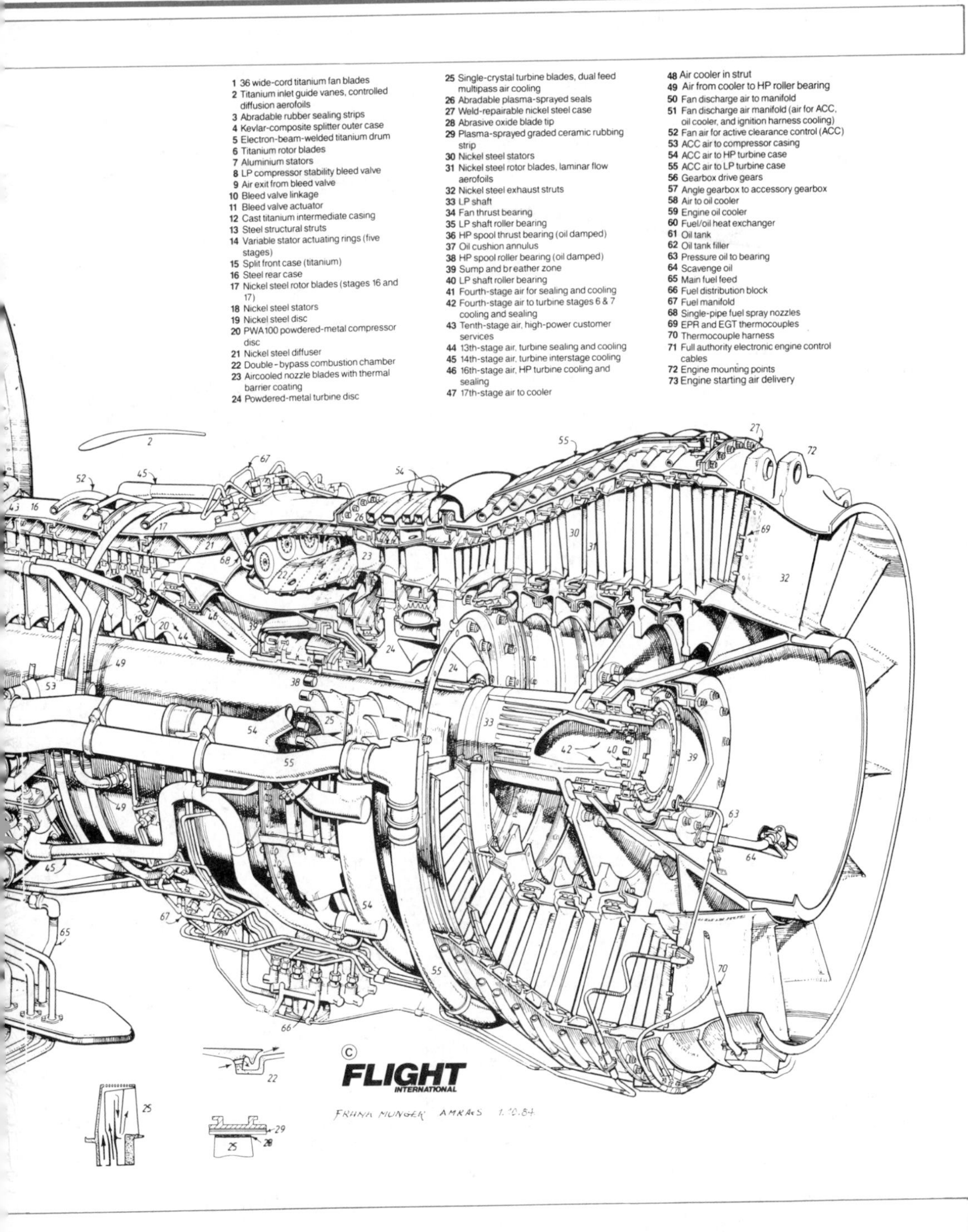
1 36 wide-cord titanium fan blades
2 Titanium inlet guide vanes, controlled diffusion aerofoils
3 Abradable rubber sealing strips
4 Kevlar-composite splitter outer case
5 Electron-beam-welded titanium drum
6 Titanium rotor blades
7 Aluminium stators
8 LP compressor stability bleed valve
9 Air exit from bleed valve
10 Bleed valve linkage
11 Bleed valve actuator
12 Cast titanium intermediate casing
13 Steel structural struts
14 Variable stator actuating rings (five stages)
15 Split front case (titanium)
16 Steel rear case
17 Nickel steel rotor blades (stages 16 and 17)
18 Nickel steel stators
19 Nickel steel disc
20 PWA100 powdered-metal compressor disc
21 Nickel steel diffuser
22 Double-bypass combustion chamber
23 Aircooled nozzle blades with thermal barrier coating
24 Powdered-metal turbine disc
25 Single-crystal turbine blades, dual feed multipass air cooling
26 Abradable plasma-sprayed seals
27 Weld-repairable nickel steel case
28 Abrasive oxide blade tip
29 Plasma-sprayed graded ceramic rubbing strip
30 Nickel steel stators
31 Nickel steel rotor blades, laminar flow aerofoils
32 Nickel steel exhaust struts
33 LP shaft
34 Fan thrust bearing
35 LP shaft roller bearing
36 HP spool thrust bearing (oil damped)
37 Oil cushion annulus
38 HP spool roller bearing (oil damped)
39 Sump and breather zone
40 LP shaft roller bearing
41 Fourth-stage air for sealing and cooling
42 Fourth-stage air to turbine stages 6 & 7 cooling and sealing
43 Tenth-stage air, high-power customer services
44 13th-stage air, turbine sealing and cooling
45 14th-stage air, turbine interstage cooling
46 16th-stage air, HP turbine cooling and sealing
47 17th-stage air to cooler
48 Air cooler in strut
49 Air from cooler to HP roller bearing
50 Fan discharge air to manifold
51 Fan discharge air manifold (air for ACC, oil cooler, and ignition harness cooling)
52 Fan air for active clearance control (ACC)
53 ACC air to compressor casing
54 ACC air to HP turbine case
55 ACC air to LP turbine case
56 Gearbox drive gears
57 Angle gearbox to accessory gearbox
58 Air to oil cooler
59 Engine oil cooler
60 Fuel/oil heat exchanger
61 Oil tank
62 Oil tank filler
63 Pressure oil to bearing
64 Scavenge oil
65 Main fuel feed
66 Fuel distribution block
67 Fuel manifold
68 Single-pipe fuel spray nozzles
69 EPR and EGT thermocouples
70 Thermocouple harness
71 Full authority electronic engine control cables
72 Engine mounting points
73 Engine starting air delivery
© FLIGHT INTERNATIONAL
FRANK MUNGER AMRAeS 1.10.84

every day. In some cases, they are succeeding; in some cases, they are not.

Health problems that result from inactive lifestyles are being overcome through diet and exercise programs. People are changing to low-fat, high-fiber diets and exercising more. Such routines are beginning to have an effect. The 1980s has had fewer heart attacks than the 1970s.

Companies that automate are setting up retraining and resettlement programs for people who lose their jobs due to technology advances.

Increased safety regulations and careful monitoring are decreasing the chances of nuclear leaks. Federal laws have been passed to ensure that industrial plants and automobiles no longer pollute the air. Special smoke scrubbers and emission devices cut down on the pollution from smokestacks and exhaust pipes.

Lower speed limits, seat belts, and strict laws about drinking and driving have helped decrease injuries from automobile accidents.

But people have not yet found a way to cope with some of the most serious problems of technology. Crime and anxiety are still serious problems in large, overcrowded cities. Drug and alcohol abuse are still on the rise. But by far the most serious technological problem that remains unsolved is the risk of nuclear destruction. The United States and the Soviet Union are trying to find ways to eliminate nuclear arsenals throughout the world. But no significant progress has been made.

CAREERS IN THE FUTURE

Each technology chapter touched on some of the most widely practiced career fields relating to the technology area in question. But the subject of careers in the future is important enough to be discussed as a topic by itself. This section explains what a career is, some emerging career fields that have a bright future, and how to select the right career field in the age of technology.

What Is a Career?

You often hear the terms "career" and "job." In fact, they are usually used interchangeably. But they are not the same thing. They differ in several ways: why we work at them, how we get into them, how we prepare for them, how long we expect to keep them, and our attitude toward them.

A **job** is something we do to earn the money to live, or it is a position within a career field. For example, if you work in a fast-food restaurant to make enough money to earn your way through college, you have a job. If you are an engineer, your career field is engineering whether or not you actually have a job in engineering at the moment. If you are working as an engineer for a company, you have a job within your career field of engineering.

A **career** is a broader concept than a job. You prepare for a career by going to school or by receiving special training of some type. When you enter your career field, it is normal to stay in it and advance. That is, in addition to doing your job in your career field, you study and keep up-to-date on developments in your field.

How to Select a Career Field for the Future

Because a career is something you will be involved with for a long time, you need to select the right one. The first step is making up your mind about what you want to do. Almost 80 percent of the people who work today got into what they do by accident or by default rather than by choice.

The next step is understanding why people work in the first place. Then you need to know how to actually select a specific career. Finally, you need to know how to determine which career fields have the best future prospects.

Why Do People Work? People work for many reasons. The most obvious, of course, is money. Some people work because they want to, and others work because they have to. Work must fulfill three broad needs:

1. Need for financial security
2. Need for personal satisfaction
3. Need to make a contribution to society

The first need is for **financial security**. People need their work to provide the money they need to live. All people need food, clothing, and shelter. These things require money. People also want material goods that extend beyond the basic necessities of life. The more material needs people have, the more money they will need to satisfy them. It is important to select a career field that will allow you to earn enough money to satisfy your basic and material needs, whatever they may be.

People need to derive **personal satisfaction** from their work. The things that give satisfaction differ from person to person. You should know what makes you happy. Is it status in the community? Is it helping people? Is it working outdoors? Whatever you need to make you happy beyond the material needs in life can come, at least in part,

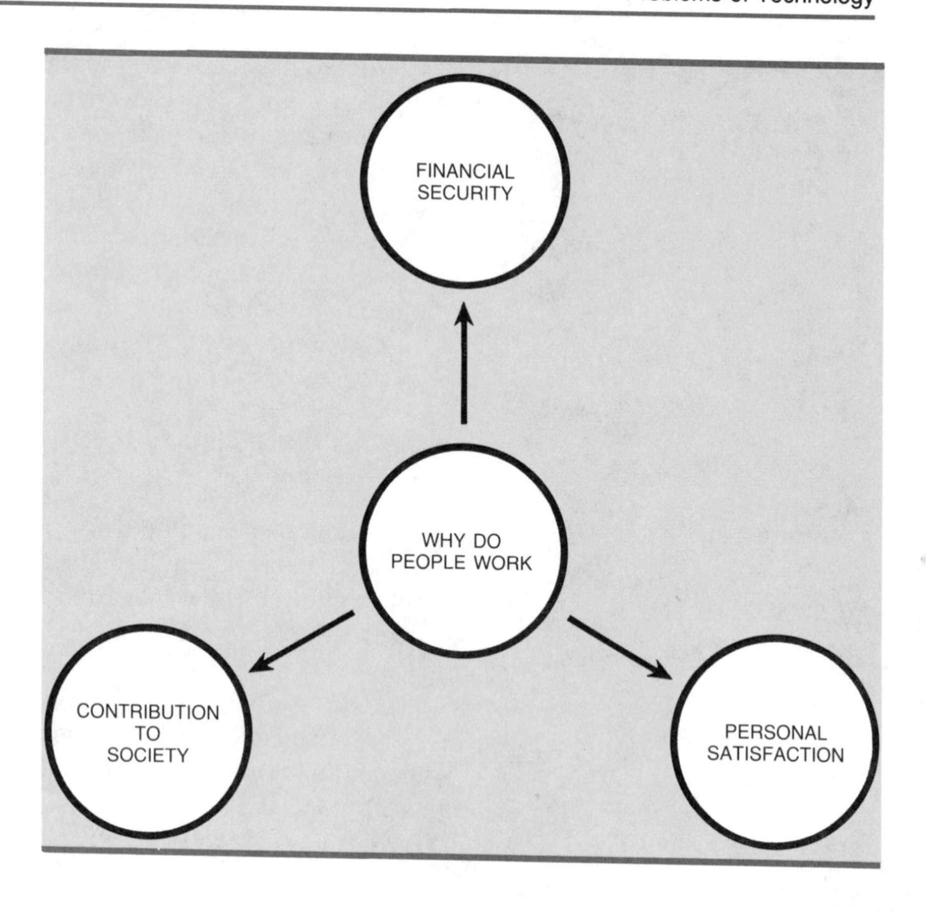

Why people work.

from your career. So you should select a career field to help supply this need.

People also need to feel that they are making some kind of **contribution to society**. In other words, they need to feel that their work is important. Everyone has his or her opinion of what is important. You should select a career field that you think is important.

Selecting a Career How can you go about selecting a career that will fulfill the basic needs we have explained? Unfortunately, there is no foolproof method because there are so many uncontrollable variables in life and people have a tendency to change. However, by using a logical, objective, systematic approach, you can cut down on your chances of selecting a career field you won't be happy with. Such an approach involves asking yourself three important questions and then getting the help necessary to find the answers.

1. Of all the career fields available, which one am I interested in?
2. Of all the career fields I am interested in, which one can I handle?

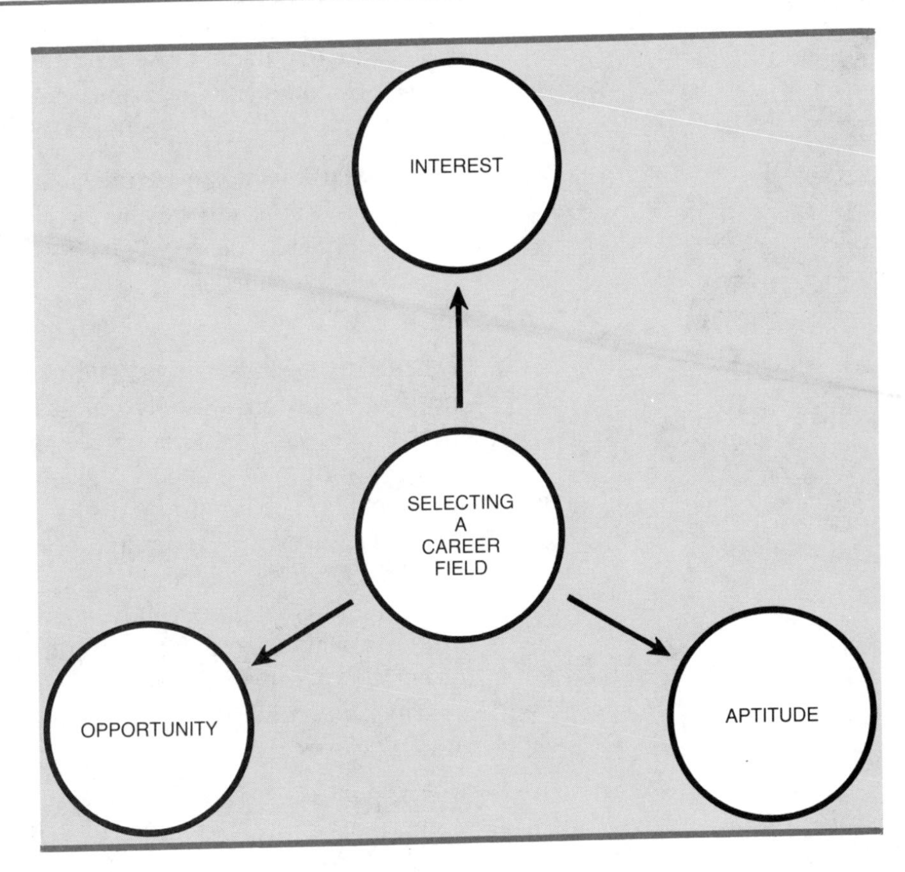

Selecting a career field.

3. Of all the career fields I can handle, which one has the best future prospects?

These three questions can be summarized as interest, aptitude, and opportunity. School counselors can help you determine where your **interests** lie and help you match them with career fields. They can also help you determine where your **aptitudes** lie and thus narrow the list of choices. Opportunity has a tendency to change as technology changes. This fact will affect what careers have the best opportunities for the future.

Some Careers

Some careers are becoming outdated, and others are just emerging. Generally, labor-intensive careers are declining. These fields involve much repetitive manual and unskilled labor. Careers tied to advances in emerging technologies and providing services are increasing. We are becoming a technology-intensive, service-oriented society rather than a do-it-yourself, labor-intensive society. This trend is leading to some new career fields. Some careers with bright prospects for the future are

- Hazardous-waste technician
- Geriatric health care worker
- Energy technician
- Robot technician
- Laser technician

Hazardous-Waste Technician You learned earlier that the disposal of hazardous waste is a difficult technological problem. Nuclear-powered electric generating plants, chemical plants, and numerous types of industrial plants produce a continuous supply of dangerous waste material that must be disposed of in a way that is not hazardous to the health of the community for now and for the future.

This need is leading to a growing industry: hazardous-waste disposal. This emerging industry will need **hazardous-waste technicians** with specialized training in dealing with hazardous-waste material. Specializations within this field will be nuclear-waste technician and chemical-waste technician. Such careers will require intensive specialized training.

Geriatric Health Care Worker One of the positive effects of technology is that the average life span of people is increasing. This means that people are living longer and this means that more and more of the world's population will fall into the elderly category in the future. The elderly require special forms of health care and medical treatment. More people will be needed to care for the elderly.

Geriatric health care workers will require specialized training similar to that received by nurses and medical technicians. However, the training will focus on the needs of the elderly.

Energy Technician You learned that fuels are available only in limited supplies, and that the world is rapidly depleting the supplies. Many different kinds of alternative energy sources are being used. Garbage is being burned to produce electricity, solar energy is being used to heat homes and water, and geothermal energy is being used to produce the steam needed in electric generating plants.

The rapidly growing need for alternative sources of energy is demanding more **energy technicians**. Such careers will require intensive specialized training in handling, extracting, processing, and distributing all of the various alternative forms of energy that are emerging; including synthetic fuels, biomass fuels, nuclear fuels, ocean thermal sources, and geothermal energy sources.

Robot Technician You learned that modern manufacturing plants are converting traditionally labor-intensive processes into automated processes through computer-aided manufacturing, including robotics. Therefore, more technicians will be needed to program, maintain, diagnose, troubleshoot, and repair robots. **Robot technicians** will need specialized training in electronics, manufacturing processes, computer programming, hydraulics, pneumatics, and mechanics.

Laser Technician The laser is a rapidly developing technological device with many applications. It has valuable uses in manufacturing, national defense, and medical technology. Laser processes in manufacturing will be one of the most widely used applications of lasers. **Laser technicians** will be needed to program, operate, maintain, troubleshoot, and repair laser equipment. Such career fields will require intensive specialized training.

These are only some of the many new and emerging careers that are growing out of technological advances. There are many more that could be listed, but these characterize the types of career fields that will be needed. In addition, many service-oriented fields will emerge. These fields will be the result of people spending less and less time doing for themselves and wanting more and more services available to them.

SUMMARY

Technology has been the vehicle through which people have improved the quality of life since the beginning of time. The contributions of technology would make a very long list. However, there are some problems that have resulted from technology. These problems can be categorized as personal, societal, and environmental.

Personal and societal problems include overcrowding, development and testing of dangerous drugs, accidents and injuries, war and disarmament, the changing nature of work, and uprooting-related problems. Environmental problems include the depletion of limited natural resources, pollution of the air, water, and ground, and nuclear-energy-related problems.

Health problems that result from inactive life styles are being overcome through diet and exercise programs. Companies that automate are setting up retraining and resettlement programs for people that lose their jobs due to technological advances. Increased safety regulations and careful monitoring are decreasing the chances of nuclear leaks. Smoke scrubbers and emission devices cut down on the pollution from smokestacks and automobile exhaust pipes. Lower speed limits, seat belts, and strict laws about drinking and driving have helped decrease injuries from automobile accidents.

However, people have not yet found a way to overcome some of the most serious problems of technology. Crime and drug and alcohol abuse are still on the rise. But by far the most serious unsolved technological problem is the risk of nuclear destruction.

There are some new and emerging career fields that are associated with advances in technology. Since a career is something you will be involved in for a long time, it is important to select a career that is right for you. To do this, you first need to know why people work. There are basically three reasons: (1) the need for financial security, (2) the need for personal satisfaction, and (3) the need to make a contribution to society.

To select the proper career field, you must ask yourself three questions concerning interest, aptitude, and opportunity and receive professional guidance and assistance in answering them. (1) Of all the careers available, which one am I interested in? (2) Of all the career fields I am interested in, which one can I handle? (3) Of all the career fields I can handle, which one has the best prospects for the future?

Some careers are becoming outdated, and others are emerging. Generally, labor-intensive careers are declining. These fields involve much repetitive manual and unskilled labor. Careers tied to advances in emerging technologies and providing services are increasing. Some careers with bright prospects for the future are hazardous-waste technician, geriatric health care worker, energy technician, robot technician, and laser technician.

KEY WORDS

Overcrowding
Accidents
War
Disarmament
Depersonalization
Submarines
Surface-to-air missiles
Mechanization
Eminent domain
Nuclear energy
Chemicals
Nuclear leak
Financial security
Contribution to society
Aptitude
Hazardous-waste technician
Energy technician
Laser technician
Drug problems
Injuries
Uprooting
Sociologists
Strategic bombers
Intercontinental ballistic missiles
Automation
Pollution
Insecticides
Nuclear wastes
Nuclear-waste disposal
Personal satisfaction
Interest
Opportunity
Geriatric health care worker
Robot technician
Job
Career

SOME QUESTIONS TO TRY

1. Name three categories of personal and societal problems that result from technological advances.
2. Explain how advances in transportation technology have caused problems for people.
3. Explain the term "automation" and the problems it can cause people.
4. Name two categories of environmental problems caused by technological advances.
5. Explain how people are solving transportation-related problems.
6. Explain how people are solving health-related problems.
7. Explain the difference between a career and a job.
8. List three reasons why people work.
9. What are the three questions a person should ask when selecting a career field?

10. Name three emerging career fields with bright prospects for the future.

TECHNOLOGY LEARNING ACTIVITIES

1. Help solve the problem of pollution by organizing a community cleanup program. Select a section of the community and work through your school and local government officials to clean it up.
2. Work through your instructor to invite your local congressman or congresswoman to your class and present him or her with a case for passing a mandatory seat belt law.
3. Obtain a list of all the state senators and congressmen and congresswomen for your district and write a letter to each, urging the passage of stricter laws for drinking and driving.
4. Divide your class into two teams: the Soviet Union and the United States. Debate the issue of nuclear disarmament. Develop a comprehensive plan for disarming the nuclear arsenals of your two countries. Write up the plan and mail it to the president of the United States and the premier of the Soviet Union.
5. Develop and implement a "say-no-to-drugs" campaign for your school.

CHAPTER 14

CHANGING TECHNOLOGY

OBJECTIVES

After completing this chapter, you should be able to

- list and describe some of the most important future developments of technology.
- explain the individual's role in changing technology.
- explain how decisions will be made in the future.

By now you know that technology changes constantly—sometimes slowly and sometimes rapidly, but it always changes. Imagine what people who lived in the last century would have thought if they could have looked ahead to the world of today. What would a person driving a horse-drawn buggy think of our automobiles and aircraft? What would a person living in a log cabin think of skyscrapers? What would an Indian sending smoke signals think of telephones, radios, and televisions? What would people who in wonder watched birds flying think of jet aircraft or the space shuttle?

Our world would no doubt amaze or even overwhelm people who lived during the last century. What do you think the world will be like in the next century? Some of the changes of tomorrow will be even more drastic than those of today when compared with those of yesterday. In every area of technology, developments will happen much faster between now and the turn of the century than they have happened in the past.

Examine the differences between the 1911 truck and the more modern 1983 aero astro truck.
Courtesy of General Motors Corporation.

FUTURE DEVELOPMENTS IN TECHNOLOGY

In construction technology, you can expect to see self-contained communities built in space and at sea. This, of course, will mean drastic changes in the tools, resources, and processes used in construction. On land, the home of the future will be vastly different from the homes of today. The home of the future will produce most of its own energy by burning garbage, using solar energy, and using wind turbines. Mundane household chores will be accomplished by household robots.

Shopping, banking, and even education will be accomplished without leaving the house. Home-based microcomputers will be linked to mainframes in banks and large shopping complexes. The computers will be used to enter and record banking transactions and to update records instantly. They will also be used to shop for groceries and other goods. By entering a command such as "MILK," a person will be able to view a list of various brands, sizes of containers, and prices available. When all grocery items have been selected in this way, a robot will collect the groceries, transfer them to a central

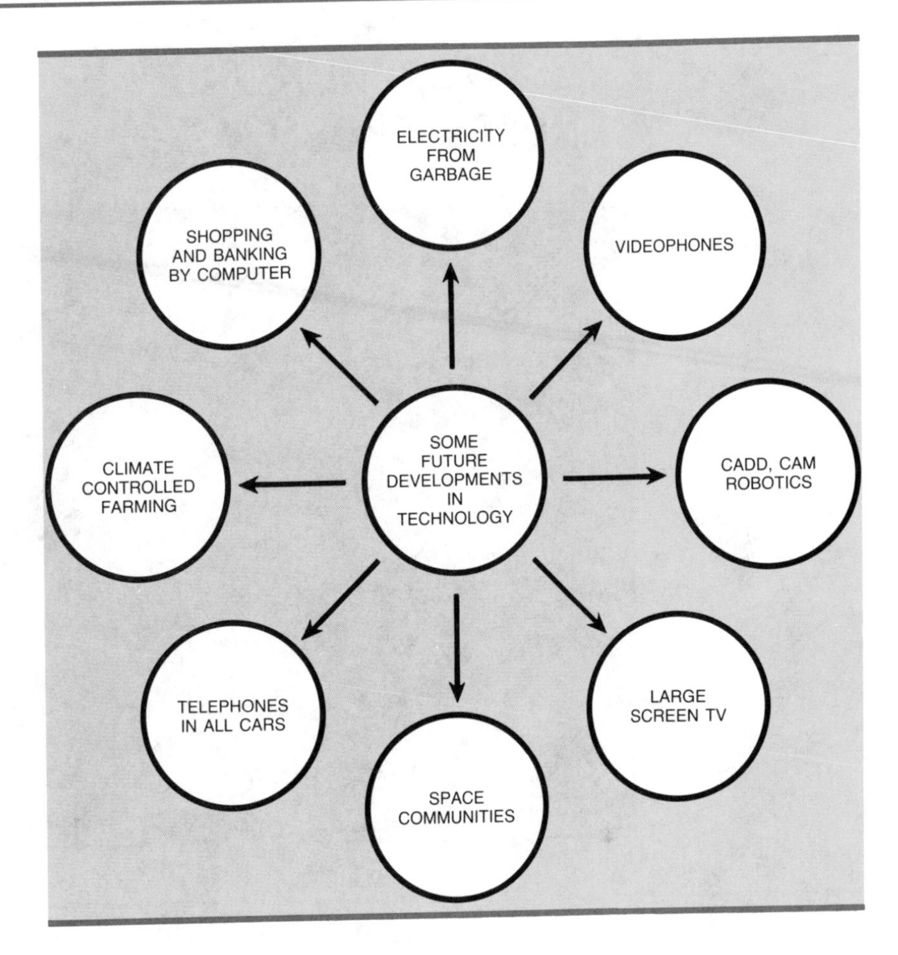

Some future developments in technology. The Promise of Technology.

location for packaging, and place them on a shipping dock for pickup and delivery.

The school of the future may be in your living room. Certainly, for students who are at home sick this will be the case. Microcomputers, televisions, and videophones in the home will be linked to centrally located schools. In this way, students can interact with teachers and other students and keep up with their work without leaving the home.

In the area of energy and power, you will see fossil fuels replaced as the primary energy source. More and more you will see garbage and waste products burned to produce electricity. You will also see solar energy electric generating plants in space where they will be closer to the sun and, hence, more efficient.

Manufacturing plants will become almost completely automated. Although CADD systems will automate the design process, people will remain the most important element in design. However, fewer people will be needed in the actual manufacturing of products. Direct

WE WILL SOLVE OUR ENERGY PROBLEMS......

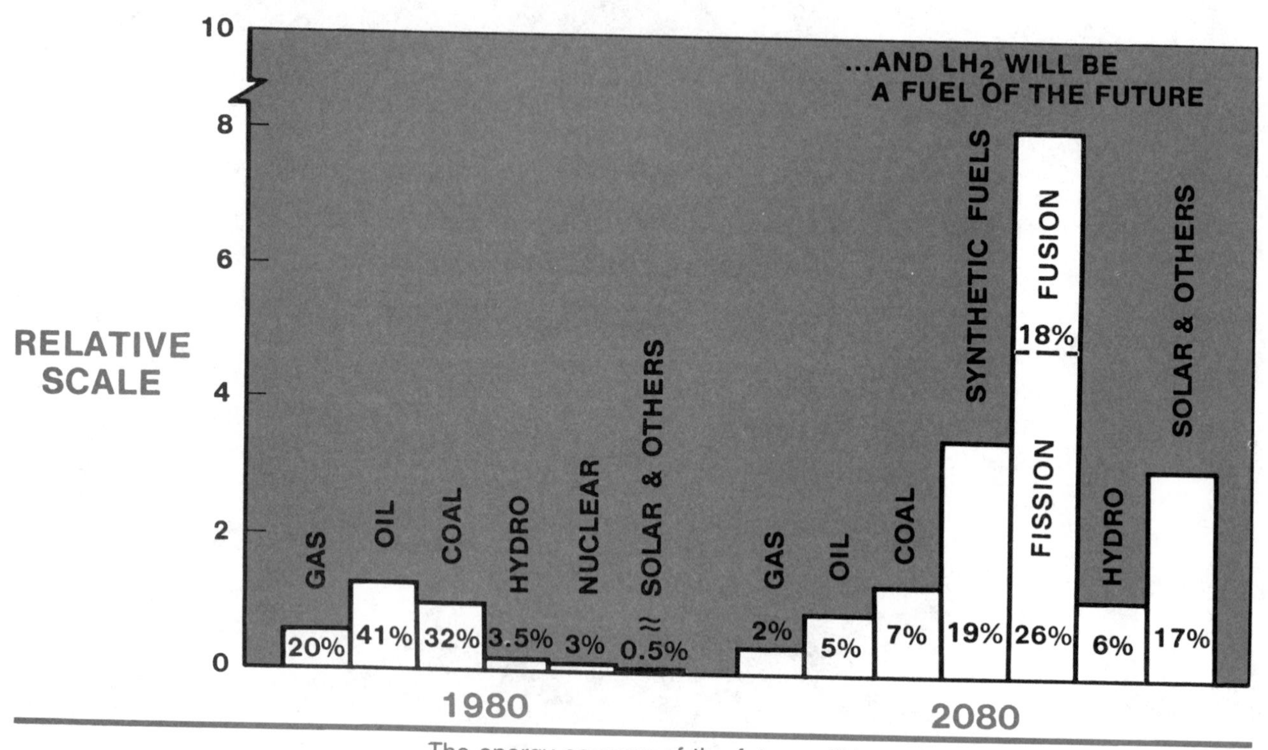

The energy sources of the future will be solar energy, nuclear energy, and synthetic fuels. *Courtesy of Lockheed Corporation.*

numerical control (DNC) machines and robots will be the norm on the shop floor. These machines will receive their instructions directly from the data base of the CADD systems. Robots will load and unload machines, assemble products, package them, and place them in storage or on loading docks for shipping. Microcomputers will be programmed to monitor all operations and signal when a problem exists. Microcomputers will also diagnose electronic and mechanical problems with DNC machines and robots.

In transportation, improved mass transit systems will be the next frontier. These systems will use extremely fast, silent monorails for earth transportation. Over water, travel will be improved by vessels that skim lightly over the water on a cushion of air. Airliners will no longer need long runways. They will take off and land by hovering like helicopters. Travel into space and back will become as common as the air travel.

CADD systems like this by Mentor Graphics Corporation will be one of the principal tools of design and manufacturing in the future. *Courtesy of Mentor Graphics Corporation.*

Fast, silent monorails will be the mainstay of mass transportation on the ground in the future.

The photograph at the top projects a subway system that will meet every incoming transport and stop at the aircraft parking apron where passenger and luggage transfer takes place. Passengers are then transported to an underground hub terminal for further dispersal. Package handling will be mechanized and almost completely automatic. The bottom picture illustrates the airport of the future. *Courtesy of Lockheed-California Company, a division of Lockheed Corporation.*

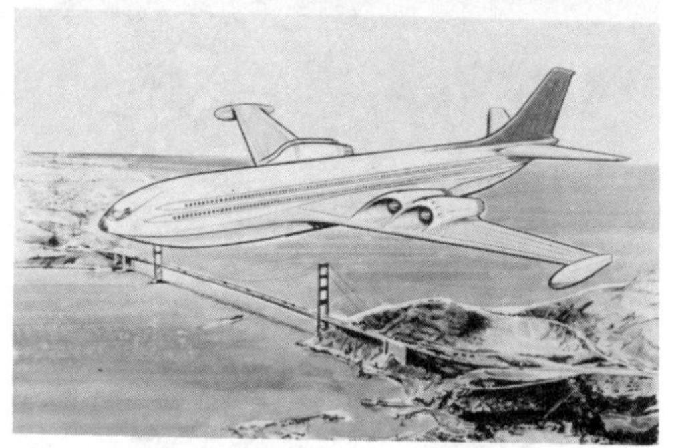

The top photograph shows an example of the return of the propeller as a partial solution to energy shortages. The prop fan concept incorporates an 8-to-10-blade swept propeller that offers high fuel efficiency and speed comparable to today's jet airliners. Large seaplanes like those at the bottom will solve the problems of landing oversized aircraft on today's runways. Such seaplanes will land on natural water bodies, human-made ponds, or powered cradles. *Courtesy of Lockheed-California Company, a division of Lockheed Corporation.*

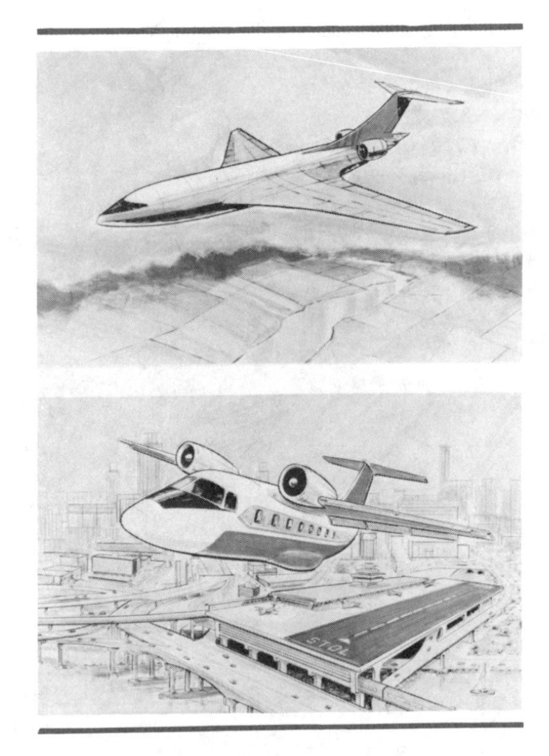

Energy-efficient transport aircraft (top) making extensive use of advanced technology should appear in the early 1990s. Short takeoff and landing aircraft (bottom) will allow airport locations closer to centers of cities, thus utilizing limited space.
Courtesy of Lockheed-California Company, a division of Lockheed Corporation.

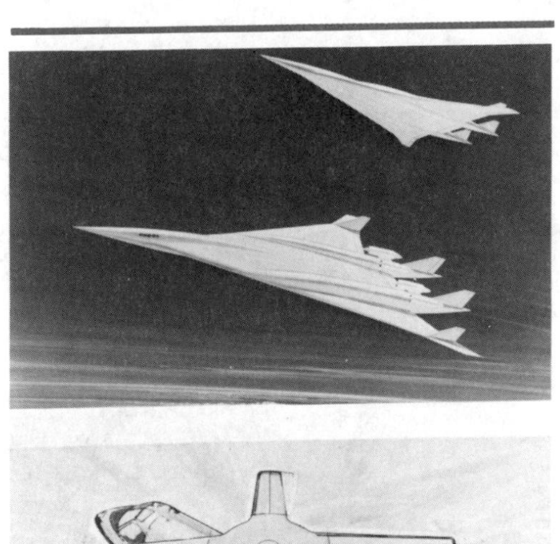

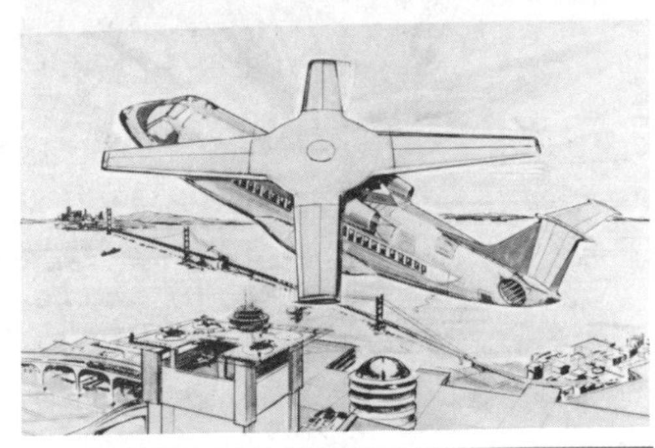

Supersonic transports (SSTs) of the future will significantly differ from those flying today. SSTs (top) will be capable of a Mach 2.7 cruise speed, a passenger capacity of over 400, and a range of at least 4,000 nautical miles. Vertical takeoff and landing aircraft (bottom) will be available after the year 2000.
Courtesy of Lockheed-California, a division of Lockheed Corporation.

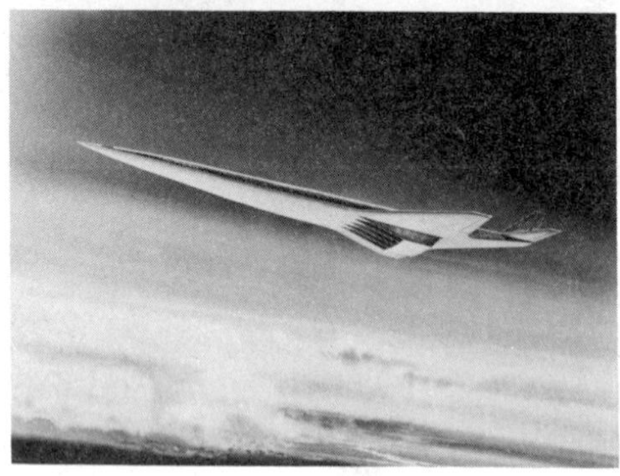

Air cargo aircraft (top) will play a major role in future air transportation. Thick wings of this aircraft will be able to accommodate two rows of standard cargo containers along the entire wing span in addition to the normal fuselage capacity. The hypersonic transport pictured at the bottom will double the speed of its predecessor, the supersonic transport or SST. *Courtesy of Lockheed-California, a division of Lockheed Corporation.*

New technologies with regard to both materials and manufacturing processes will emerge in the future. *Courtesy of Lockheed-California, a division of Lockheed Corporation.*

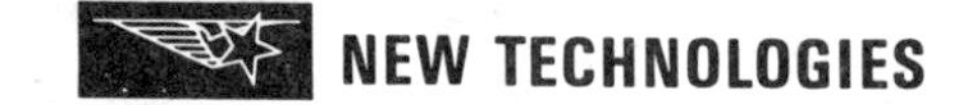

NEW TECHNOLOGIES	BENEFITS
MATERIAL	
• COMPOSITES	LIGHTER AIRCRAFT
• FIBER REINFORCED METALLICS	LOWER COSTS
• ADVANCED ALUMINUM ALLOYS	
• CERAMICS/METALLICS	LOWER FUEL CONSUMPTION
MANUFACTURING PROCESSES	
• SUPER PLASTIC FORMING	LOWER COSTS
• ADHESIVE BONDING	LONGER LIFE CYCLES
• AUTOMATED LARGE PIECE MANUFACTURING	

Energy-efficient transport aircraft will appear in the next century. *Courtesy of Lockheed-California, a division of Lockheed Corporation.*

ENERGY EFFICIENT TRANSPORTS WILL APPEAR

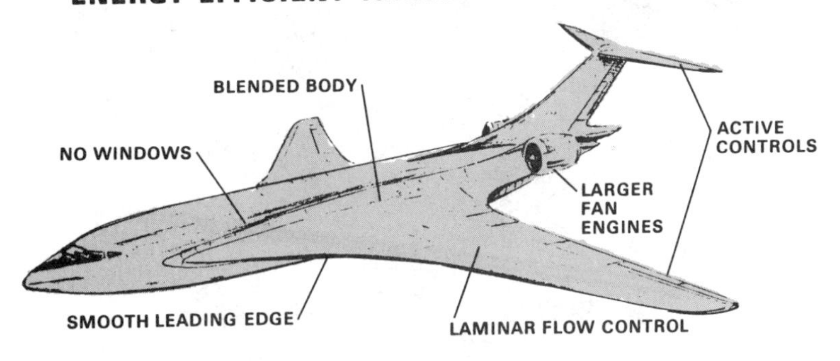

TECHNOLOGY	IMPROVEMENT
• COMPOSITES & NEW ALUMINUM ALLOYS	10%
• PROPULSION	15%
• AERODYNAMICS	30%
• FLIGHT CONTROLS	10%
TOTAL	65%

TRAVEL WILL INCREASE
NEW MARKET AND MORE LEISURE
YEAR 2080

Worldwide travel will increase significantly by the year 2080. *Courtesy of Lockheed-California, a division of Lockheed Corp.*

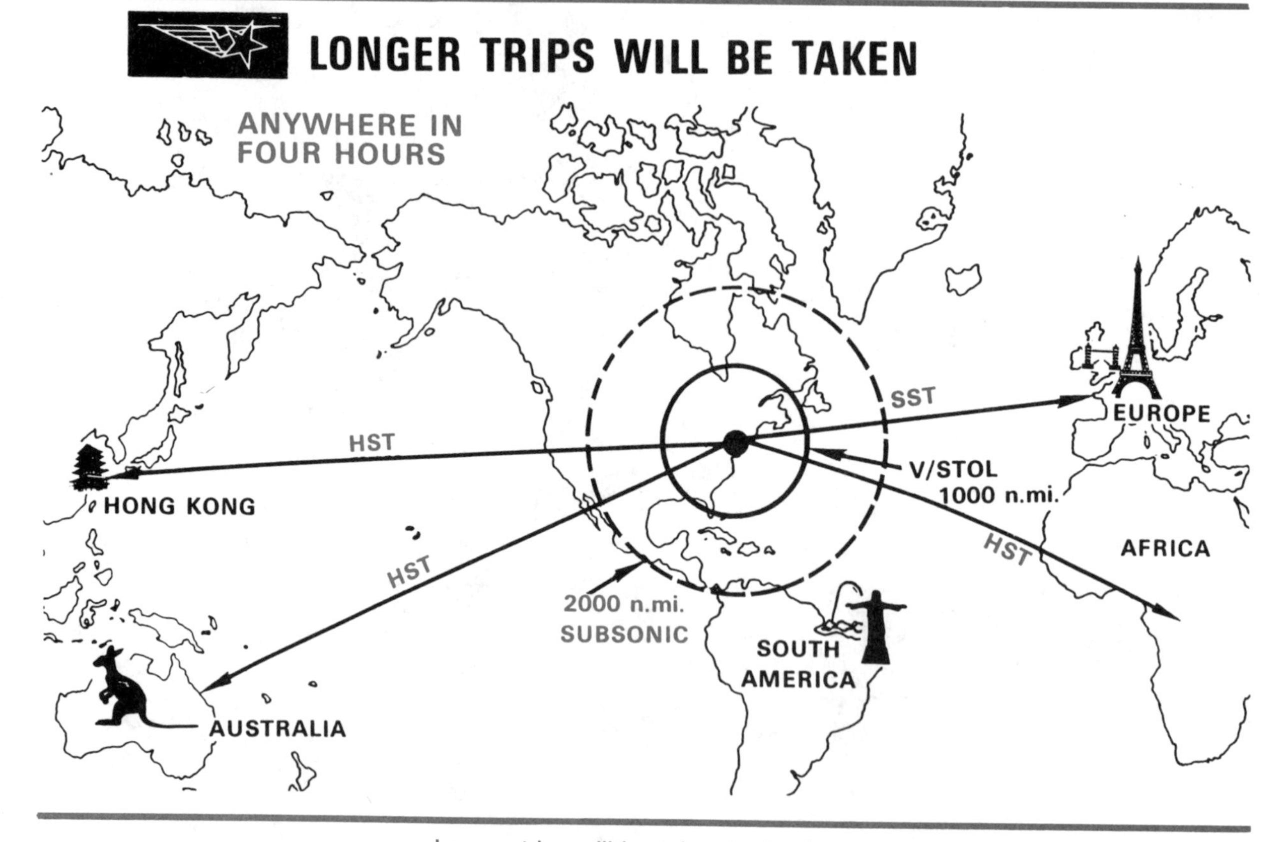

Longer trips will be taken in the future. People will be able to travel anywhere in the world in 4 hours or less. *Courtesy of Lockheed-California, a division of Lockheed Corporation.*

Information processing will undergo continual changes. Advances in microelectronic circuitry will allow for the production of increasingly powerful microcomputers that will continue to become smaller and less expensive. The microcomputer will replace the wheel as the most widely used, most important tool of technology.

The list of new and expanding applications of microcomputers is long. Companies with offices scattered nationwide will be networked. All activities that take place at one location will be instantly communicated to all locations, and records at all locations will be updated instantly.

Sales representatives who travel on the road will be linked into the mainframe computers of their home offices by microcomputer/telephone hookups that can be activated almost anywhere. This will give them instant access to an inventory listing, so they don't place

Continuous advances in microcomputer technology will expand significantly the number of microcomputer applications. *Courtesy of IBM Corporation.*

Telephone-microcomputer linkups, such as those made possible by the AT & T Unix PC, will allow immediate communication among people at many different locations across the country and the world. *Courtesy of AT & T.*

orders for items that are out of inventory. In addition, it will allow them to place an order, update the inventory, and activate the billing process all from remote locations.

People will spend more time working at home instead of traveling to the office. Microcomputer and videophone networks will allow them to have complete access to files, records, and people at the office. Doctors and lawyers will no longer spend long hours in medical and law libraries. Instead they will tie into libraries contained in centralized mainframes. The information they need will be instantly available to them.

In graphic communication technology, microcomputers will be produced that are capable of interfacing with phototypesetting equipment and even replacing it. Authors will begin to do their own typesetting, or they will send diskettes instead of manuscripts to publishers.

The future will see a tremendous increase in the number and types of electronic communication options. The telephone will be replaced by the *videophone*, a combination of a telephone and a television. It allows callers to see who they are talking to. In addition, telephones will become standard features on all automobiles. In airlines, each seat will be equipped with its own telephone, radio, and television system.

In the future, agricultural technology will be less and less dependent on nature. Much of our food will be grown in huge climate- and environment-controlled domes. All phases of agricultural technology, from planting through food processing, will be accomplished in the dome.

Some of the most fascinating developments will take place in health care and medical technology. Many homes will be equipped with special computerized diagnostic machines that will give instant feedback on vital signs and other measures of health. Long-term-care patients will be returned to the home but linked to the hospital by a computer and videophone linkup. Artificial hearts and other artificial organs will become commonplace, and the life expectancy may reach as high as 100 years in the next century.

As overwhelming as some of these developments seem now, they will be common in the future as the automobile and television are today. Every person can play a role in the technological changes that occur in the future. In fact, every person will have a role.

YOUR ROLE IN CHANGING TECHNOLOGY

People have the opportunity to play two separate roles with regard to changing technology. Both roles carry important responsibilities. The first role is that of the change agent. A **change agent** is a person who

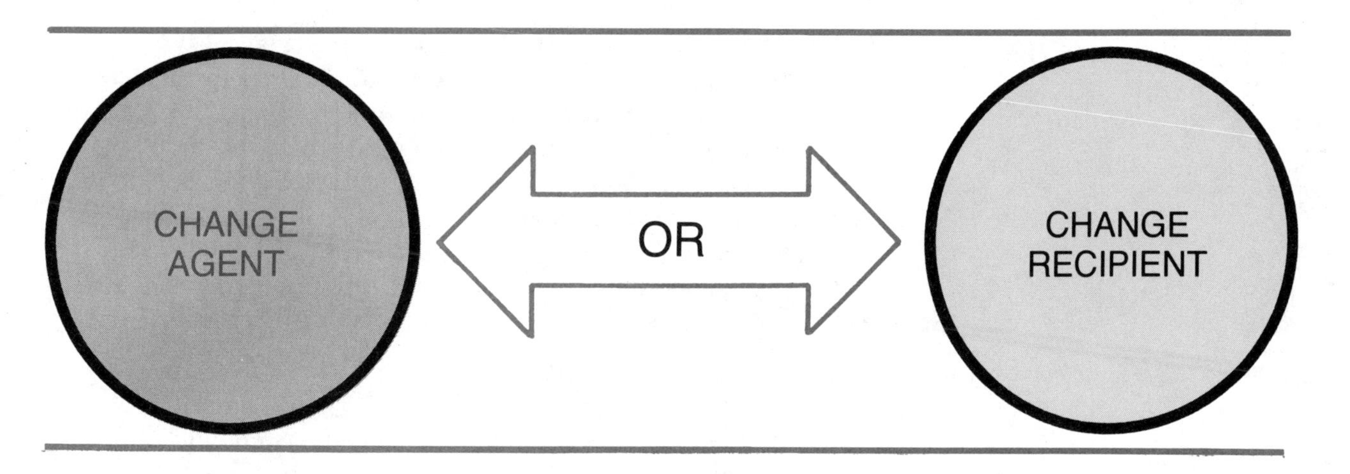

Roles people play with regard to change.

causes change to happen. Some famous change agents have been Alexander Graham Bell and Eli Whitney. Any person can be a change agent. You don't have to invent something as important as the telephone or wheat reaper to be a change agent. In fact, you don't have to invent anything at all. Change agents are people who work at any level to bring about change of any kind.

If you and your friends walk to school every day and find a shorter route that saves time, you are acting as a change agent. The change would be from the old route to the new. If you work through the political system to get new laws passed or old ones changed, you are acting as a change agent. In fact, the political system is one of the most effective ways for individuals to work as change agents. It is the one system in which we can all participate. Therefore it is important to register and vote.

Anybody can be a change agent. But anyone who chooses to be must be willing to accept responsibility for the changes. Thus, we must be aware of the consequences of the change and be willing to take responsibility for them.

Determining the Consequences of Change

To determine what the consequences of change will be, people must ask questions. This process of change is important. Suppose you are the manager of a manufacturing plant that is losing business to foreign competition. If someone told you they could improve production in your plant by 200 percent by installing robots, you would probably like the idea. But how would you feel if you learned that installing the robots would mean 50 people who had worked for you a long time would lose their jobs? Are you willing to accept responsibility for that consequence?

In order to determine what the **consequences** of a technological change will be, people have to ask a number of questions such as:

1. Who will this change help?
2. Who will this change hurt?
3. Can the negative effects of the change be overcome?
4. Will the change do more harm than good?

Answering such questions takes time. Consequently, change agents must be patient. Rushing changes, even the best of changes, can lead to problems. The reason is that it is human nature to resist change. Even change for the better makes people uncomfortable at first. It is important to understand this, because changes usually will not succeed without the support of people. Being a change agent requires patience and persistence.

The second role people play with regard to change is change recipient. **Change recipients** are the people affected by the change. At one time or another, we are all change recipients. As recipients, we might view the change as good or bad. In either case, change recipients have responsibilities too.

The first responsibility is to be open minded and objective about the change. Those recipients who benefit from a change have a responsibility to help those who do not. In the previous example, as the manager of the manufacturing plant that is considering robots, you stand to benefit from the improved productivity they will bring. So does the public that will be able to purchase your products at a lower price. But what about the 50 people who will lose their jobs? You have a responsibility to help them get retrained and resettled in new jobs.

DECISION MAKING IN THE FUTURE

In the earlier example, you were asked to suppose that you are the manager of a manufacturing plant. You are forced to make a decision in which the options are to continue your operation as usual or to change and install robots. The consequences of continuing as usual are that you will continue to lose business to foreign competition and, eventually, go out of business. The consequences of changing and installing robots are that 50 loyal employees will lose their jobs. A

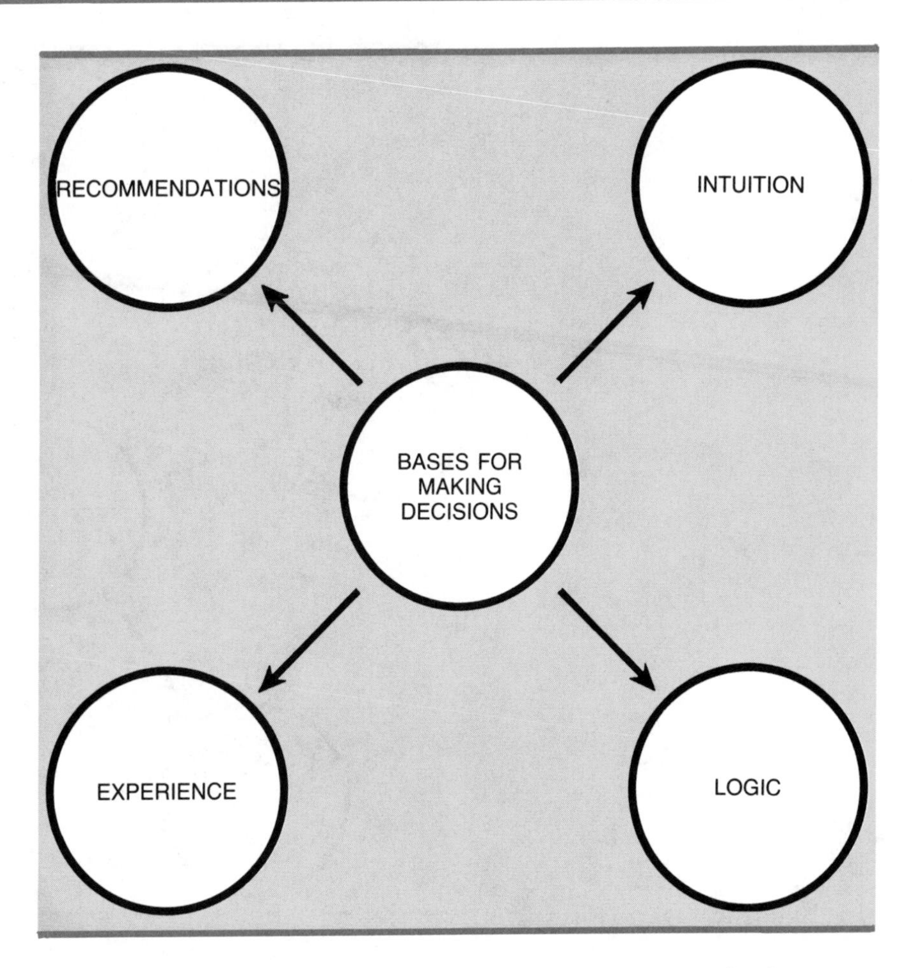

Bases for making decisions.

decision must be made, but how do you do it? Such decisions characterize the types that will be an everyday part of life in the world of the future. Consequently, it is important for you to know how decisions are made.

Decision making involves selecting from among a number of alternatives. A decision is simply a choice that you make. People make hundreds of decisions every day. When you wake up in the morning, the first thing you do is make a decision. "Do I get out of bed or stay in?" You decide what to wear and what to eat, and you decide whether to walk or to ride. How do you make these decisions?

How Do People Make Decisions?

People must have some basis for selecting alternative A instead of alternatives B, C, or D in a given situation. Let's say that in the morning, while dressing for school, you decide to wear a raincoat. You do this

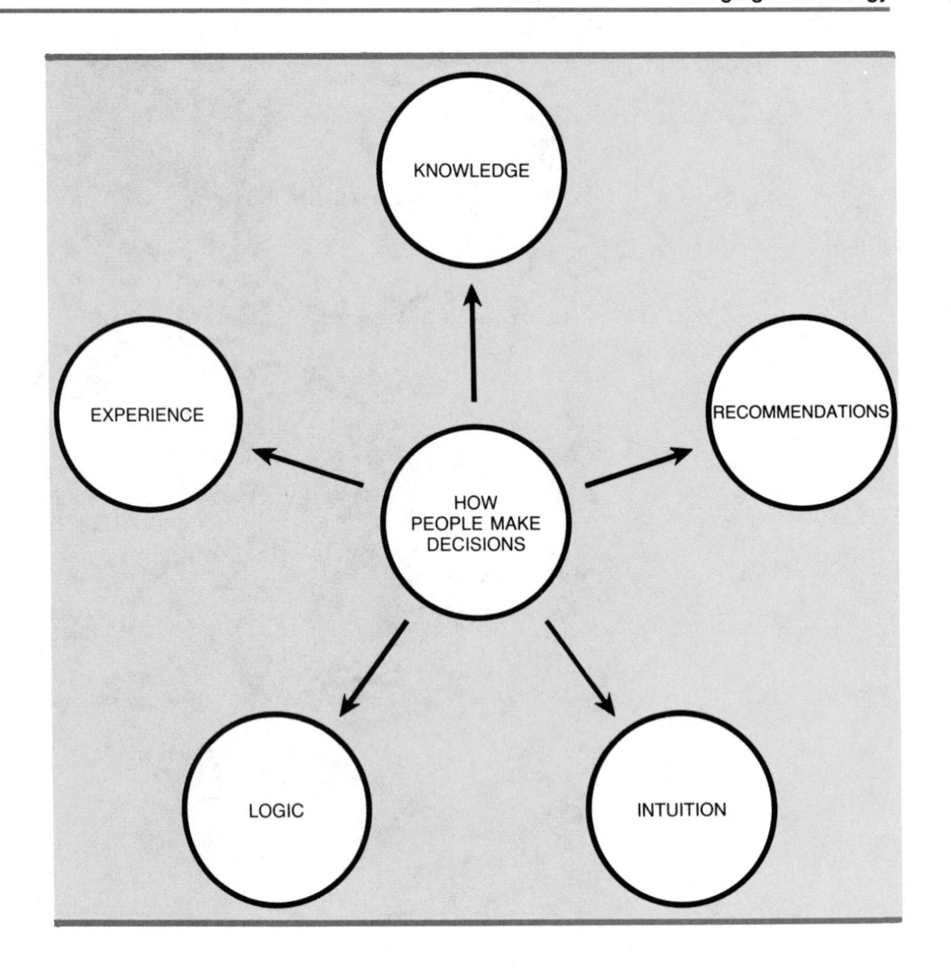

How people make decisions.

because you know that it is raining outside. You have knowledge. **Knowledge** is an important factor in making the right decision. To make the best decision possible, we need to have as much knowledge as possible about all the alternatives. Knowledge can form the basis for decisions we make. So can

1. The recommendations of others
2. Intuition
3. Logic
4. Experience

Recommendations of Others Even when we have all of the information available on the various options we are trying to decide on, it can still be difficult to make a decision. One of the problems with decisions is that, in some cases, not making a decision or not making one in time is the same thing as making one. For example, if your friend has fallen through thin ice while skating and is in danger of drowning, you must decide what to do. Doing nothing is the same thing as deciding to let your friend drown.

Not all decisions must be made in a split second. In some cases, there is time to give consideration to all of the various courses of action available to you. When this is the case, one of the things people do is ask for **recommendations** from others. What to do in this case or how to handle this problem are the types of questions that are sometimes asked when making a decision. The importance attached to a recommendation from someone depends on the amount of respect you have for the person's opinion. For this reason, it is important to ask for the recommendations of only those people who are knowledgeable and experienced with regard to the problem in question. If you are trying to decide whether or not to see a particular movie, you would want to ask someone who has seen it.

Intuition **Intuition** is the ability to know without having any concrete evidence. Have you ever said, "I don't know why I am right, I just know that I am." That is intuition. Intuition is a feeling that causes you to make a certain choice without knowing exactly why. Generally speaking, the more knowledgeable and experienced you are in a certain situation, the better your intuition will be with regard to making decisions in that situation.

Logic **Logic** involves taking a step-by-step, carefully sequenced approach in weighing all of the consequences of each alternative when making decisions. In applying logic, you ask yourself, "If I do this, then what will happen?" This is called an if-then question. Such questions are asked about every alternative, and the consequences are weighed. The opposite of logic is random choice. A **random choice** is the equivalent of closing your eyes and making a choice without weighing the consequences.

Experience **Experience** can be a valuable tool in making decisions, if you have experience relative to the decision. For example, if the gas gauge in your car indicates empty but you are only two miles from home, you have to decide whether to stop for gas or hope you will be able to make it home. If you have been in this situation with this car before, your experience will help you decide. What did you decide last time and what were the consequences? Did you make it home, or did you run out of gas? If you have always made it before, experience says you will make it this time. If not, you had better find a gas station.

Making the right decision is often difficult. In the world of the future, this will be increasingly true. Technological innovations, such as the computer, help. With a computer, we can store and retrieve

An example of an if-then logic test.

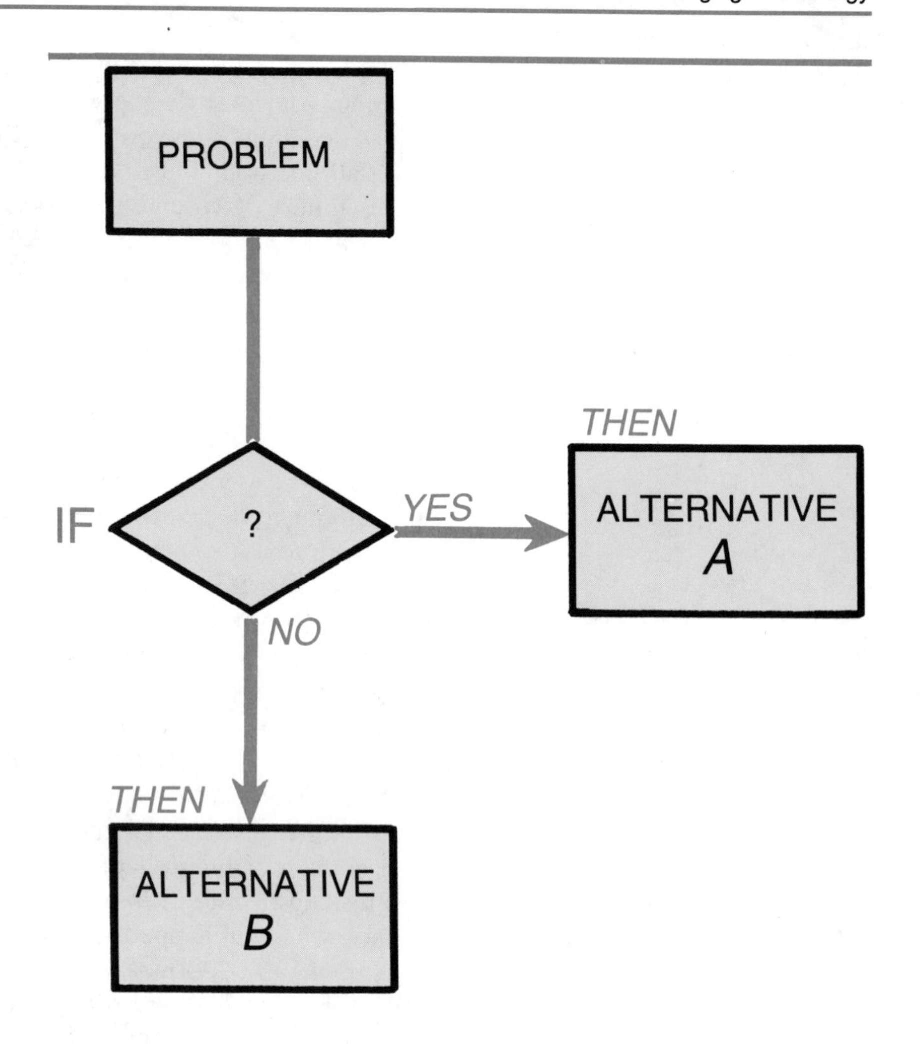

volumes of information that will help us make decisions. Through telecommunications technology, we can seek the advice of others almost anywhere on earth. By storing records of our experiences in computers, we can run a series of if-then tests and run a printout of recommendations.

But in the long run, as helpful as technological innovations can be in making decisions, people must ultimately make them. Therefore, in spite of all the amazing technological developments we can expect over the next 100 years, people will still control their own destinies. Because, in the end, people make the decisions.

SUMMARY

Technology never stands still. It is always changing. In the years to come, there will be numerous technological developments. Self-contained communities built at sea and in space, energy self-sufficient homes, and household robots are in our future. Banking, shopping, and learning at home by computer and telecommunications linkups are also in the future.

Garbage and waste will be converted into an important energy source by burning to produce the steam to generate electricity. Solar energy electric generating plants in space will also be seen.

The number of people working in manufacturing plants will decrease substantially as more and more plants automate with DNC machines and robots. Troubleshooting and diagnosing mechanical and electronic problems will be accomplished by microcomputers.

Fast, silent monorails and everyday space travel will become common in transportation technology. Increasingly powerful micro-computers that are smaller and less expensive will be the norm. This development will allow the list of microcomputer applications to grow rapidly.

Videophones will replace the telephones. Telephones will become standard features on all automobiles. Climate- and environment-controlled farming will become common, and human life expectancy will continue to increase.

All people play a role with regard to technological change. People can be change agents, change recipients, or both. Change agents make change happen. Change recipients are the people affected by change. Change agents must be aware of the consequences of the changes and be willing to accept responsibility for them.

To determine what the consequences of a technological change will be, people must ask the questions: (1) Who will the change help? (2) Who will the change hurt? (3) Can any negative effects be overcome? (4) Will the change do more harm than good?

Change recipients have responsibilities too. Those who are affected in a negative way must be open minded and objective about the change. Those who are affected in a positive way must be willing to help those hurt by the change.

There are five bases for making decisions. People rely on one or more of them each time they make a decision. The bases for decision making are knowledge, recommendations of others, intuition, logic, and experience. Technological innovations such as the computer can help people make decisions, but they cannot make decisions for them.

KEY WORDS

Change agent
Change recipient
Consequences
Recommendations
Knowledge
Intuition
Logic
Experience

SOME QUESTIONS TO TRY

1. How many future developments can you name?
2. What is a change agent?
3. What is a change recipient?
4. What is a decision?
5. What do people base decisions on?

TECHNOLOGY LEARNING ACTIVITIES

1. Make a poster showing the future developments expected in one or more of these technology areas: construction, manufacturing, transportation, health care and medical, electronic communication, graphic communication, agricultural, and energy/power.
2. Practice decision making by completing the following activity: Form a group of three students as a committee. Collect information from the library, the community, and fellow students. Then make the following hypothetical decisions and explain why you made each decision. Rotate the committee for each decision.
   a. Should a power company be allowed to build a nuclear reactor near your school?
   b. Should NASA's space shuttle program be scrapped because of the accidental deaths that have occurred?

c. Should the speed limit be raised to more than 55?
d. Should the drinking age be lowered to 18?
e. Should drunk drivers be given mandatory jail sentences?

Remember that you are assuming that your committee has the power to make these decisions and that the decisions will be binding. Make sure that you are willing to accept the consequences of each decision.

3. To learn how to become a change agent through the political system, complete each step in the following activity:
   a. Find out where people register to vote in your community. Learn what the requirements are to register.
   b. Learn where people in your community vote and where you would vote if you were old enough.
   c. Record the steps you would go through if you wanted to get a law passed in your state.
   d. Record the steps you would go through if you wanted to get an old law amended in your state.

TECHNOLOGY TRIVIA

What follows is some technology trivia to test your knowledge of technology. Answers to some of the questions can be found in this book, but not all of them. You may have to go to a library and do some reading to answer many of these questions. This can also be used as a trivia game. Divide the class into two groups and have a moderator ask the questions. See which groups can answer more questions.

TECHNOLOGY TRIVIA TEST NUMBER 1

1. What was the name given to the great wagons that hauled cargo for settlers going west during the 1700s and 1800s?
2. What was the name given to the horsedrawn coaches of the 1800s that carried passengers west?
3. One of the most elegant models of the horsedrawn carriage was named after the Queen of England. It was well balanced, had springs, and used rubber tires. What was this model called?
4. The first streetcars were pulled by horses along rails. What were they called?
5. What famous mode of ground transportation started in San Francisco in the late 1800s?
6. One of the first trucks looked like a cross between a covered wagon and a horseless carriage. What was it called?
7. The famous inventor Buckminster Fuller designed a car in 1933 that would go up to 120 miles per hour. What was it called?
8. What was the name of America's first locomotive?

9. One of the first successful steam powered locomotives won a contest in Liverpool, England in 1829 against horsedrawn and cable powered locomotives. What was it called?
10. What is the name of the sleek, fast locomotive that broke the international speed record in 1893 by running 112.5 miles per hour?
11. What is the name of America's first diesel powered stream-lined train?

TECHNOLOGY TRIVIA TEST NUMBER 2

1. What is the name of the ship many feel was the earliest known seagoing ship?
2. Approximately how many people manned the early viking ships?
3. What is the name given to the Chinese ships which travel the Yangtze and other rivers?
4. What was the name of Christopher Columbus's flagship?
5. What are the names of the other two ships that sailed with Columbus?
6. Which of the three ships of Columbus's fleet was the smallest?
7. What is the real name of the American ship nicknamed "Old Ironsides?"
8. What is the nickname given to the paddle steamer riverboats that were so popular on the Mississippi River during the 1800s?
9. From 1919 to 1935 the largest ship afloat was a German ocean liner. What was its name?
10. Today's modern supertankers can carry a huge capacity of over 250,000 tons. What was the approximate capacity of the first diesel powered tankers?

TECHNOLOGY TRIVIA TEST NUMBER 3

1. What was the name of the famous Wright brothers' airplane?
2. Which of the Wright brothers flew the first engine powered aircraft at Kitty Hawk on December 17, 1903?
3. What is the real name of the World War I biplane nicknamed the "Jenny?"
4. What was the first airplane to fly transcontinental airmail routes commercially?
5. What was the nickname of the airplane that Charles A. Lindbergh flew in making the first solo transatlantic flight in May, 1927?
6. What type of plane did Amelia Earhart fly in becoming the first woman to fly across the "Atlantic?"
7. What is the airplane that is credited with the birth of widescale commercial passenger airlines?
8. One of the most famous airplanes on the German side in World War II was called the "Stuka" divebomber. What is the Stuka's real name?
9. The most popular and versatile American bomber in World War II was the B-17. What was the B-17's nickname?
10. What was the name of the small, high powered fighter plane the British developed to fight off German aerial attacks in World War II?

TECHNOLOGY TRIVIA TEST NUMBER 4

1. The modern Mercedes Benz is an expensive automobile. So was the first model. What did a Benz originally cost in the late 1800s?
2. What was the real name of the steam powered car of the early 1900s known as the "Stanley Steamer"?
3. What famous model Ford car is credited with putting America behind the wheel?

4. What famous automobile body maker made bodies out of wood instead of metal?
5. What was the name of the first model of Cadillac built?
6. What kind of car won the first Indianapolis 500 race?
7. Who developed the original idea for the Lincoln Continental?
8. What was the first all metric car built in the United States?
9. What company manufactured the first Rambler?
10. What widely sold automobile was named after a wild horse?

TECHNOLOGY TRIVIA TEST NUMBER 5

1. What was the first airplane to fly faster than the speed of sound?
2. What was the first airplane to go faster than twice the speed of sound?
3. What was the name of the capsule flown by Alan B. Shephard in America's first manned space flight in 1961?
4. What was the name of the spacecraft flown by John Glenn when he became the first person to orbit the earth in 1962?
5. What was the name of the first spacecraft to carry two men in an orbital flight?
6. What is the real name of America's "Blackbird" reconaissance airplane?
7. What was the airplane that consistently set new speed and altitude records in the 1950s? Hint: It had to be launched from under the wing of a B-52 bomber.
8. What is the name of the spacecraft that carried three astronauts through 10 orbits of the moon and back in 1968?

10. What is the full name of the spacecraft which Neil Armstrong landed on the moon in July 1969?
11. What was the first space shuttle to orbit the earth?

TECHNOLOGY TRIVIA TEST NUMBER 6

1. Who was the designer of the first liquid fuel rocket?
2. What was the name of the first Russian satellite?
3. What was the first American satellite to orbit the earth?
4. What was the name of the rocket that launched America's first satellite?
5. What was the name of the rocket that launched America's first manned space flight?
6. What was the name of the spacecraft that undertook the first exploratory mission to another planet?
7. What is the name of the generation of unmanned spacecraft that landed on the moon between 1966 and 1968?
8. What was the name of the mission when Neil Armstrong landed on the moon?
9. What was the name of the USSR-USA joint program in space?
10. What was the name of the manned spacecraft put into orbit in May 1973 to study the effects of long-term exposure to weightless conditions?

GLOSSARY

Abrading A process that involves wearing away the surface through scraping or rubbing.

Advertise To get the public's attention through newspapers, handbills, radio, television, billboards, and so forth.

Aerobic exercise Physical activity designed to strengthen the cardiovascular system. For example, running, walking, roller skating, dancing, bicycling, swimming, and cross-country skiing are forms of aerobic exercise.

Agriculture The science and art of farming for the purpose of producing vegetable and animal foods.

Alloys A mixture of two or more types of metals. The alloy has the positive characteristics of the metals used to make it.

Aluminum A lightweight, easily worked metal that resists corrosion and is abundantly available in nature.

AM Amplitude modulation.

Amplifier An electronic device used to increase the strength of an electrical signal.

Architect A person who designs and plans houses, buildings, and other structures.

Automation A method in which any or all of the processes of a system are automatically performed or controlled by self-operating machinery or electronic devices.

Auto mechanic A person who diagnoses, troubleshoots, and repairs automobile engines and systems.

Bacteria Simple one-celled microorganisms that can multiply by simple division and can only be seen through a microscope. Bacteria are often called germs.

BASIC A simple computer programming language. BASIC is the acronym for *B*eginners *A*ll-purpose *S*ymbolic *I*nstruction *C*ode.

Blood pressure The amount of pressure exerted against the walls of the arteries as blood is pushed through them with each beat of the heart.

Boring Making a hole in an object. Boring is sometimes called drilling.

Breeding The producing of plants or animals, usually for the purpose of developing new or better types.

Brittleness The characteristic of being easily broken or shattered.

Broadcasting Transmitting by radio or television.

C A machine-independent language developed for system programming.

CADD *C*omputer *A*ided *D*esign and *D*rafting. Accomplishing design and drafting tasks with the aid of a computer.

Carbohydrates A class of foods, including sugars, starches, and celluloses, that supplies energy to the body.

Cardiovascular system The system comprising the heart, lungs, arteries, and veins.

Carpentry An occupation that involves building with wood materials. A manufacturing process that involves pouring molten metal into a mold to form a particular shape.

Cathode ray tube A vacuum tube in which the electrons streaming from the cathode are directed to strike a fluorescent screen and produce illuminated traces visible from outside the tube. A television screen or a computer screen is a cathode ray tube.

Chemical analysis A process in which chemicals are used to analyze blood or specific human excretions.

Chemical dependency An unnatural addiction to some chemical or drug. Alcoholics and drug addicts have chemical dependencies.

Cholesterol A crystalline fatty alcohol found in animal fats, thought to be a factor in heart and artery health problems.

Clip art Graphic images that have already been created and are clipped out to be used to form another graphic image.

CNC *C*omputer *N*umerical *C*ontrol; a computer manufacturing process that involves controlling machine operation with a computer.

COBOL A high-level programming language designed primarily for business applications. COBOL is the acronym for *C*ommon *B*usiness *O*riented *L*anguage.

Cold composition A classification of composition not involved with metal type. Cold composition does not include photographic techniques.

Commercial transportation Various means of transportation in which the person being transported pays. Commercial transportation includes commercial buses, taxi cabs, airlines, and ships.

Communicable disease Any type of disease that can be spread from one person to another or from one source to another.

Communication Sending and receiving information, signals, and messages.

Compact disk A small, clear plastic disk of less than 5 inches in diameter used to record and play back audio and visual signals. A compact disk resembles a tiny long-playing record album.

Computer A device capable of solving problems by accepting data, performing prescribed operations on the data, and supplying the results of these operations.

Computer-aided manufacturing A broad term encompassing all of the many different ways that manufacturing processes are aided by computers.

Computer graphics Creating graphic images with the assistance of a computer. A computer-generated drawing is a form of computer graphics.

Computer program A highly structured set of instructions that guide the operation of a computer.

Construction The act of building or forming by fitting parts or elements together systematically.

Continuous-path movement Movement from location to location along smooth unbroken pathways.

CPU The control section in the arithmetic logic unit of a computer.

Crop management An agricultural technique in which crops are planted, cultivated, harvested, and rotated annually to ensure the highest possible yield without depleting the soil.

Cultivating Breaking up the surface soil around plants in order to destroy weeds, prevent crusting, preserve moisture, and promote healthy crops.

Cutting A manufacturing process involving sawing through any type of material.

Data Raw, unprocessed facts and figures.

Dehydration The removal of water from a compound, substance, or tissue.

Depersonalization A state in which people are deprived of their individuality or identity. It often occurs in overcrowded conditions, in large cities, for example.

Design A problem-solving process used by architects, engineers, and designers.

Development The stage that accompanies research and involves bringing new ideas into being.

Diagnosis The act or process of determining the nature of a disease, condition, or any other type of problem with an animal or machine by examining the symptoms.

Diesel engine A type of internal combustion engine that burns fuel oil. Ignition occurs through heat from air compression.

Distribution The process in which consumer products are moved from where they are produced to where they are needed.

DNC *D*irect *N*umerical *C*ontrol; a form of computer manufacturing in which instructions go directly from a computer to the machine being controlled.

Drilling The act of boring holes in wood, metal, stone, or other material.

Drug A chemical substance used as an ingredient in medicine or as some type of narcotic.

Ductility The ability of a material to be stretched or drawn without breaking.

Duplicating Reproducing exact copies of graphic images.

Elasticity The ability of a material to stretch and return to its original shape.

Electronic communication Sending and receiving messages by electronic devices, machines, or systems.

Eminent domain The right of the government to take an individual's property for the good of the general public.

Energy The capacity for work or vigorous physical activity.

Engineer A person who works in an occupation which involves the science concerned with putting knowledge to practical uses. Engineering is divided into civil, electrical, mechanical, chemical and other branches.

ENIAC *E*lectronic *N*umerical *I*ntegrator *A*nd *C*alculator developed in the 1940s was an early version of the computer.

Facility Building or special room that makes possible some activity.

Fastening The act or means of making something secure.

Ferrous metals Metals containing iron. Ferrous metals will rust.

Finishing A process involving making final refinements to a product before it is distributed to the market.

Fixed-wing aircraft Any type of air transportation vehicle with static wings, such as airplanes.

Fluid transmission Movement of matter from one location to another through liquids or gases.

FM Frequency modulation.

Food preservation Processing food so that it can be stored for a long time.

Food processing Converting food from its natural form to the form we see in a grocery store.

Food production All of the various processes involved in providing food: planting, harvesting, processing, and storing crops and breeding animals.

FORTRAN A high-level programming language designed primarily for mathematical and scientific applications. FORTRAN is the acronym for *For*mula *Tran*slator.

Fossil fuel Fuels derived from the remains of plant or animal life of some previous geological period and now preserved in the earth's crust.

Gasoline A volatile, highly flammable, colorless liquid mixture of hydrocarbons produced by distilling petroleum; used principally as fuel in internal combustion engines.

Gasoline engine An internal combustion engine that uses gasoline as its fuel.

Gas turbine An engine driven by the pressure of gas against the curved veins of a wheel or a set of wheels fastened to a driving shaft.

Gear A toothed wheel or disk designed to mesh with another so that the motion of one is passed on to the other.

General-purpose transportation Any noncommercial form of transportation without a specific purpose. An example is the family car.

Generator A device or machine for changing mechanical energy into electrical energy.

Genetic problems Health problems inherited at birth and passed along to future generations.

Geothermal energy Energy derived from steam escaping from the earth's core.

Grains A variety of small, hard, seed or seedlike fruits, including plants such as wheat, rice, corn, rye, and barley.

Graphic communication Communication through static graphic images such as letters, numbers, and pictures.

Graphic image An imitation or representation of anything produced by any one of a variety of methods.

Hardness The quality of a material that resists indentation.

Hardware A computer-related term describing the electromechanical devices and machines that go into a computer system.

Hazardous-waste technician A person who is involved in the safe disposal of hazardous-waste products, such as nuclear materials.

Hieroglyphics A pictorial or symbolic representation of a word, syllable, or sound used by the ancient Egyptians.

Horsepower A unit for measuring the power of automobile engines. One horsepower is equal to a rate of 33,000 foot-pounds per minute.

Hot composition Composition where type metal is used to create graphic images.

Immunization An inoculation that keeps a person from getting a certain disease.

Inclination The slope or slant of a surface.

Information Data that has been processed.

Information processing Collecting, reading, manipulating, sorting, and performing other operations on data according to a specific set of instructions.

Input The first phase of a computer operation. Input can also mean data that is to be processed by a computer.

Input/output devices Peripheral devices that allow people to put data into and take data out of a computer system. A keyboard is an input device. A printer is an output device. A printer with a keyboard is an input/output device.

Insecticide Any chemical substance used to kill insects.

Intaglio printing An image-creation process in which the image areas are cut into the surface of a plate. Also called gravure printing.

Integrated circuit A circuit module with many transistors, resistors, capacitors, and connecting leads. Integrated circuits can be miniaturized and placed on tiny wafers of silicon smaller than a child's fingernail.

Interchangeable parts Parts that can replace identical or nearly identical parts in manufactured items.

Intercontinental ballistic missiles (ICBMs) Long-range missiles used by several countries for national defense. They usually carry nuclear warheads.

Internal combustion engine A type of engine in which the fuel is ignited either by heat or by spark plug inside an enclosed chamber. Gasoline and diesel engines are internal combustion engines.

Intuition The direct knowing or learning of something without the conscious use of reasoning or without any conscious attempt to learn or understand.

Iron A malleable, ductile metal that can be readily magnetized. It rusts rapidly in moist or salty air. Iron is one of the principal ingredients in steel.

Irrigation Supplying land with water by artificial ditches, channels, or sprinklers.

Isometric exercise Exercise designed to build, tone, or firm the skeletal muscles by pushing against heavy or immovable weights.

Knowledge Information that has been learned and understood.

Laminate To cover with or bond to one or more thin layers of clear plastic.

Layout In manufacturing and construction, the initial preparation for a job or task.

Legumes A family of nitrogen-fixing trees, shrubs, plants, and vegetables that are often used for replenishing the soil in agricultural systems. Soybeans and peanuts are examples of legumes.

Leisure Free or unoccupied time for a person to engage in any activity he or she likes.

Leverage Increased mechanical power, or the means of accomplishing some purpose through the use of a lever.

Lithographic printing A major printing process where the image and nonimage areas fall on the same plane but are separated by a chemical action. The nonimage areas are water receptive and ink repellent; the image areas are ink receptive and water repellent.

Logic The science of correct reasoning that describes relationships among propositions.

Machinability Characteristic of a material that allows it to be easily machined by common machine tools.

Machine The highest level of tool. Machines are made of moving parts that all work together to do a specific job.

Mainframe computer A large, powerful computer into which a number of computer terminals may be tied.

Manufacturing The making of goods and products by machinery on a large scale and with divisions of labor.

Mass communication Giving a message to many people at the same time. Television and radio broadcasts are forms of mass communication.

Mass production Producing goods and materials in large quantities.

Material Any matter or substance that occupies space.

Measurement An extent, quantity, or size as determined by measuring.

Mechanical transmission Transmitting power by mechanical devices like pulleys and gears.

Mechanism Mechanical device to produce power or energy or to extend the power or energy of humans beyond their basic abilities.

Mechanization To convert from manual processes to machine-operated processes.

Microcomputer The smallest type of computer. It has a complete computer on a single miniature circuit board.

Microscopic organisms Organisms that can only be seen with a microscope.

Milling A machining process that uses rotating cutters to cut, grind, or shape metal parts.

Modulator Device that varies the power of a radio frequency.

Morse code A special code invented by Samuel F. B. Morse that uses dots and dashes to transmit and receive messages by telegraph systems.

Movable-type printing A printing process in which individual letters, symbols, and characters can be placed on a carrying frame and interchanged to create and print graphic messages.

Natural materials Materials found in and provided by nature.

NC *N*umerical *C*ontrol. A system of controlling manufacturing machines and equipment.

Nitrogen A colorless, tasteless, odorless, gaseous chemical found in the atmosphere. It is an important ingredient in growing plants, trees, and agricultural products.

Nonferrous metals Metals that do not contain iron and therefore are highly resistant to rusting.

Nuclear energy Energy produced by splitting the nuclei of atoms into two fragments of approximately equal mass accompanied by conversion of part of the mass into energy.

Nuclear-waste disposal The process of eliminating and disposing of hazardous nuclear-waste materials that are the by-products of nuclear energy.

Oscillator A device that generates alternating current by the exchange of energy between an inductance and a capacitance connected together as a resonant circuit.

Output The last phase in the operation of a computer, or that which comes out of a computer.

Oxygen A colorless, odorless, tasteless, gaseous chemical element. Oxygen is the most abundant of all elements. It occurs freely in the atmosphere.

Packaging The final physical preparation of a consumer good before it is distributed to the market.

PASCAL A computer language named after the French physicist and mathematician Blaise Pascal.

Pathway The routes used to move from one location to another.

Petroleum An oily, flammable, liquid solution of hydrocarbons occurring naturally in the rock strata of certain geological formations. When distilled, petroleum can yield paraffin, fuel oil, kerosene, and gasoline.

Photography The process of producing images of objects on a photosensitive surface, such as camera film, by the chemical action of light or some other type of radiant energy.

Phototypesetting A text composition process that involves producing an image from a master film matrix. The matrix is a type font containing a negative image of the style of type being used.

PL/1 Programming Language 1. A high-level programming language designed to make it easier to prepare computer programs for business and scientific applications.

Planing The process in which an object is made smooth or level.

Pollution Contamination of a substance. Automobile exhaust fumes are a form of pollution.

Power The ability to convert energy into work or force.

Preservatives Substances added to foods to keep them from spoiling.

Principle A basic truth, law, or doctrine on which other truths, laws, or doctrines are based.

Process A set of step-by-step procedures that lead to a specific result.

Processing To read, manipulate, or perform operations according to a specific set of instructions.

Production The act or process through which goods and materials are manufactured.

Pulley A mechanism consisting of a small wheel (or wheels) with a grooved rim, turning in a block, in which a rope or chain runs so as to raise a weight attached to one end by pulling on the other end.

Pulse rate The number of beats per minute of the human heart.

Radio A device for communicating over long distance by converting sounds or signals into electromagnetic waves and transmitting them through space.

Radio waves Electromagnetic waves sent through space without connecting wires.

Raw materials Materials from nature or some other source that have not yet undergone a conversion process.

Receiver The person or device who receives messages that have been transmitted through any mechanism. When you listen to another person talk, you are a receiver. A radio is a receiver.

Refrigeration The process that preserves food and other perishable materials by keeping them cold or frozen.

Rehabilitation The process through which people are restored to a normal or close to normal state of health and physical activity.

Relief printing The printing process where the nonimage area is cut below the image area. During reproduction, ink is applied to the raised image area to produce the actual image.

Research A careful, systematic study or investigation into some field of knowledge to discover or establish facts or principles.

Resource Anything that is available to help us survive and improve the quality of our lives.

Respiration The act of breathing, inhaling, and exhaling air.

Robotics The design, manufacture, and effective use of robots.

Roots Parts of a plant that are usually found below the ground, such as nodes, shoots, and leaves. They hold the plant in position and draw water and nourishment from the soil.

Rotary-wing aircraft Helicopters.

Sawing To cut or divide with a special tool, usually called a saw.

Screen process printing A process that transfers an image by forcing ink or paint through a screen image carrier.

Sea-lane A pathway followed by oceangoing vessels.

Secondary storage Storage that supplements primary storage. Secondary storage is also referred to as auxillary storage.

Sender The person or device that originally transmits a message.

Shelter Any type of structure that covers, protects, defends, or provides safe refuge from the elements.

Shipping lane An ocean pathway specifically set aside for vessels moving cargo from one place to another.

Sickness A weakened or unhealthy condition.

Software The nonelectromechanical parts of a computer system, such as computer programs, program documentation, user manuals, instructions, and technical manuals.

Solar energy Energy derived directly from the sun.

Soldering Joining metal parts by melting an alloy of tin and lead.

Steam engine An engine that uses steam under pressure to supply mechanical energy, usually to drive a piston in a cylinder.

Steam turbine A special turbine turned by steam moving under great pressure.

Steel A hard, tough metal made of iron and other elements such as carbon, nickle, and chromium.

Stethoscope An instrument used by doctors to listen to the heart and lungs.

Structure Anything that is built or constructed, such as a dam, a tower, or a large building.

Submarine A military vessel, armed with torpedos and guided missiles, that can operate under water.

Supercomputer The largest type of computer.

Surface-to-air missiles Special military weapons fired from the ground at targets in the air.

Synthesizer An electronic device containing filters, oscillators, and voltage-controlled amplifiers to produce different types of sounds.

Synthetic materials Materials made by people.

System A set or arrangement of related parts connected to form a unified whole.

Technology People using tools, resources, and processes to solve problems or extend their capabilities.

Telegraph A device for transmitting messages by electric impulses through a wire or by radio waves.

Telephone A device for converting sound into electrical impulses for transmitting and receiving over long distances.

Tensile strength Resistance of a material to lengthwise stress.

Textiles Fabrics made by weaving or knitting cloth.

Tool A device or object that allows people to perform tasks they couldn't perform with their bare hands.

Transistor A solid-state electronic device composed of semiconductor materials.

Transmitter A person or device that conveys a message.

Transportation Moving passengers, goods, and materials.

Treatment Medical or health care aimed at curing a particular illness.

Triglycerides Any of a group of substances derived from glycerol containing one, two, or three specific fatty acids.

Turbine An engine driven by the pressure of steam, water, or air against the curved veins of a wheel fastened to a driving shaft.

Turbofan A jet engine in which additional thrust is obtained from the part of the air that bypasses the engine and is accelerated by a fan.

Turbojet A jet engine using a turbine to drive an air compressor that takes in and compresses air for fuel combustion.

Turboprop A jet engine whose turbine shaft, through reduction gears, drives a propeller that develops most of the thrust.

UNIVAC An early computer produced by Sperry-Rand that was installed at the U. S. Census Bureau in 1951. UNIVAC was the first commercially produced computer.

Uranium A hard radioactive metallic element that is important in the work of atomic energy.

VCR Video cassette recorder.

Vital signs The pulse rate, blood pressure, and respiration rate. Vital signs are used as initial indicators of potential health problems.

Vitamin A complex organic substance found in food, and sometimes made in the body. It is essential for regulating metabolism and normal growth and functioning of the body.

Welding The process of uniting two or more pieces of metal by heating until they fuse.

Wheel A disk or circular frame connected by spokes to a central hub capable of turning on a central axis and used to move vehicles or transmit power in machinery.

Windmill A structure used to capture wind energy and convert it to mechanical energy.

Wireless telegraph A later version of the telegraph that transmitted with radio waves instead of telegraph wires.

Word processor A microcomputer-based system for composing, storing, editing, and printing written materials.

Xerography A process for copying printed materials and other graphic images. The latent image of the original is transferred by the action of light to an electrically charged surface. The image attracts oppositely charged dry ink particles to this surface. The particles are then fused in place on the copy paper.

INDEX

D

E

F

U

V

W

X